Library of
Davidson College

Bäcklund Transformations
and
Their Applications

This is Volume 161 in
MATHEMATICS IN SCIENCE AND ENGINEERING
A Series of Monographs and Textbooks
Edited by RICHARD BELLMAN, *University of Southern California*

The complete listing of books in this series is available from the Publisher upon request.

Bäcklund Transformations
and
Their Applications

C. Rogers and W. F. Shadwick
University of Waterloo
Waterloo, Ontario, Canada

1982

ACADEMIC PRESS
A Subsidiary of Harcourt Brace Jovanovich, Publishers
New York   London
Paris   San Diego   San Francisco   São Paulo   Sydney   Tokyo   Toronto

COPYRIGHT © 1982, BY ACADEMIC PRESS, INC.
ALL RIGHTS RESERVED.
NO PART OF THIS PUBLICATION MAY BE REPRODUCED OR
TRANSMITTED IN ANY FORM OR BY ANY MEANS, ELECTRONIC
OR MECHANICAL, INCLUDING PHOTOCOPY, RECORDING, OR ANY
INFORMATION STORAGE AND RETRIEVAL SYSTEM, WITHOUT
PERMISSION IN WRITING FROM THE PUBLISHER.

ACADEMIC PRESS, INC.
111 Fifth Avenue, New York, New York 10003

United Kingdom Edition published by
ACADEMIC PRESS, INC. (LONDON) LTD.
24/28 Oval Road, London NW1 7DX

Library of Congress Cataloging in Publication Data

Rogers, C.
  Bäcklund transformations and their applications.

  (Mathematics in science and engineering)
  Includes bibliographical references and index.
  1. Differential equations, Partial.   2. Bäcklund
transformations.   3. Jet bundles (Mathematics)
4. Mathematical physics.   I. Shadwick, W. F., Date
II. Title.   III. Series.
QA374.R64   515.3'53   81-22783
ISBN 0-12-592850-5   AACR2

PRINTED IN THE UNITED STATES OF AMERICA

82 83 84 85   9 8 7 6 5 4 3 2 1

To

DR. DAVID CLEMENTS
Reader in Applied Mathematics
University of Adelaide

and

DR. GEOFFREY POWER
Reader in Mathematics
University of Nottingham

Contents

Preface   xi
Acknowledgments   xiii

General Introduction and Outline   1

CHAPTER **1**   **Bäcklund Transformations and Their Application to Nonlinear Equations of Mathematical Physics**

1.1   The Origin and Importance of Bäcklund Transformations   12
1.2   The $1 + 1$ Sine-Gordon Equation. The Permutability Theorem. Bianchi Diagrams   16
1.3   Ultrashort Optical Pulse Propagation in a Resonant Medium. $2N\pi$ Light Pulses   22
1.4   Propagation of Magnetic Flux through a Long Josephson Junction. An Elliptic Analog of the $1 + 1$ Sine-Gordon Equation. Generation of Solutions via a Bäcklund Transformation   29
1.5   The Cole–Hopf Reduction of Burgers' Equation to the Heat Equation. Solution of Initial Value Problems   38
1.6   Finite Amplitude Dispersive Waves and the Korteweg–deVries Equation   43
1.7   The Wahlquist–Estabrook Invariant Transformation of the Korteweg–deVries Equation. The Soliton Ladder   51

1.8	An Auto-Bäcklund Transformation of the Modified Korteweg–deVries Equation. The Permutability Theorem and Generation of Solutions	57
1.9	The Miura Transformation	60
1.10	A Nonlinear Schrödinger Equation	62
1.11	Bäcklund Transformations and the AKNS System. An Invariant Transformation of the Cubic Schrödinger Equation	65
1.12	The Hirota Bilinear Operator Formulation of Bäcklund Transformations. The Boussinesq Equation. Kaup's Higher Order Water Wave Equation	69
1.13	Bäcklund Transformations for Higher Order Korteweg–deVries Equations	79
1.14	Bäcklund Transformations in Higher Dimensions. The Sine-Gordon Equation in 3 + 1 Dimensions. The Kadomtsev–Petviashvili Equation. The Yang Equations	82
1.15	The Benjamin–Ono Equation for Internal Deep Water Waves. The Nakamura Transformation	92
1.16	A Bäcklund Transformation for Joseph's Equation for Internal Waves in a Stratified Fluid with Finite Depth	101
1.17	Bäcklund Transformations of Nonlinear Lattice Equations in Their Continuum Approximation. The Konno–Sanuki Transformation	105
1.18	Bäcklund Transformations of Nonlinear Differential-Difference Equations. The Toda and Associated Lattices	110
1.19	Bäcklund Transformations in General Relativity. The Ernst Equations. Neugebauer's Permutability Theorem	115

CHAPTER 2 A Local Jet-Bundle Formulation of Bäcklund Transformations

2.1	Preliminaries	123
2.2	Contact Structures	126
2.3	Partial Differential Equations on Jet Bundles	129
2.4	Fibered Products of Jet Bundles	132
2.5	Bäcklund Maps	134
2.6	Bäcklund Transformations Determined by Bäcklund Maps	139
2.7	Symmetries of Differential Equations and One-Parameter Families of Bäcklund Maps	143
2.8	The Wahlquist–Estabrook Procedure	145

CHAPTER 3 Bäcklund Transformations in Gasdynamics, Nonlinear Heat Conduction, and Magnetogasdynamics

3.1	The Reciprocal Relations and the Haar Transformation	151
3.2	Properties of the Reciprocal Relations. Invariance of the Equation of State	155
3.3	Reciprocal Relations in Subsonic Gasdynamics	160
3.4	Reciprocal-Type Transformations in Steady Magnetogasdynamics	162

	3.5	A Bäcklund Transformation in Nonlinear Heat Conduction	168
	3.6	Bäcklund Transformations of the Loewner Type	171
	3.7	Reduction of the Hodograph Equations to Canonical Form in Subsonic, Transsonic, and Supersonic Gasdynamics	176
	3.8	Bäcklund Transformations in Aligned Nondissipative Magnetogasdynamics	183
	3.9	Invariant Transformations in Nonsteady Gasdynamics	185
	3.10	Bäcklund Transformations in Lagrangian Gasdynamics. Reflection of a Centered Wave in a Shock Tube	189

CHAPTER 4 **Bäcklund Transformations and Wave Propagation in Nonlinear Elastic and Nonlinear Dielectric Media**

4.1	Propagation of Large Amplitude Waves in Nonlinear Elastic Media. Reduction to Canonical Form of the Riemann Representation	197
4.2	The Model Constitutive Laws	202
4.3	Reflection and Transmission of a Large Amplitude Shockless Pulse at a Boundary	211
4.4	Electromagnetic Wave Propagation in Nonlinear Dielectric Media. Reduction to Canonical Form via Bäcklund Transformations	225
4.5	Evolution of a Large Amplitude Centered Fan in a Nonlinear Dielectric Slab	230

CHAPTER 5 **Bäcklund Transformations and Stress Distribution Theory in Elastostatics**

5.1	Weinstein's Correspondence Principle in the Context of Bäcklund Transformations of the Generalized Stokes–Beltrami System	238
5.2	Application of Weinstein's Correspondence Principle in Elastostatics. Associated Axially Symmetric Punch, Crack, and Torsion Boundary-Value Problems	242
5.3	Antiplane Crack and Contact Problems in Layered Elastic Media. Bäcklund Transformations and the Bergman Series Method	253
5.4	Stress Concentration for Shear-Strained Prismatic Bodies with a Nonlinear Stress–Strain Law	262

APPENDIX I **Properties of the Hirota Bilinear Operators**   274

APPENDIX II **Differential Forms**

II.1	Tangent Spaces and Vector Fields on $\mathbb{R}^k$	276
II.2	Differential p-Forms on $\mathbb{R}^k$	277
II.3	The Exterior Derivative	279

	II.4 Pull-Back Maps	282
	II.5 The Interior Product of Vector Fields and Differential p-Forms	283
	II.6 The Lie Derivative of Differential Forms	284

APPENDIX III **Differential Forms on Jet Bundles**

III.1 Preliminaries	286
III.2 Exterior Differential Systems on Jet Bundles	288
III.3 The System of Differential Equations Associated with an Exterior Differential System	289
III.4 An Exterior Differential System of m-Forms on $J^k(M, N)$ Associated with a Quasi-Linear Equation on $J^{k+1}(M, N)$	290

APPENDIX IV **The Derivation of the Equations that Define $B^{k+1,1}(\psi)$** 292

APPENDIX V **Symmetries of Differential Equations and Exterior Systems**

V.1 Point Transformations	294
V.2 Symmetries	297

APPENDIX VI **Composition of Symmetries and Bäcklund Maps** 299

References 302

Author Index 323
Subject Index 329

Preface

Bäcklund transformations have emerged over the past decade as an important tool in the study of a wide range of nonlinear partial differential equations in mathematical physics. A need has emerged for a single text that presents a compendium of the diverse applications of Bäcklund transformations and provides an acceptable underlying theory for the subject. This monograph attempts to meet that need.

The main significance of Bäcklund transformations in connection with nonlinear equations is that typically they have associated nonlinear superposition principles whereby infinite sequences of solutions to nonlinear equations may be generated by *purely algebraic procedures*. Multi-soliton solutions of many important nonlinear evolution equations have been thereby constructed. An independent development involves the use of Bäcklund transformations in general relativity to construct exact solutions of the stationary axisymmetric vacuum Einstein field equations. Again, a nonlinear superposition principle is available for the generation of such solutions. These and other applications of Bäcklund transformations in mathematical physics are surveyed in Chapter 1.

Whereas the classical notion of Bäcklund transformations has proved too restrictive for contemporary needs, a universally accepted modern theory is yet to be established. In Chapter 2, we present a unifying geometric framework for Bäcklund transformations based on a jet-bundle formalism. This approach meets many of the requirements of a modern theory and, in particular, allows for a natural derivation of associated linear scattering problems.

The remaining chapters of the monograph are devoted to applications of Bäcklund transformations in areas of continuum mechanics, such as gasdynamics, magnetogasdynamics, and elasticity. This represents a development in Bäcklund transformation theory that has proceeded quite independently of the work described in Chapters 1 and 2. Much of the material contained in these chapters may be new to theoretical physicists.

Our aim has been to produce an introductory monograph that is reasonably self-contained. To this end, appendixes have been included on Hirota bilinear operators, differential forms, and aspects of jet-bundle theory needed in the main text. The range of application of Bäcklund transformations, however, is broad indeed; and such has been the rapid growth of the subject that it has not been practical here to cover all of the most recent advances. Certain topics have been omitted as being more appropriate to an advanced treatise. Thus we have not discussed the relationship of Bäcklund transformations to canonical transformations, the use of Bäcklund transformations in the generation of conservation laws [310, 447–450], or the novel application of these transformations to reveal hidden symmetry properties in similarity solutions of nonlinear equations [433, 435, 437]. Likewise, details of the generalized Wronskian technique and its relationship to Bäcklund transformations must be sought elsewhere [42–46]. Nevertheless, an attempt has been made to produce a comprehensive bibliography, and the purpose of this monograph will have been fulfilled if the reader is led to inquire into the original literature for details of these and other exciting recent developments in the expanding subject of Bäcklund transformations. The book is appropriate for use as a graduate level text for applied mathematicians or theoretical physicists. Indeed, it has grown out of a graduate course given at the University of Waterloo. In this connection, it should be noted that an applications-oriented course may be taught based on Chapters 1 and 3–5, together with Appendix I. On the other hand, a course with emphasis on the theoretical aspects of the subject might consist of Chapter 2, together with Appendixes II–VI.

Acknowledgments

The authors are indebted to their students and colleagues who have scrutinized parts of the manuscript of this book, and whose comments at various stages have led to improvements in the final version. Special thanks in this regard are due to Drs. H. M. Cekirge, J. Isenberg, D. C. Robinson, and Professors K. Dunn, R. B. Gardner, G. M. L. Gladwell, and F. A. E. Pirani. For his initial suggestion that this work be undertaken, the authors are indebted to Professor Richard Bellman.

One of the authors (C. R.) wishes to express his gratitude to the Department of Applied Mathematics and Theoretical Physics, University of Cambridge, and to the Technical University of Denmark for the kind hospitality shown him during development of this project. His support under a National Science and Engineering Research Council of Canada Grant A0879 and NATO grant R. G. 113.80 is also gratefully acknowledged.

It is a pleasure to express our gratitude to Helen Warren, who typed the manuscript.

General Introduction and Outline

In recent years, a class of transformations having their origin in work by Bäcklund [1-5] in the late nineteenth century has provided a basis for remarkable advances in the study of nonlinear partial differential equations. The importance of Bäcklund transformations and their generalizations is basically twofold. Thus, on the one hand, invariance under a Bäcklund transformation may be used to generate an infinite sequence of solutions of certain nonlinear equations by purely algebraic superposition principles. On the other hand, Bäcklund transformations may also be used to link certain nonlinear equations to canonical forms whose properties are well known. Both kinds of Bäcklund transformations have been exploited extensively not only in mathematical physics but also in continuum mechanics. However, for the most part, the developments in these two fields have proceeded independently. Here, the results in both of the areas are brought together in a single account. Furthermore, a jet-bundle formulation of Bäcklund transformations is presented which, in the current absence of a universally accepted theory, provides, in the opinion of the authors, the best framework for a unified treatment of the subject.

It was Bäcklund who, in 1880, introduced generalized Bianchi–Lie transformations between pairs of surfaces Σ, Σ' embedded in $\mathbb{R}^3$ such that the surface element $\{x^1, x^2, u, \partial u/\partial x^1, \partial u/\partial x^2\}$ of Σ is connected to the surface element $\{x'^1, x'^2, u', \partial u'/\partial x'^1, \partial u'/\partial x'^2\}$ of Σ' by four relations of the type

$$\mathbb{B}_i(x^1, x^2, u, \partial u/\partial x^1, \partial u/\partial x^2; x'^1, x'^2, u', \partial u'/\partial x'^1, \partial u'/\partial x'^2) = 0,$$
$$i = 1, \ldots, 4. \qquad (0.1)$$

Application of the integrability condition

$$\frac{\partial^2 u'}{\partial x'^1 \partial x'^2} - \frac{\partial^2 u'}{\partial x'^2 \partial x'^1} = 0, \qquad (0.2)$$

may be shown to lead, under certain circumstances, either to a pair of third-order equations or to a single second-order equation for u [107]. The analogous integrability condition on the unprimed quantities leads, again under appropriate conditions, to a pair of third-order equations or to a single second-order equation for u'. In this situation, implicit in the Bäcklund relations (0.1) is a mapping between the solutions $u(x^a)$, $u'(x'^a)$ of generally distinct systems of partial differential equations. This fact may be used to advantage in one of two ways. Thus, on the one hand, if the solution of the transformed equation or pair of equations is known, then the Bäcklund relations (0.1) may be used to generate the solution of the original equation or pair of equations. On the other hand, if the equations are invariant under the Bäcklund transformation (0.1), then the latter may be used to construct an infinite sequence of new solutions from a known trivial solution. Both types of Bäcklund transformations have important applications. Accordingly, both will be discussed extensively in this monograph.

Chapter 1 presents a comprehensive account of those Bäcklund transformations which have been found to have application in mathematical physics. It opens with a description of the classical notion of a Bäcklund transformation together with a simple illustration that leads to the solution of Liouville's equation. There follows a derivation of the celebrated auto-Bäcklund transformation for the 1 + 1 sine-Gordon equation

$$u_{12} = \sin u, \qquad (0.3)$$

together with its associated permutability theorem. The latter constitutes a nonlinear superposability principle whereby an infinite sequence of solutions of (0.3) may be constructed by purely algebraic means. The Bianchi diagrams descriptive of this procedure are also introduced.

The nonlinear evolution equation (0.3) occurs in a diversity of areas of physical importance. In particular, it was the work of Lamb [17–21] connected with ultrashort optical pulse propagation phenomena modeled by the 1 + 1 sine-Gordon equation which, in some measure, led to the renaissance of the subject of Bäcklund transformations. Thus, Lamb employed the permutability theorem associated with an auto-Bäcklund transformation of (0.3) to generate analytical expressions descriptive of the evolution of ultrashort light pulses. An account of this work, together with its extension to consideration of $2N\pi$ pulses by Barnard [22], is given in Section 1.3.

The analysis of the steady propagation of magnetic flux through a long Josephson junction leads to an elliptic analog of the sine-Gordon equation. In this connection, Leibbrandt [25–27] established a Bäcklund transformation which links the equations

$$\alpha_{11} + \alpha_{22} = \sin\{\alpha + v^1\}, \tag{0.4}$$

and

$$\beta_{11} + \beta_{22} = \sinh\{\beta + v^2\}, \tag{0.5}$$

where v^i, $i = 1, 2$, are arbitrary plane harmonic functions. This result, together with associated permutability theorems whereby infinite sequences of α and β solutions may be constructed, is derived in Section 1.4.

In the next section, as a further illustration of a Bäcklund transformation of a Monge–Ampère equation, the Cole–Hopf transformation of Burgers' equation

$$u_2 + uu_1 - vu_{11} = 0, \tag{0.6}$$

is presented. This equation is of considerable significance, since it represents the simplest model which incorporates both amplitude dispersion and diffusion effects. The Cole–Hopf transformation has the remarkable property that it reduces the *nonlinear* equation (0.6) to the *linear* heat equation. Here it is shown how this result may be used to solve a variety of initial value problems for Burgers' equation, including those with initial step data and δ function initial conditions.

Apart from the intrinsic importance of the Cole–Hopf transformation in the solution of initial value problems for (0.6), there is no doubt that its discovery also served to stimulate the search for Bäcklund transformations for other nonlinear evolution equations. One such equation was that derived in 1895 by Korteweg and deVries [30] in a study of the evolution of long water waves along a canal of rectangular cross section. The Korteweg–deVries equation

$$u_2 + 6uu_1 + u_{111} = 0, \tag{0.7}$$

was rederived in 1960 by Gardner and Morikawa [31] in an analysis of collision-free hydromagnetic waves. Subsequently, it has been shown to model a rich diversity of finite amplitude dispersive wave phenomena in the theory of solids, liquids, gases, and plasmas.

In Section 1.6, the Korteweg–deVries equation is derived in the context of shallow water wave theory. In the following section, a generalization of the classical Bäcklund transformation (0.1) is introduced wherein second-order derivatives are included. Thus, transformations of the type

$$\mathbb{B}_i(x^a, u, u_a, u_{ab}; x'^a, u', u'_a, u'_{ab}) = 0, \qquad i = 1, \ldots, 4, \quad a, b = 1, 2, \qquad (0.8)$$

are introduced, and following Lamb's adaption of Clairin's procedure, an auto-Bäcklund transformation is obtained for the Korteweg–deVries equation. This Bäcklund transformation, originally derived in 1973 by Wahlquist and Estabrook [36], is then used to obtain a permutability theorem whereby multi-soliton solutions may be constructed.

The next two sections of Chapter 1 concern the modified Korteweg–deVries equation

$$u_2 + 6u^2 u_1 + u_{111} = 0, \qquad (0.9)$$

which arises both in the theory of nonlinear Alfvén waves in a collisionless plasma and in the analysis of acoustic wave propagation in anharmonic lattices [203, 33]. In Section 1.8, a Bäcklund transformation is introduced which leaves invariant the equation governing an integral over a pulse profile v, where v satisifies the modified Korteweg–deVries equation. The permutability theorem associated with the auto-Bäcklund transformation is then used to generate soliton solutions. In Section 1.9, Miura's celebrated Bäcklund transformation, linking the Korteweg–deVries and modified Korteweg–deVries equations, is developed, using Clairin's procedure as adapted by Lamb [35]. It is remarked that Miura's transformation may be exploited to construct a variety of constants of motion for the Korteweg–deVries equation [37].

Lamb also used Clairin's procedure to obtain an auto-Bäcklund transformation for another important nonlinear evolution model, namely, the cubic Schrödinger equation

$$iu_2 + u_{11} + vu^2 \bar{u} = 0. \qquad (0.10)$$

In Section 1.10, the relevance of this equation to the evolution of weakly nonlinear deep water gravity wave trains is summarized and its role as a canonical form noted. In the following section, the AKNS system of nonlinear evolution equations amenable to the inverse scattering method is introduced. It is shown to include as particular cases not only the sine-Gordon, Korteweg–deVries, and modified Korteweg–deVries equations but also the cubic Schrödinger equation. A link with the construction of Bäcklund transformations is established via an invariance property contained in a paper by Crum [311] on associated Sturm–Liouville systems. The section concludes with the statement of an auto-Bäcklund transformation of (0.10), together with an associated permutability theorem. The latter has recently been used to reveal, via the language of Painlevé transcendents, hidden symmetries of the similarity solutions of (0.10).

Hirota and Satsuma [201] have recently demonstrated that the permutability theorems for a wide class of nonlinear evolution equations have a particularly simple generic form when written in terms of certain bilinear operators. The latter formulation was originally introduced by Hirota [118] in connection with a direct expansion method for the construction of multi-soliton solutions of certain nonlinear evolution equations. The basic properties of the Hirota bilinear operators are set forth in Appendix I. In Section 1.12, this formalism is used to derive a Bäcklund transformation and associated permutability theorem for both the Boussinesq equation

$$u_{tt} - u_{xx} - 3(u^2)_{xx} - u_{xxxx} = 0, \qquad (0.11)$$

and a higher order water wave equation due to Kaup [236]. In Section 1.13 the same method is used to construct an auto-Bäcklund transformation in bilinear form for the higher order Korteweg–deVries equation

$$u_t + 180u^2 u_x + 30(uu_{xxx} + u_x u_{xx}) + u_{xxxxx} = 0, \qquad (0.12)$$

(Sawada and Kotera [245]; Caudrey *et al.* [242]). Miura-type Bäcklund transformations exist that link (0.12) to another important higher order Korteweg–deVries equation [246, 248]. Moreover, bilinear representations of Bäcklund transformations for the entire Lax hierarchy of higher order Korteweg–deVries equations have been very recently developed by Matsuno [413].

Whereas there has been notable progress in the construction of Bäcklund transformations for higher order equations in $1 + 1$ dimensions, particularly via the Hirota bilinear representation, there has been no corresponding rapid advance in the development of Bäcklund transformations for nonlinear evolution equations in higher dimensions. A review of the limited progress that has been made is presented in Section 1.14. The section starts with a summary of Leibbrandt's results on Bäcklund transformations and permutability theorems for higher dimensional sine-Gordon equations. This work has recently been extended by Popowicz [254], who constructed Bäcklund transformations and nonlinear superposition principles for the $O(3)$ nonlinear σ model in both $2 + 1$ and $3 + 1$ dimensions. There follows a derivation, in terms of the Hirota bilinear formalism, of a Bäcklund transformation for the two-dimensional Korteweg–deVries equation

$$u_{xt} + \alpha u_{yy} + (3u^2)_{xx} + u_{xxxx} = 0, \qquad (0.13)$$

originally developed by Kadomtsev and Petviashvili [256] in connection with the propagation of disturbances in weakly dispersive media. Finally, an interesting Bäcklund transformation of the four-dimensional Yang equations is recorded. This result is returned to in Chapter 2 when it is viewed in a jet-bundle context.

The Benjamin–Ono equation

$$u_t + 2uu_x + H[u_{xx}] = 0, \tag{0.14}$$

where H is the Hilbert transform operator defined by

$$Hf(x) := \frac{1}{\pi} P \int_{-\infty}^{\infty} \frac{f(z)}{z - x} dz, \tag{0.15}$$

models internal water wave propagation in a deep stratified fluid [274, 276]. In Section 1.15, an auto-Bäcklund transformation and associated superposition theorem are established for (0.14) whereby N-soliton solutions may be constructed. In the following section, the results are extended to Joseph's nonlinear integrodifferential equation, which is descriptive of internal wave propagation in a stratified fluid of finite depth [282, 283]. Bäcklund transformations for Joseph's equation have been recently obtained by Chen et al. [287] and Satsuma et al. [288] employing differential-difference operators.

In the next two sections, the subject of Bäcklund transformations as applied to nonlinear difference systems is discussed. It was Toda in 1967 who, in an analysis of the longitudinal vibration of a chain of masses interconnected by nonlinear springs, introduced a nonlinear differential-difference equation which may be shown to admit stable N-soliton solutions [290, 291]. Subsequently, a Bäcklund transformation and nonlinear superposition principle were obtained for this Toda lattice equation and analytical expressions for the multi-soliton solutions thereby generated [292, 293]. Bäcklund transformations for a number of other nonlinear differential-difference equations have since appeared. In Section 1.17, the Konno–Sanuki transformation for a continuum approximation to a nonlinear lattice equation is recorded. In Section 1.18, the Bäcklund transformation for the nonlinear differential-difference equation associated with the discrete Toda lattice is described. It is noted that the Hirota bilinear operator representation of this Bäcklund transformation links the Toda lattice equation to other important nonlinear lattice equations.

The final section of Chapter 1 deals with a quite independent development, involving the application of Bäcklund transformations in general relativity. Thus, in recent years much research has been directed toward the construction of exact solutions of the stationary axisymmetric vacuum Einstein field equations that could represent the gravitational field of a spinning mass. Bäcklund transformations have been discovered by Harrison [376], Belinskii and Zakharov [377], and Neugebauer [378], whereby solutions of the axially symmetric stationary Einstein equations can be generated. Moreover, recently Cosgrove [380] has shown that the Neugebauer–Bäcklund transformations provide an important framework for the unification of diverse solution-generating techniques for the Einstein equations. In Section 1.19, the Neugebauer–Bäcklund transformations of the Ernst equations are introduced together with a permutability theorem. The latter is then used to derive the Kerr-NUT metric from Minkowski space–time.

The rich diversity of areas of application of Bäcklund transformations is apparent in the compendium presented in Chapter 1. However, it has been pointed out that such has been the rapid advance of the subject in recent years that, whereas the classical notion of Bäcklund transformation has proved too limited, a universally accepted modern theory has yet to be established. Current work on Bäcklund transformations is primarily concerned with an attempt to, on the one hand, extend application to other nonlinear equations of physical interest, and on the other, to embody the extant results in a comprehensive theory. The progress in the area of applications in mathematical physics has been reviewed in Chapter 1. In Chapter 2, we present a jet-bundle formulation of Bäcklund transformations based in the main on the work of Pirani et al. [47, 48, 407]. The jet-bundle approach is seen to provide an appropriate geometric setting for the study of Bäcklund transformations and their connection with the inverse scattering formalism, the prolongation structure of Wahlquist and Estabrook, and symmetries of differential equations. The development of the classical Bäcklund transformation (0.1) not only incorporates higher derivatives but also allows the introduction of several dependent variables. Detailed applications of a particular class of the latter type of Bäcklund transformation form the basis for Chapters 3–5.

Chapter 2 commences with a reasonably self-contained introduction to the jet-bundle formalism. Thus, Sections 2.1–2.4 deal, in turn, with k-jet notation, contact structures, prolongation of partial differential equations, and fibered products of jet bundles. This account is augmented by Appendixes II and III, which deal with background material on differential forms. The remaining sections (2.5–2.8) show how jet-bundle transformations known as Bäcklund maps reproduce features common to many of the Bäcklund transformations described in Chapter 1. Thus, in Section 2.5 and Appendix IV, the concept of Bäcklund map is developed and is illustrated both for the sine-Gordon equation and for a class of Bäcklund transformations originally introduced by Loewner [64] in connection with the reduction of hodograph systems to canonical form.

In Section 2.6, it is shown how Bäcklund maps may be used to determine Bäcklund transformations. Illustrations are given both for the Korteweg–deVries equation and for the AKNS system associated with the sine-Gordon equation. Furthermore, the notion of Bäcklund map is extended to incorporate constraint equations. Particular reference is made to Pohlmeyer's Bäcklund transformation for the Yang equations.

In Section 2.7, the subject of symmetries of systems of differential equations is addressed and the manner in which they may be combined with Bäcklund maps is described. This leads to a generalization of the classical "theorem of Lie" for the sine-Gordon equation whereby a parameter is inserted in the Bäcklund relations by conjugation of a parameter-free Bianchi transformation with a Lie transformation (Eisenhart [168]). This construction is an important one since the parameter so intruded adopts the role of the eigenvalue in associated inverse scattering problems and, furthermore, is intrinsic to both the permutability theorems and a method for the generation of conservation laws [310, 407]. Background results on symmetries of differential equations and exterior systems, together with detailed calculations related to the composition of symmetries and Bäcklund maps are incorporated in Appendixes V and VI.

Chapter 2 concludes with a discussion of the Wahlquist–Estabrook procedure for the construction of Bäcklund maps. This technique has the advantage that it reveals the roles of the various Lie groups associated with the differential equations which admit Bäcklund transformations. Furthermore, the method admits a natural formulation within the jet-bundle framework. Here a generalization of the Wahlquist–Estabrook procedure to the case of n independent variables is presented [48, 407]. The application of the method to the sine-Gordon equation is presented in detail.

The application of Bäcklund transformations in continuum mechanics as opposed to mathematical physics has experienced a quite independent growth. Indeed, the theory underlying the separate developments has only recently been brought together in the jet-bundle formulation of Pirani et al. [48] discussed in Chapter 2. Applications in continuum mechanics up to 1973 have been summarized in [50]: it was seen that the range of application is as diverse in continuum mechanics as in mathematical physics. Developments since that time have reinforced this view. The remaining chapters of this monograph are devoted, therefore, to an up-to-date and comprehensive treatment of applications of Bäcklund transformations to areas of continuum mechanics such as gasdynamics, magnetogasdynamics, and elasticity.

It was Haar [51], in 1928, who first introduced a class of transformations that leave invariant, up to the equation of state, the governing equations of plane potential gasdynamics. However, as will be seen, the pressure–density approximation to the adiabatic gas law, developed as early as 1904 by Chaplygin [52] in his now classical work on gas jets, may be set in the context of a broad class of Bäcklund transformations of importance not only in subsonic, but also in transsonic and supersonic flow. Later, in 1938, Bateman [53] constructed a further class of transformations that leave invariant the gasdynamic equations. These have been termed the "reciprocal relations" and a specialization of these was used by Tsien [54] in connection with the approximation of certain adiabatic gas flows. This specialization was shown to lead to results equivalent to those derivable from the von Kármán–Tsien approximation [55].

Bateman [56] subsequently observed that both the Haar transformation and the reciprocal relations are of the Bäcklund type. Since that time, reciprocal and other invariant transformations in both steady gasdynamics and magnetogasdynamics have been the subject of extensive inquiry [57–63]. The reciprocal and adjoint transformations are presented in the introductory section of Chapter 3, while in Section 3.2, certain invariance properties of the reciprocal relations are recorded. In general, the equation of state is not invariant under the reciprocal transformations. Indeed, if the original gas is incompressible, then the reciprocal gas has a constitutive law of the von Kármán–Tsien type. This allows the application of the reciprocal relations to the approximation of subsonic flows of an adiabatic gas. The subject of reciprocal relations in subsonic gasdynamics is described in Section 3.3, while in Section 3.4, reciprocal transformations are constructed for steady two-dimensional nondissipative magnetogasdynamics [63].

In Section 3.5, a Bäcklund transformation which leaves invariant the nonlinear heat equation

$$u_t - [k(u)u_x]_x = 0, \tag{0.16}$$

is presented. Equations of this type arise notably in plasma physics [423, 424], boundary layer theory [425], and the theory of Darcian filtration [426–428]. It is shown here that, on appropriate specialization, a result originally due to Rosen [430] is retrieved whereby an important nonlinear conduction equation which governs the temperature distribution in solid crystalline hydrogen may be reduced to the linear heat conduction equation. This transformation is used to solve an initial value problem for the temperature distribution in a semi-infinite block of such a material subjected to surface cooling or warming. Further, it is shown that the Rosen transformation may be used to link fixed noninsulated boundary problems for a nonlinear heat conduction equation to moving noninsulated boundary problems for the linear heat conduction equation. This work has recently been extended by Berryman [432].

In 1950, Loewner [64] introduced a generalization of the classical finite Bäcklund transformations (0.1) to systems involving pairs of dependent variables. Loewner's investigation was concerned with the reduction to canonical form of the hodograph equations of gasdynamics (specifically, to the Cauchy–Riemann, Tricomi, and classical wave equations in subsonic, transsonic, and supersonic flow, respectively). Such was achieved for certain multiparameter pressure–density relationships which may be used to approximate the prevailing empirical equation of state. Power et al. [66] later showed that Loewner's formulation may be adopted as a unifying framework for earlier transformation theory of the hodograph equations due to von Kármán [67, 68], Pérès [69], Sauer [70], and Dombrovskii [71]. In Dombrovskii's monograph, what are essentially Bäcklund transformations were used extensively in the solution of boundary value problems in compressible adiabatic flow. A detailed account of Loewner's work and its application to gasdynamics is given in Sections 3.6 and 3.7.

Aligned field nondissipative magnetogasdynamics has been the subject of much attention in view of its potential relevance to astrophysics. Reference may be made, for instance, to the work of Grad [72] on reducibility; to that of Jeffrey and Taniuti [73] on the characteristic theory associated with aligned field magnetogasdynamic wave propagation; and for general background material to the treatise by Dragos [74]. In Section 3.8, it is shown how Loewner's class of Bäcklund transformations may readily be applied to obtain elliptic and hyperbolic canonical forms in both sub-Alfvénic aligned magnetogasdynamic regimes.

Invariant transformations in nonsteady gasdynamics and magnetogasdynamics have been developed and utilized considerably in recent years, notably by the Russian school [75–80]. On the other hand, transformations of the reciprocal and Haar type for $(\varepsilon + 1)$-dimensional spherically symmetric nonsteady gasdynamics were first constructed by Rogers [81, 82] and were subsequently applied in [83] to solve certain nonuniform shock-wave problems by the generation of new gasdynamic solutions based on one discovered by Sedov [84]. In Section 3.9, the reciprocal relations are developed and are then shown to constitute a Bäcklund transformation of a Monge–Ampère formulation of the governing nonsteady gasdynamic equations.

Finally in Chapter 3, it is shown that Bäcklund transformations of the Loewner-type may be used to reduce the Lagrangian equations of one-dimensional nonsteady gasdynamics to the classical linear wave equation. This reduction is available for certain multiparameter gas laws that may be used for both local and global approximation to the adiabatic equation of state. A summary is presented of the application by Cekirge and Varley [85] of a model pressure–density relation to the analysis of the pressure variation at the closed end of a shock tube during the reflection of a centered, simple wave.

The use of model constitutive laws in the study of the one-dimensional propagation of large amplitude longitudinal disturbances in other nonlinear media of finite extent was developed not only by Cekirge and Varley [85], but also in subsequent papers by Kazakia and Varley [141, 142], Cekirge [143], and Mortell and Seymour [144]. Thus certain multiparameter stress–strain laws may be used, in particular, to approximate the response of saturated soil, dry sand, clay silt, and certain hard materials under dynamic compression. In Sections 4.1–4.3 we present a review of recent work on the application of these laws to the analysis of reflection and transmission phenomena. It is also shown that for the model constitutive laws under consideration the Riemann representation of the Lagrangian equations of motion may be reduced to the linear wave equation by Loewner-type Bäcklund transformations.

Kazakia and Venkataraman [94], in a study of electromagnetic wave propagation through nonlinear dielectric media, introduced a variant of the Cekirge–Varley procedure wherein the Riemann characteristic equations were shown to be integrable for a certain class of **B-H** and **D-E** constitutive laws. An exact representation of a centered fan which describes the interaction of this wave with an oncoming signal was constructed. In Section 4.4, the work of Kazakia and Venkataraman is presented and set in the broader context of Bäcklund transformation theory.

Bäcklund transformations have also been recently applied to solve a variety of boundary value problems in elastostatics. Clements and Rogers [96] used Loewner-type transformations in the analysis of the stress distribution due to sharply curved notches in a class of shear-strained nonlinear elastic materials. In [97], the same authors noted that Weinstein's correspondence principle may be regarded as a simple Bäcklund transformation of the Stokes–Beltrami equations. The importance of Stokes–Beltrami systems in the theory of torsion of shafts of revolution had been first pointed out by Arndt [316] in 1915. Subsequently, Weinstein [98] noted that in fact such systems also arise in hydrodynamics, transsonic gasdynamics, and electrostatics. A correspondence principle was developed, which, in particular, allows the solution of certain boundary value problems involving cracks and dislocations in linear elastic media [328].

In Chapter 5, the first two sections are concerned, in turn, with the derivation of Weinstein's correspondence principle as a Bäcklund transformation and its application to axially symmetric punch, crack, and torsion problems in elasticity. In Section 5.3, a link is established between Loewner's Bäcklund transformations and the Bergman integral operator method. Application is made to crack and contact problems that involve layered elastic media under mode III displacement. Finally, in Section 5.4, Bäcklund transformations of the Loewner type are employed in the analysis of stress concentration and displacement in a notched half-space of nonlinear elastic material subject to antiplane deformation.

CHAPTER 1

Bäcklund Transformations and Their Application to Nonlinear Equations of Mathematical Physics

1.1 THE ORIGIN AND IMPORTANCE OF BÄCKLUND TRANSFORMATIONS

The classical treatment of the surface transformations which provide the origins of Bäcklund theory is to be found in the work of Lie [103, 104] and Bäcklund [1–5]. The subject was subsequently developed by both Goursat [105] and Clairin [106]. The modern interest in Bäcklund transformations lies in that they may be used in one of two important ways in connection with integral surfaces of certain nonlinear partial differential equations. Thus, invariance under a Bäcklund transformation may be used, under appropriate circumstances, to generate an infinite sequence of solutions of such equations by a *purely algebraic superposition principle*. On the other hand, Bäcklund transformations may sometimes be adduced to link nonlinear equations to *canonical forms* whose properties are well established. Both kinds of Bäcklund

1.1 THE ORIGIN AND IMPORTANCE OF BÄCKLUND TRANSFORMATIONS

transformation have important applications in mathematical physics and continuum mechanics and, accordingly, each will be treated in detail in this monograph.

An extensive bibliography of the early literature on Bäcklund transformations is contained in Goursat's treatise cited above. In this connection, of particular interest is Clairin's method of derivation of Bäcklund transformations. Indeed, Lamb [35] adopted this classical procedure in the construction of Bäcklund transformations for the Korteweg–deVries, modified Korteweg–deVries, and nonlinear Schrödinger equations. However, in the main, classical theory has been subsumed by a jet-bundle formalism to be described at length in the next chapter. Thus, whereas we shall illustrate Clairin's procedure as it applies to the Miura transformation, generally the emphasis will be on modern methods for the construction of Bäcklund transformations. The reader interested in the classical theory may consult a recent monograph by Anderson and Ibragimov [110]. The latter work is devoted to a generalization of Lie's classical theory of contact transformations based on what are termed *Lie–Bäcklund tangent transformations*. An interesting discussion of the early literature on surface transformations in general is contained therein.

Bäcklund, in a study of possible extensions of Lie contact transformations, was led to introduce the important class of surface transformations which bear his name and which, together with their modern extensions, form the basis of the present monograph. The historical progression to their introduction is marked by the key work of Lie [103, 104, 108] and Bäcklund [1–5].[†]

A simple illustration of a Bäcklund transformation and its application is provided by a well-known reduction of Liouville's equation

$$\frac{\partial^2 u}{\partial x^1 \partial x^2} = e^u. \tag{1.1}$$

Thus, consider the relations

$$\frac{\partial u'}{\partial x'^1} = \frac{\partial u}{\partial x^1} + \beta e^{1/2(u+u')} := \mathbb{B}'_1(u, u_1; u'),$$

$$\frac{\partial u'}{\partial x'^2} = -\frac{\partial u}{\partial x^2} - \frac{2}{\beta} e^{1/2(u-u')} := \mathbb{B}'_2(u, u_2; u'), \tag{1.2}$$

$$x'^a = x^a, \quad a = 1, 2,$$

where $\beta \in \mathbb{R}$ is a nonzero constant known as a "Bäcklund parameter."

[†] Additional background material to be found in Bianchi [113–114] and Darboux [115].

Application of the integrability condition[†]

$$\frac{\partial \mathbb{B}'_1}{\partial x'^2} - \frac{\partial \mathbb{B}'_2}{\partial x'^1} = 0, \qquad (1.3)$$

to $(1.2)_{1,2}$ produces Liouville's equation (1.1). On the other hand, if $(1.2)_{1,2}$ are rewritten as

$$\frac{\partial u}{\partial x^1} = \frac{\partial u'}{\partial x'^1} - \beta e^{1/2(u+u')} := \mathbb{B}_1(u', u'_1; u),$$

$$\frac{\partial u}{\partial x^2} = -\frac{\partial u'}{\partial x'^2} - \frac{2}{\beta} e^{1/2(u-u')} := \mathbb{B}_2(u', u'_2; u), \qquad (1.4)$$

then the integrability requirement

$$\frac{\partial \mathbb{B}_1}{\partial x^2} - \frac{\partial \mathbb{B}_2}{\partial x^1} = 0 \qquad (1.5)$$

produces

$$\frac{\partial^2 u'}{\partial x^1 \partial x^2} = 0. \qquad (1.6)$$

Thus, implicit in the set of relations (1.2) is a link between the *nonlinear* equation (1.1) and the *linear* equation (1.6). This connection may be exploited to solve (1.1) in full generality. Thus, insertion of the general solution

$$u' = X^{(1)}(x^1) + X^{(2)}(x^2) \qquad (1.7)$$

of (1.6) into the "Bäcklund relations" $(1.2)_{1,2}$ and subsequent integration produces the general solution of Liouville's equation in the form

$$u = 2 \ln \left[\frac{\exp[(X^{(1)}(x^1) - X^{(2)}(x^2))/2]}{(\beta/2) \int_{x_0^1}^{x^1} \exp[X^{(1)}(\sigma)] \, d\sigma + (1/\beta) \int_{x_0^2}^{x^2} \exp[-X^{(2)}(\tau)] \, d\tau} \right], \qquad (1.8)$$

where x_0^a, $a = 1, 2$, are constants.

The encouraging nature of the above result has led to an extensive search for such transformations for other nonlinear equations of mathematical physics: this with a view to their use in the generation of solutions. It has emerged in recent years that, in fact, a remarkable diversity of important nonlinear equations admit these transformations which, it turns out, are of a type originally introduced by Bäcklund.

[†] It is assumed that u_{12}, u_{21}, u'_{12}, and u'_{21} are continuous.

1.1 THE ORIGIN AND IMPORTANCE OF BÄCKLUND TRANSFORMATIONS

Thus, let $u = u(x^a)$, $u' = u'(x'^a)$, $a = 1, 2$, represent two surfaces Σ and Σ', respectively, in $\mathbb{R}^3$. A set of four relations

$$\mathbb{B}_i^*(x^a, u, u_a; x'^a, u', u'_a) = 0, \quad i = 1, \ldots, 4, \quad a = 1, 2, \tag{1.9}$$

which connect the surface elements $\{x^a, u, u_a\}$ and $\{x'^a, u', u'_a\}$ of Σ and Σ', respectively, is termed a *Bäcklund transformation* in the classical literature.[†] In contemporary applications, the particular interest is when $u = u(x^a)$, $u' = u'(x'^a)$ represent integral surfaces of partial differential equations of significance in mathematical physics and continuum mechanics.

If (1.9) admits the explicit resolutions

$$u'_i = \mathbb{B}'_i(x^a, u, u_a; u'), \quad i = 1, 2, \tag{1.10}$$

and

$$u_i = \mathbb{B}_i(x'^a, u', u'_a; u), \quad i = 1, 2, \tag{1.11}$$

together with

$$x'^j = X^j(x^a, u, u_a; u'), \quad j = 1, 2, \tag{1.12}$$

then, in order that these relations transform a surface $u = u(x^a)$ with surface element $\{x^a, u, u_a\}$ to a surface $u' = u'(x'^a)$ with surface element $\{x'^a, u', u'_a\}$, it is required that the relations

$$du - \mathbb{B}_a\, dx^a = 0, \tag{1.13}$$

$$du' - \mathbb{B}'_a\, dx'^a = 0, \tag{1.14}$$

be integrable. Hence, we obtain the conditions

$$\frac{\partial \mathbb{B}_1}{\partial x^2} - \frac{\partial \mathbb{B}_2}{\partial x^1} = 0, \tag{1.15}$$

$$\frac{\partial \mathbb{B}'_1}{\partial x'^2} - \frac{\partial \mathbb{B}'_2}{\partial x'^1} = 0. \tag{1.16}$$

Application of (1.16) to the Bäcklund relations (1.10) and (1.12) is shown by Forsyth [107] to lead to a nonlinear equation of the form

$$U\{u_{11}u_{22} - u_{12}^2\} + Ru_{11} + 2Su_{21} + Tu_{22} + V = 0, \tag{1.17}$$

where U, R, S, T, and V, in general, involve the quantities x^a, u, u_i, and u' but not partial derivatives in u of second or higher order. Thus, if u' is absent in (1.17), the well-known Monge–Ampère form is obtained.

[†] The modern notion of a Bäcklund transformation in terms of a jet-bundle formalism is developed in Chapter 2.

In a similar manner, application of (1.15) to the Bäcklund relations (1.11) and (1.12) leads to a primed counterpart of (1.17). In particular, if the equations for u and u' so derived are both of the Monge–Ampère form, then the Bäcklund transformation may be regarded as a mapping between their integral surfaces.

The question as to what types of Monge–Ampère equations admit Bäcklund transformations was addressed in [107], where it was shown for the case in which the $\mathbb{B}_i^j$ or X^j in (1.10) and (1.12) are independent of u' that Bäcklund transformations are not available for the most general such equation. However, there remain certain Monge–Ampère equations of significance which do possess Bäcklund transformations, notably Liouville's equation, the sine-Gordon equation, and Burgers' equation. The Bäcklund transformation for Liouville's equation has already been exhibited while those for the sine-Gordon and Burgers' equations will be dealt with in subsequent sections of this chapter.

While the classical procedures may be extended in a number of ways (see, for example, Goursat [109] and Loewner [64, 65]), in the main, alternative methods for the construction of Bäcklund transformations have been adopted in recent developments. It has already been pointed out that an exception is the work by Lamb [35], wherein Clairin's method is used to generate Bäcklund transformations for higher order nonlinear evolution equations. As an illustration of this classical approach, in Section 1.9 we follow Lamb in our derivation of Miura's transformation for the Korteweg–deVries equation. Additional aspects of classical theory have been discussed by Dodd and Bullough [112] in connection with the possibility of the existence of Bäcklund transformations for higher dimensional sine-Gordon equations.

In the present monograph, we shall be concerned mainly with the modern methods of construction of Bäcklund transformations. These are associated with the inverse scattering method, Hirota's technique, and the Estabrook–Wahlquist procedure.

1.2 THE 1 + 1 SINE-GORDON EQUATION. THE PERMUTABILITY THEOREM. BIANCHI DIAGRAMS

The sine-Gordon equation

$$u_{12} = \sin u, \qquad (1.18)$$

models transmission in a wide variety of areas of physical interest, notably, the propagation of quantized flux in Josephson junctions [6–9, 157–159], crystal dislocation theory [10, 160–162], elementary particle theory [11–14, 163], splay wave propagation along lipid membranes [164], the analysis of mechanical modes of nonlinear wave propagation [15], the motion of a Bloch wall between ferromagnetic domains [16, 165, 166], nuclear magnetic resonance [167], and massive Thirring model theory [23, 24]. In an interesting recent development, the occurrence of the $1 + 1$ sine-Gordon equation in the description of the nonlinear evolution of wave packets in rotating baroclinic shear flow has been documented by Gibbon et al. [173]. However, it was the relevance of (1.18) to the propagation of ultrashort optical pulses in resonant laser media that provided the mainspring for a renewal of interest in Bäcklund transformation theory. Thus, Lamb [17] reintroduced a classical invariant transformation of (1.18) of a Bäcklund type and employed it in an elegant manner to generate multi-soliton solutions by an iterative procedure.

Here the classical auto-Bäcklund transformation of (1.18) is presented together with important consequences of this invariance. Thus, a nonlinear superposability theorem is established whereby a sequence of soliton solutions of the $1 + 1$ sine-Gordon equation may be constructed by *purely algebraic means*.

The basic invariance property is embodied in the following result:

THEOREM 1.1   The sine-Gordon equation

$$u_{12} = \sin u$$

is invariant under the Bäcklund transformation

$$u'_1 = \mathbb{B}'_1(u, u_1; u') = u_1 - 2\beta \sin\left\{\frac{u + u'}{2}\right\},$$

$$u'_2 = \mathbb{B}'_2(u, u_2; u') = -u_2 + \frac{2}{\beta} \sin\left\{\frac{u - u'}{2}\right\}, \qquad (1.19)$$

$$x'^a = x^a, \qquad a = 1, 2,$$

where $\beta \in \mathbb{R}$ is a nonzero Bäcklund parameter. ■

The above result is readily demonstrated. Thus, application of the integrability condition

$$\frac{\partial \mathbb{B}'_1}{\partial x^2} - \frac{\partial \mathbb{B}'_2}{\partial x^1} = 0$$

in (1.19) shows that

$$u_{12} - \beta(u_2 + u'_2)\cos\left\{\frac{u + u'}{2}\right\} - \left[-u_{12} + \frac{1}{\beta}(u_1 - u'_1)\cos\left\{\frac{u - u'}{2}\right\}\right] = 0,$$

whence

$$u_{12} = \sin u.$$

1. BÄCKLUND TRANSFORMATIONS AND NONLINEAR EQUATIONS

Similarly, application of the integrability requirement

$$\frac{\partial \mathbb{B}_1}{\partial x^2} - \frac{\partial \mathbb{B}_2}{\partial x^1} = 0,$$

where $(1.19)_{1,2}$ are rewritten as

$$u_1 = \mathbb{B}_1(u', u'_1; u) = u'_1 + 2\beta \sin\left\{\frac{u+u'}{2}\right\},$$

$$u_2 = \mathbb{B}_2(u', u'_2; u) = -u'_2 + \frac{2}{\beta} \sin\left\{\frac{u-u'}{2}\right\},$$

yields

$$u'_{12} = \sin u'. \tag{1.20}$$

Hence, relations (1.19) leave (1.18) invariant, that is, they constitute an auto-Bäcklund transformation of the sine-Gordon equation. ∎

The relevance of the above Bäcklund transformation to the geometry of pseudospherical surfaces is described by Eisenhart [168] while the derivation of the Bäcklund transformation by classical procedures is given by Ames [169].

Note that both the Liouville and sine-Gordon equations are of the hyperbolic form[†]

$$u_{12} = \Phi(u). \tag{1.21}$$

Accordingly, it is natural to inquire as to what restrictions must be placed on Φ in order that (1.21) admit a Bäcklund transformation. McLaughlin and Scott [170] investigated this question via a classical approach. A treatment based on the jet bundle formalism of Chapter 2, wherein the Bäcklund transformations for both (1.1) and (1.18) arise in a natural manner, is given by Shadwick [171].

We now turn to the application of the Bäcklund transformation (1.19). Its power lies in that it may be used to generate additional solutions of the second-order nonlinear equation (1.18) via the pair of first-order equations $(1.19)_{1,2}$ on insertion of a known solution. Thus, for instance, $u' = 0$ is a trivial "vacuum" solution of (1.20). Substitution of this solution into $(1.19)_{1,2}$ indicates that a second but nontrivial solution may be constructed by integration of the pair of equations

$$u_1 = 2\beta \sin\left\{\frac{u}{2}\right\}, \qquad u_2 = \frac{2}{\beta} \sin\left\{\frac{u}{2}\right\}, \tag{1.22}$$

[†] There are, of course, elliptic analogs of both (1.1) and (1.18). A Bäcklund transformation for an elliptic version of the sine-Gordon equation is presented in Section 1.4.

1.2 SINE-GORDON EQUATION; PERMUTABILITY THEOREMS

leading to

$$u = 4\tan^{-1}\{\exp(\beta x^1 + \beta^{-1} x^2 + \bar{\alpha})\}, \qquad (1.23)$$

where $\bar{\alpha}$ is a constant of integration.

Let ϕ_n, ψ_n be solutions of (1.18) generated by application of the Bäcklund transformation (1.19) to a known solution ϕ_{n-1} with, in turn, the Bäcklund parameters β_1 and β_2. Further, let ϕ_{n+1} and ψ_{n+1} denote solutions of (1.18) obtained by application of the Bäcklund transformation with parameter β_2 to ϕ_n and with parameter β_1 to ψ_n, respectively. The situation may be represented schematically by a "Bianchi diagram" (Fig. 1.1). In the latter, the b_i, b_i', $i = 1, 2$, represent arbitrary constants of integration.

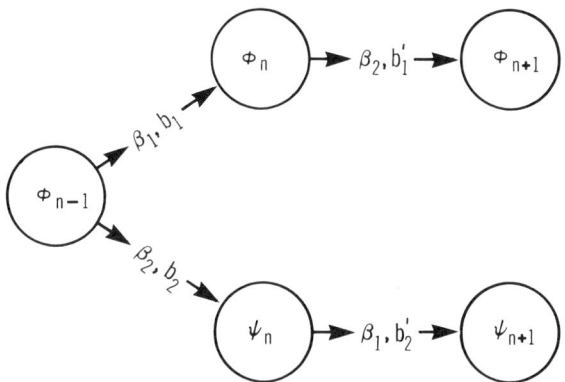

Fig. 1.1 A Bianchi diagram. (McLaughlin and Scott [170].)

Once a second solution of (1.18) such as (1.23) has been derived, that an infinite sequence of additional solutions may be generated without further recourse to integration is a consequence of the following remarkable result:

THEOREM 1.2 There exist ϕ_{n+1} and ψ_{n+1} such that

$$\phi_{n+1} = \psi_{n+1} = \phi,$$

where

$$\phi = 4\tan^{-1}\left[\left(\frac{\beta_1 + \beta_2}{\beta_1 - \beta_2}\right)\tan\left\{\frac{\phi_n - \psi_n}{4}\right\}\right] + \phi_{n-1}. \quad \blacksquare$$

In order to establish the above result, we follow the argument developed by McLaughlin and Scott [170]. Thus, on appropriate use of the Bäcklund relations (1.19), the Bianchi diagram (Fig. 1.1) indicates that [†]

$$\phi_{n,x^1} = \phi_{n-1,x^1} + 2\beta_1 \sin\left\{\frac{\phi_n + \phi_{n-1}}{2}\right\}, \tag{1.24}$$

$$\phi_{n,x^2} = -\phi_{n-1,x^2} + \frac{2}{\beta_1}\sin\left\{\frac{\phi_n - \phi_{n-1}}{2}\right\}, \tag{1.25}$$

$$\phi_{n+1,x^1} = \phi_{n,x^1} + 2\beta_2 \sin\left\{\frac{\phi_{n+1} + \phi_n}{2}\right\}, \tag{1.26}$$

$$\phi_{n+1,x^2} = -\phi_{n,x^2} + \frac{2}{\beta_2}\sin\left\{\frac{\phi_{n+1} - \phi_n}{2}\right\}, \tag{1.27}$$

$$\psi_{n,x^1} = \phi_{n-1,x^1} + 2\beta_2 \sin\left\{\frac{\psi_n + \phi_{n-1}}{2}\right\}, \tag{1.28}$$

$$\psi_{n,x^2} = -\phi_{n-1,x^2} + \frac{2}{\beta_2}\sin\left\{\frac{\psi_n - \phi_{n-1}}{2}\right\}, \tag{1.29}$$

$$\psi_{n+1,x^1} = \psi_{n,x^1} + 2\beta_1 \sin\left\{\frac{\psi_{n+1} + \psi_n}{2}\right\}, \tag{1.30}$$

$$\psi_{n+1,x^2} = -\psi_{n,x^2} + \frac{2}{\beta_1}\sin\left\{\frac{\psi_{n+1} - \psi_n}{2}\right\}. \tag{1.31}$$

Now, (1.24) + (1.26) and (1.27) − (1.25) yield, in turn

$$\phi_{n+1,x^1} = \phi_{n-1,x^1} + 2\beta_1 \sin\left\{\frac{\phi_n + \phi_{n-1}}{2}\right\} + 2\beta_2 \sin\left\{\frac{\phi_{n+1} + \phi_n}{2}\right\} \tag{1.32}$$

and

$$\phi_{n+1,x^2} = \phi_{n-1,x^2} - \frac{2}{\beta_1}\sin\left\{\frac{\phi_n - \phi_{n-1}}{2}\right\} + \frac{2}{\beta_2}\sin\left\{\frac{\phi_{n+1} - \phi_n}{2}\right\}, \tag{1.33}$$

while (1.28) + (1.30) and (1.31) − (1.29) show that

$$\psi_{n+1,x^1} = \phi_{n-1,x^1} + 2\beta_1 \sin\left\{\frac{\psi_{n+1} + \psi_n}{2}\right\} + 2\beta_2 \sin\left\{\frac{\psi_n + \phi_{n-1}}{2}\right\} \tag{1.34}$$

and

$$\psi_{n+1,x^2} = \phi_{n-1,x^2} - \frac{2}{\beta_2}\sin\left\{\frac{\psi_n - \phi_{n-1}}{2}\right\} + \frac{2}{\beta_1}\sin\left\{\frac{\psi_{n+1} - \psi_n}{2}\right\}, \tag{1.35}$$

respectively.

[†] Thus, for example, to obtain (1.24) and (1.25) we set $u' = \phi_{n-1}$, $u = \phi_n$, and $\beta = \beta_1$ in (1.19)$_{1,2}$.

1.2 SINE-GORDON EQUATION; PERMUTABILITY THEOREMS

Thus, if we require

$$\phi_{n+1} = \psi_{n+1} = \phi, \tag{1.36}$$

we see that relations (1.32)–(1.35) provide two necessary consistency requirements, namely,

$$\beta_1 \sin\left\{\frac{\phi_n + \phi_{n-1}}{2}\right\} + \beta_2 \sin\left\{\frac{\phi + \phi_n}{2}\right\}$$
$$= \beta_1 \sin\left\{\frac{\phi + \psi_n}{2}\right\} + \beta_2 \sin\left\{\frac{\psi_n + \phi_{n-1}}{2}\right\} \tag{1.37}$$

and

$$\beta_1 \sin\left\{\frac{\phi - \phi_n}{2}\right\} - \beta_2 \sin\left\{\frac{\phi_n - \phi_{n-1}}{2}\right\}$$
$$= -\beta_1 \sin\left\{\frac{\psi_n - \phi_{n-1}}{2}\right\} + \beta_2 \sin\left\{\frac{\phi - \psi_n}{2}\right\}. \tag{1.38}$$

If we now define

$$\Sigma_\pm := \tfrac{1}{4}[\phi + \phi_{n-1} \pm (\psi_n + \phi_n)], \tag{1.39}$$

$$\theta_\pm := \tfrac{1}{4}[\phi - \phi_{n-1} \pm (\psi_n - \phi_n)], \tag{1.40}$$

then the conditions (1.37) and (1.38) may be written as

$$\beta_1 \sin(\Sigma_+ - \theta_+) + \beta_2 \sin(\Sigma_+ + \theta_-) = \beta_1 \sin(\Sigma_+ + \theta_+) + \beta_2 \sin(\Sigma_+ - \theta_-), \tag{1.41}$$
$$\beta_1 \sin(\Sigma_- + \theta_+) + \beta_2 \sin(\Sigma_- - \theta_-) = \beta_1 \sin(\Sigma_- - \theta_+) + \beta_2 \sin(\Sigma_- + \theta_-), \tag{1.42}$$

whence, on reduction,

$$\{\beta_1 \sin \theta_+ - \beta_2 \sin \theta_-\} \cos \Sigma_+ = 0, \tag{1.43}$$
$$\{\beta_1 \sin \theta_+ - \beta_2 \sin \theta_-\} \cos \Sigma_- = 0. \tag{1.44}$$

Hence the consistency conditions (1.37) and (1.38) are satisfied if

$$\beta_1 \sin \theta_+ = \beta_2 \sin \theta_-. \tag{1.45}$$

The relationship (1.45) may be solved explicitly for ϕ in terms of ϕ_{n-1}, ϕ_n, and ψ_n. Thus, (1.45) implies that

$$\beta_1 \sin \tfrac{1}{4}[\phi - \phi_{n-1} + (\psi_n - \phi_n)] = \beta_2 \sin \tfrac{1}{4}[\phi - \phi_{n-1} - (\psi_n - \phi_n)],$$

whence,

$$\beta_1[\sin \tfrac{1}{4}(\phi - \phi_{n-1}) \cos \tfrac{1}{4}(\psi_n - \phi_n) + \cos \tfrac{1}{4}(\phi - \phi_{n-1}) \sin \tfrac{1}{4}(\psi_n - \phi_n)]$$
$$= \beta_2[\sin \tfrac{1}{4}(\phi - \phi_{n-1}) \cos \tfrac{1}{4}(\psi_n - \phi_n) - \cos \tfrac{1}{4}(\phi - \phi_{n-1}) \sin \tfrac{1}{4}(\psi_n - \phi_n)],$$

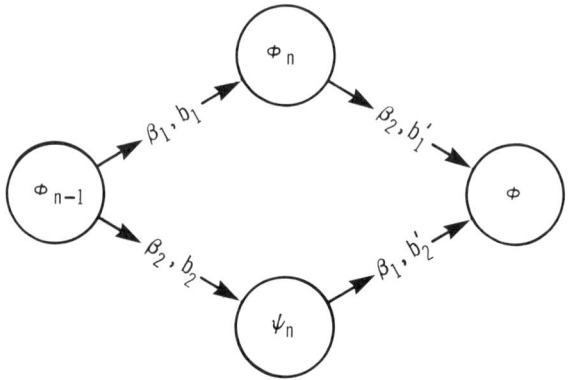

Fig. 1.2 A commutative Bianchi diagram. (McLaughlin and Scott [170].)

or

$$\beta_1[\tan\tfrac{1}{4}(\phi - \phi_{n-1}) + \tan\tfrac{1}{4}(\psi_n - \phi_n)]$$
$$= \beta_2[\tan\tfrac{1}{4}(\phi - \phi_{n-1}) - \tan\tfrac{1}{4}(\psi_n - \phi_n)].$$

Accordingly,

$$\phi = 4\tan^{-1}\left[\left(\frac{\beta_1 + \beta_2}{\beta_1 - \beta_2}\right)\tan\left\{\frac{\phi_n - \psi_n}{4}\right\}\right] + \phi_{n-1}. \qquad (1.46)$$

It may now readily be verified that the insertion of $\phi_{n+1} = \phi$ in (1.26) and (1.27) and $\psi_{n+1} = \phi$ in (1.30) and (1.31) satisfies these Bäcklund relations. This completes the demonstration. ∎

Theorem 1.2 has the important consequence that a commutative Bianchi diagram may now be constructed as in Fig. 1.2. Iteration of the Bäcklund transformation may be used to generate a Bianchi lattice whereby an infinite sequence of additional solutions of the 1 + 1 sine-Gordon equation may be generated without further integration. This procedure is adopted in the next section to construct analytical expressions for $2N\pi$ pulses in an unbroadened resonant medium.

1.3 ULTRASHORT OPTICAL PULSE PROPAGATION IN A RESONANT MEDIUM. $2N\pi$ LIGHT PULSES

The equations of ultrashort pulse propagation under the assumption of vanishing bandwidth and with neglect of nonresonant losses reduce to consideration of (Lamb [18], Barnard [22])

$$\frac{\partial \bar{\varepsilon}}{\partial t} + C\frac{\partial \bar{\varepsilon}}{\partial x} = \frac{2\pi N_0 \omega p^2}{\hbar}\mathbb{P}, \qquad (1.47)$$

1.3 OPTICAL PULSE PROPAGATION; $2N\pi$ LIGHT PULSES

$$\frac{\partial \mathbb{P}}{\partial t} = \bar{\varepsilon}\mathbb{N}, \tag{1.48}$$

$$\frac{\partial \mathbb{N}}{\partial t} = -\bar{\varepsilon}\mathbb{P}, \tag{1.49}$$

where the optical field $E(x, t)$ and the macroscopic polarization density $P(x, t)$ are expressed in terms of their envelopes and carriers as

$$E(x, t) = (\bar{\varepsilon}\hbar/p)\cos(kx - \omega t), \tag{1.50}$$

$$P(x, t) = N_0 p \mathbb{P}(x, t)\sin(kx - \omega t), \tag{1.51}$$

where N_0 is the number of active atoms per unit volume; p is the dipole matrix element; $\mathbb{N} = N/N_0$, where N is the population inversion, and c is the phase velocity of light in the medium.

Integration of (1.48), (1.49) for uniform initial population of the active medium yields

$$\mathbb{N}(x, t) = \pm\cos\sigma, \tag{1.52}$$

$$\mathbb{P}(x, t) = \pm\sin\sigma, \tag{1.53}$$

where

$$\bar{\varepsilon} = \partial\sigma/\partial t. \tag{1.54}$$

Substitution of (1.54) into (1.47) now gives the $1 + 1$ sine-Gordon equation

$$\frac{\partial^2 \sigma}{\partial \xi \partial \tau} = \pm\sin\sigma, \tag{1.55}$$

where $\xi = \alpha x/c$, $\tau = \alpha(t - x/c)$, and

$$\alpha = (2\pi N_0 \omega p^2/\hbar)^{1/2}. \tag{1.56}$$

Thus, if we set $x^1 = \tau$, $x^2 = \pm\xi$, $u = \sigma$ in (1.55), we obtain (1.18).

It is a remarkable property of the Bäcklund transformation, established in the previous section, that it may be used to construct solutions of (1.55) which correspond to ultrashort light pulses that propagate in a resonant medium with *conservation of pulse area*. Specifically, a pulse which travels in such a manner that the total area under the pulse envelope remains at $2N\pi$, so that

$$\int_{-\infty}^{\infty} \bar{\varepsilon}(x, t) \, dt = 2N\pi, \tag{1.57}$$

is called a $2N\pi$ *pulse*. The permutability theorem generates analytic expressions for such pulses by an iterative procedure. Furthermore, these solutions exhibit a distinctive property for such pulses that has been observed experimentally, namely, their decomposition into N stable 2π pulses. The reader interested in experimental evidence for this phenomenon should consult, for example, the work by Gibbs and Slusher [116] concerning the observed decomposition of a 6π pulse into three 2π pulses in a Rb vapor.

We now proceed to describe the generation of $2N\pi$ pulse solutions via the Bianchi diagrams associated with the Bäcklund method.

2π Pulse

The best known solution of (1.55) is that associated with the propagation characteristic of self-induced transparency in an attenuator. This may be obtained as a specialization of the solution (1.23) generated by the Bäcklund transformation (1.19) out of the vacuum solution. Thus, if the negative sign is chosen in (1.55) as is required for propagation in an attenuator, on omission of the translation constant $\bar{\alpha}$, we obtain

$$\sigma = 4\tan^{-1}\{e^{a\tau - (\xi/a)}\}, \tag{1.58}$$

where a is the Bäcklund parameter. The associated electric field envelope is given by

$$\bar{\varepsilon} = 2a\alpha\,\text{sech}[a\alpha(t - x/V_p)], \tag{1.59}$$

where

$$V_p = c/[1 + a^{-2}], \tag{1.60}$$

is the envelope velocity.

4π Pulse

According to both experimental and numerical evidence, a 4π pulse does not propagate as a single pulse but rather, separates into two distinct 2π pulses. Such decomposition is also a property of the associated solution as derived by the Bäcklund method. Thus, use of the permutability theorem with starting solution $\sigma_0 = 0$ and first-generation solutions

$$\sigma_i = 4\tan^{-1}\{e^{v_i}\}, \quad i = 1, 2 \quad (v_i = a_i\tau - \xi/a_i), \tag{1.61}$$

corresponding to a single application of the Bäcklund transformation with parameters $a = a_i$, $i = 1, 2$, produces the second-generation solution

$$\begin{aligned}\sigma_3 &= 4\tan^{-1}\left[\left(\frac{a_1 + a_2}{a_1 - a_2}\right)\tan\tfrac{1}{4}(\sigma_1 - \sigma_2)\right], \\ &= 4\tan^{-1}\left[\left(\frac{a_1 + a_2}{a_1 - a_2}\right)\frac{\sinh\tfrac{1}{2}(v_1 - v_2)}{\cosh\tfrac{1}{2}(v_1 + v_2)}\right].\end{aligned} \tag{1.62}$$

The procedure is illustrated by a Bianchi diagram (Fig. 1.3).

1.3 OPTICAL PULSE PROPAGATION; $2N\pi$ LIGHT PULSES

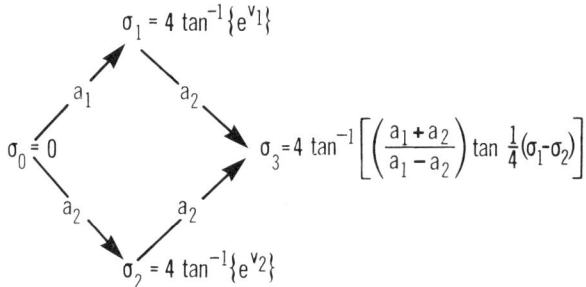

Fig. 1.3 Generation of the 4π pulse solution via the Bianchi diagram.

For $a_1 > -a_2 > 0$, σ_3 varies between -2π and 2π as τ proceeds from $-\infty$ to $+\infty$. Accordingly, the area under the associated pulse envelope is given by

$$\int_{-\infty}^{\infty} \bar{\varepsilon}\, dt = \sigma_3(\infty) - \sigma_3(-\infty) = 4\pi,$$

as is required for a 4π pulse.

Use of (1.54) produces the electric field

$$\bar{\varepsilon} = \left(\frac{a_1^2 - a_2^2}{a_1^2 + a_2^2}\right)\left[\frac{2a_1\alpha \operatorname{sech} X_1 - 2a_2\alpha \operatorname{sech} X_2}{1 - A(\tanh X_1 \tanh X_2 - \operatorname{sech} X_1 \operatorname{sech} X_2)}\right], \quad (1.63)$$

where

$$X_1 = a_1\alpha(t - x/V_1), \quad (1.64)$$

$$X_2 = -a_2\alpha(t - x/V_2), \quad (1.65)$$

$$A = -2a_1 a_2/(a_1^2 + a_2^2), \quad (1.66)$$

while the velocities V_i are given by

$$V_i = C_i/(1 + a_i^{-2}), \quad i = 1, 2. \quad (1.67)$$

A graph of $\alpha^{-1}\bar{\varepsilon}$ against $a_1^{-1} X_1$ is shown in Fig. 1.4. The solution is seen to evolve into two steady 2π pulses according to

$$\bar{\varepsilon} = 2a_1\alpha \operatorname{sech}(X_1 \pm \beta) - 2a_2\alpha \operatorname{sech}(X_2 \mp \beta), \quad (1.68)$$

where

$$\beta = \tanh^{-1} A, \quad (1.69)$$

and the upper sign in (1.68) is to be taken when $1/a_1 < -1/a_2$ and the lower sign when $1/a_1 > -1/a_2$.

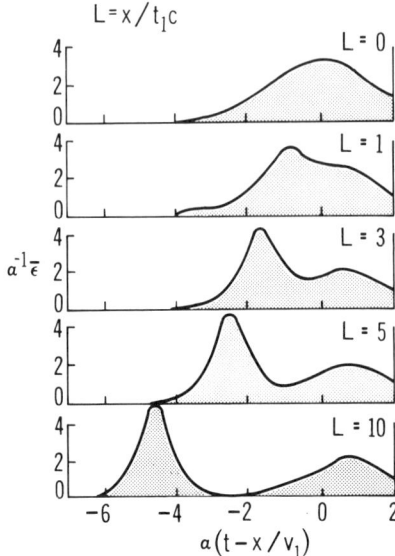

Fig. 1.4 Breakup of 4π pulse. (Lamb [21].)

6π Pulse

A 6π pulse solution may be obtained as a third-generation solution $\sigma_3^{(1)}$ of the Bäcklund procedure as indicated in Fig. 1.5. According to the permutability theorem,

$$\sigma_3^{(1)} = \sigma_1^{(2)} + 4\tan^{-1}\left[\left(\frac{a_1 + a_3}{a_1 - a_3}\right)\tan\frac{1}{4}(\sigma_2^{(1)} - \sigma_2^{(2)})\right], \quad (1.70)$$

where

$$\sigma_1^{(2)} = 4\tan^{-1} e^{v_2}, \quad (1.71)$$

$$\sigma_2^{(1)} = 4\tan^{-1}\left[\left(\frac{a_1 + a_2}{a_1 - a_2}\right)\frac{\sinh\frac{1}{2}(v_1 - v_2)}{\cosh\frac{1}{2}(v_1 + v_2)}\right], \quad (1.72)$$

$$\sigma_2^{(2)} = 4\tan^{-1}\left[\left(\frac{a_2 + a_3}{a_2 - a_3}\right)\frac{\sinh\frac{1}{2}(v_2 - v_3)}{\cosh\frac{1}{2}(v_2 + v_3)}\right]. \quad (1.73)$$

The constraints $a_1 < 0$, $a_2 > 0$, $0 < a_3 < a_2$ on the Bäcklund parameters a_i, $i = 1, 2, 3$, may be shown to generate a 6π pulse (Lamb [21]). The decomposition of such a pulse is illustrated in Fig. 1.6.

As noted earlier, the decomposition of a 6π pulse into three 2π pulses has been observed in Rb vapor by Gibbs and Slusher [116], who obtained pulse profiles similar to those shown in Fig. 1.6.

1.3 OPTICAL PULSE PROPAGATION; $2N\pi$ LIGHT PULSES

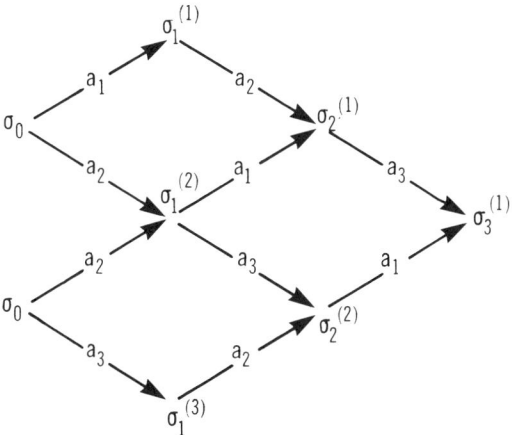

Fig. 1.5 The Bianchi diagram associated with the 6π pulse solution.

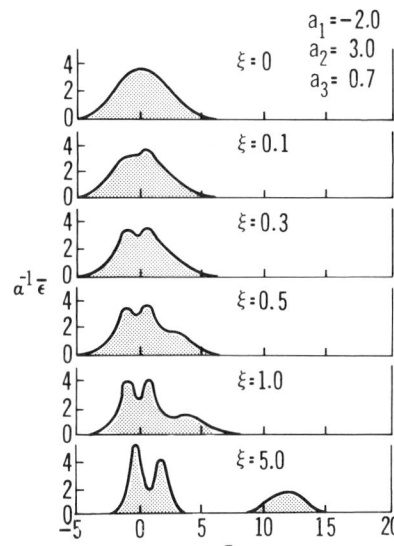

Fig. 1.6 Decomposition of a 6π pulse. (Lamb [21].)

$2N\pi$ Pulse

On introduction of the notation

$$\psi_{k,j+1} = \mathbb{B}_{a_j}\psi_{k,j}, \tag{1.74}$$

$$\psi_{k+1,j} = \mathbb{B}_{a_k}\psi_{k,j} \tag{1.75}$$

where $\mathbb{B}_{a_j}$ is the Bäcklund transformation associated with the parameter a_j, iteration of the procedure indicated above may be used to generate

28 1. BÄCKLUND TRANSFORMATIONS AND NONLINEAR EQUATIONS

expressions $\sigma = \psi_{i,j}$ for $2N\pi$ pulses via the following relations (Barnard [22]):

$$\psi_{i+1,i} = 0, \quad \psi_{i,i} = 4\tan^{-1} e^{v_i + \gamma_i}, \quad (-1)^i a_i < 0, \quad i > 0, \quad (1.76)$$

$$\begin{aligned}\psi_{i,j} &= \psi_{i+1,j-1} + 4\tan^{-1}[k_{ij}\tan\tfrac{1}{4}(\psi_{i,j-1} - \psi_{i+1,j})], \\ k_{ij} &= (a_i + a_j)/(a_i - a_j), \quad (-1)^i a_i < (-1)^j a_j, \quad j > i,\end{aligned} \quad (1.77)$$

while the associated electric field amplitude is given by

$$\bar{\varepsilon}_{ij} = \bar{\varepsilon}_{i+1,j-1} + k_{ij}\frac{1+g_{ij}^2}{1+k_{ij}^2 g_{ij}^2}\{\bar{\varepsilon}_{i,j-1} - \bar{\varepsilon}_{i+1,j}\}, \quad j > i, \quad (1.78)$$

where

$$g_{ij} = \tan[\tfrac{1}{4}(\psi_{i,j-1} - \psi_{i+1,j})]. \quad (1.79)$$

In the above, γ_i is an arbitrary constant of integration associated with the initial location of the first generation solution $\psi_{i,i}$.

The Bianchi diagram descriptive of the generation of the 12π pulse solution is given in Fig. 1.7. The areas of the pulse envelopes associated with the intermediate solutions are indicated. The decomposition of the 12π pulse solution as given by the relations (1.76)–(1.79) is exhibited in Fig. 1.8.

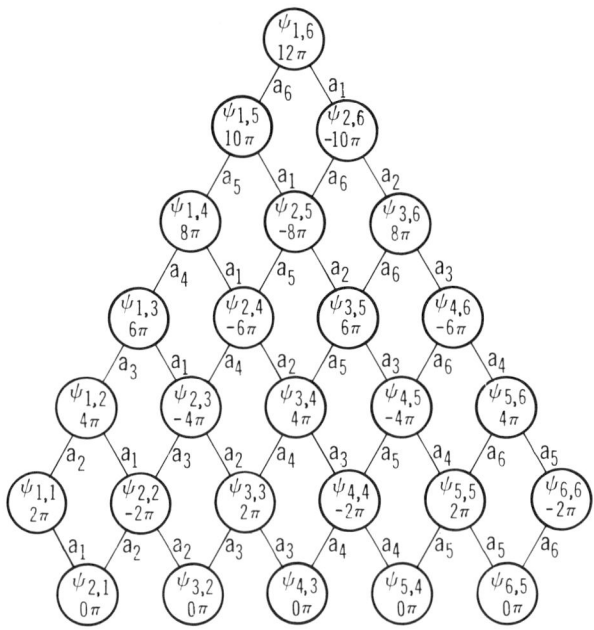

Fig. 1.7 Sequence of Bäcklund transformations for generation of 12π pulse. (Barnard [22].)

1.4 LONG JOSEPHSON JUNCTION: ELLIPTIC SINE-GORDON EQUATION

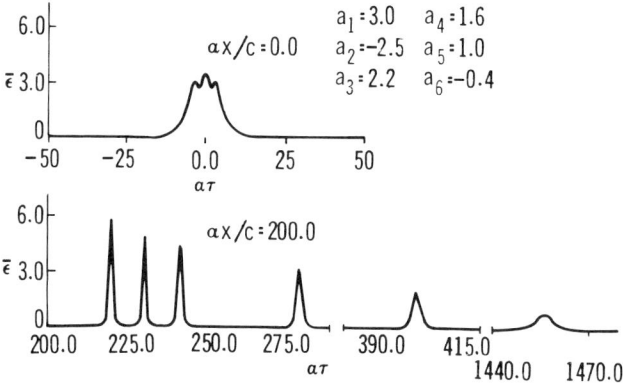

Fig. 1.8 12π pulse decomposition. (Barnard [22].)

1.4 PROPAGATION OF MAGNETIC FLUX THROUGH A LONG JOSEPHSON JUNCTION. AN ELLIPTIC ANALOG OF THE 1 + 1 SINE-GORDON EQUATION. GENERATION OF SOLUTIONS VIA A BÄCKLUND TRANSFORMATION

The Bäcklund transformation treated in the two preceding sections is such that it leaves the 1 + 1 sine-Gordon equation invariant. A Bäcklund transformation has been introduced recently by Leibbrandt [25–27] for an elliptic analog of this 1 + 1 equation. This Bäcklund transformation however, rather than leaving the nonlinear equation invariant, maps it to an associated nonlinear equation. A permutability theorem is available for the construction of solutions to both nonlinear equations.

We commence our discussion of the two-dimensional analog of (1.18) with its derivation in connection with the steady propagation of magnetic flux along a long Josephson junction (Leibbrandt [25]).

A Josephson junction consists of two superconducting metals separated by a very thin nonsuperconducting barrier, here taken to occupy the region

$$\{-\infty < x^1 < +\infty\} \times \{-\infty < x^2 < +\infty\} \times \{-b/2 \leq x^3 \leq b/2\}.$$

The two superconductors are assumed to occupy the regions $x^3 < -b/2$ and $x^3 > b/2$ (see Fig. 1.9).

The key quantity in the macroscopic description of the Josephson effect is the *relative phase* $\phi^* := \phi_1 - \phi_2$ between the superconducting metals I, II which causes the *Josephson tunneling current* to flow across the nonsuperconducting barrier.

If $\mathbf{H} = (H^1, H^2)$ is the magnetic field strength (assumed to be independent of x^3 and t) and if $V(x^a, t)$, $a = 1, 2$, is the potential difference across the

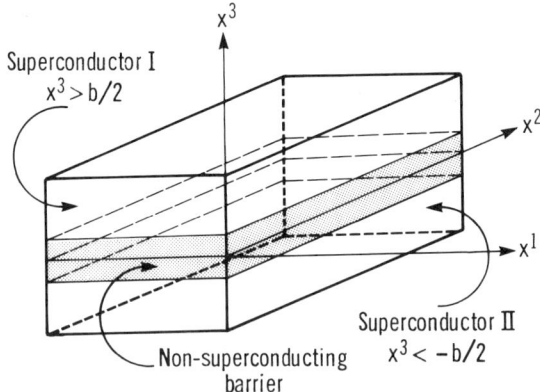

Fig. 1.9 A Josephson junction.

barrier, then the basic equations descriptive of the Josephson effect are (Josephson [6])

$$\partial \phi^*/\partial t = 2eV/h, \tag{1.80}$$

$$\partial \phi^*/\partial x^1 = (2ed/hc)H^2, \tag{1.81}$$

$$\partial \phi^*/\partial x^2 = (-2ed/hc)H^1, \tag{1.82}$$

together with

$$j_3 = j_0(x^1, x^2) \sin \phi^*, \tag{1.83}$$

where the latter gives the Josephson current per unit area across the barrier. Here, e is the electron charge, c the speed of light *in vacuo* while $d = \lambda_1 + \lambda_2 + b$, where λ_1, λ_2 denote, in turn, the London penetration depths for metals I and II; j_0 depends on the properties of the barrier.

Substitution of (1.80)–(1.83) into the Maxwell equation

$$\frac{\partial H^2}{\partial x^1} - \frac{\partial H^1}{\partial x^2} - \frac{4\pi}{c} C \frac{\partial V}{\partial t} = \frac{4\pi}{c} j_3 \tag{1.84}$$

produces the *barrier equation*

$$\nabla^2 \phi^* - \frac{1}{c_0^2} \frac{\partial^2 \phi^*}{\partial t^2} = \lambda_J^{-2} \sin \phi^*, \tag{1.85}$$

where

$$c_0 = c(4\pi Cd)^{-1/2}, \tag{1.86}$$

$$\lambda_J = c[h/(8\pi edj_0)]^{1/2} \tag{1.87}$$

denote, respectively, the speed of electromagnetic waves along the surface of the barrier in the absence of the Josephson current and a measure of the

1.4 LONG JOSEPHSON JUNCTION: ELLIPTIC SINE-GORDON EQUATION

Josephson penetration depth; C denotes the capacitance of the barrier per unit area. Here, and in the sequel, ∇^2 designates the two-dimensional Laplace operator.

Introduction of the scaling

$$\bar{x}^a = x^a/\lambda_J, \quad a = 1, 2,$$

reduces (1.85), in the static case, to the elliptic sine-Gordon equation investigated by Leibbrandt. Solutions of this nonlinear equation here generated by the Bäcklund method were termed *fluxons* by Leibbrandt since they can be shown to carry an integral number of flux quanta through the Josephson junction.

The method of generation of fluxon solutions is a consequence of the following result:

THEOREM 1.3 The equation

$$\nabla^2 \alpha = \sin\{\alpha + v^1\} \tag{1.88}$$

is mapped to the equation

$$\nabla^2 \beta = \sinh\{\beta + v^2\} \tag{1.89}$$

by the Bäcklund transformation

$$\frac{\partial \alpha}{\partial x^1} = \mathbb{B}_1(\alpha; \beta, \frac{\partial \beta}{\partial x^2})$$

$$= \frac{-\partial v^1}{\partial x^1} - \frac{\partial}{\partial x^2}\{\beta + v^2\}$$

$$+ 2\left[\cos\phi \sin\left\{\frac{\alpha + v^1}{2}\right\}\cosh\left\{\frac{\beta + v^2}{2}\right\}\right.$$

$$\left. - \sin\phi \cos\left\{\frac{\alpha + v^1}{2}\right\}\sinh\left\{\frac{\beta + v^2}{2}\right\}\right],$$

$$\frac{\partial \alpha}{\partial x^2} = \mathbb{B}_2(\alpha; \beta, \frac{\partial \beta}{\partial x^1}) \tag{1.90}$$

$$= \frac{-\partial v^1}{\partial x^2} + \frac{\partial}{\partial x^1}\{\beta + v^2\}$$

$$+ 2\left[\cos\phi \cos\left\{\frac{\alpha + v^1}{2}\right\}\sinh\left\{\frac{\beta + v^2}{2}\right\}\right.$$

$$\left. + \sin\phi \sin\left\{\frac{\alpha + v^1}{2}\right\}\cosh\left\{\frac{\beta + v^2}{2}\right\}\right],$$

$$x'^a = x^a, \quad a = 1, 2,$$

where ϕ is a Bäcklund parameter and v^i, $i = 1, 2$, are arbitrary harmonic functions in x^a, $a = 1, 2$. ∎

The above result may be readily established. Thus, the constituent Bäcklund relations of (1.90) yield

$$\left\{\frac{\partial}{\partial x^1} + \frac{i\partial}{\partial x^2}\right\}\left\{\frac{\alpha + v^1 - i(\beta + v^2)}{2}\right\} = e^{i\phi}\sin\left\{\frac{\alpha + v^1 + i(\beta + v^2)}{2}\right\}, \quad (1.91)$$

together with its complex conjugate

$$\left\{\frac{\partial}{\partial x^1} - i\frac{\partial}{\partial x^2}\right\}\left\{\frac{\alpha + v^1 + i(\beta + v^2)}{2}\right\} = e^{-i\phi}\sin\left\{\frac{\alpha + v^1 - i(\beta + v^2)}{2}\right\}. \quad (1.92)$$

Operation on (1.91) with $\partial/\partial x^1 - i\partial/\partial x^2$ and subsequent use of the relation (1.92) show that

$$\nabla^2\left\{\frac{\alpha - i\beta}{2}\right\} = \sin\left\{\frac{\alpha + v^1 - i(\beta + v^2)}{2}\right\}\cos\left\{\frac{\alpha + v^1 + i(\beta + v^2)}{2}\right\}, \quad (1.93)$$

whence, on separation into real and imaginary parts, the nonlinear equations (1.88) and (1.89) are generated as required. ∎

The Bäcklund transformation (1.90) may now be used to generate first-generation nontrivial α and β solutions. We remark that the transformation $\mathbb{B}_\phi$ with Bäcklund parameter ϕ takes a solution $\alpha + v^1$ of (1.88) to a new solution $i(\beta + v^2)$. This may be represented by

$$\mathbb{B}_\phi(\alpha + v^1) = i(\beta + v^2). \quad (1.94)$$

Thus, in this sense, the mapping $\mathbb{B}_\phi$ may be regarded as being an auto-Bäcklund transformation.

First-Generation α Solutions

In order to construct a first-generation α solution $\underset{1}{\alpha}$ of (1.88), we start with a "vacuum" solution $\underset{0}{\beta} + v^2 = 0$ of (1.89). Substitution of $\underset{0}{\beta} = -v^2$ into (1.90)$_{1,2}$ yields

$$\frac{\partial}{\partial x^1}\left\{\underset{1}{\alpha} + v^1\right\} = 2\cos\phi\sin\left\{\frac{\underset{1}{\alpha} + v^1}{2}\right\},$$

$$\frac{\partial}{\partial x^2}\left\{\underset{1}{\alpha} + v^1\right\} = 2\sin\phi\sin\left\{\frac{\underset{1}{\alpha} + v^1}{2}\right\},$$

1.4 LONG JOSEPHSON JUNCTION: ELLIPTIC SINE-GORDON EQUATION

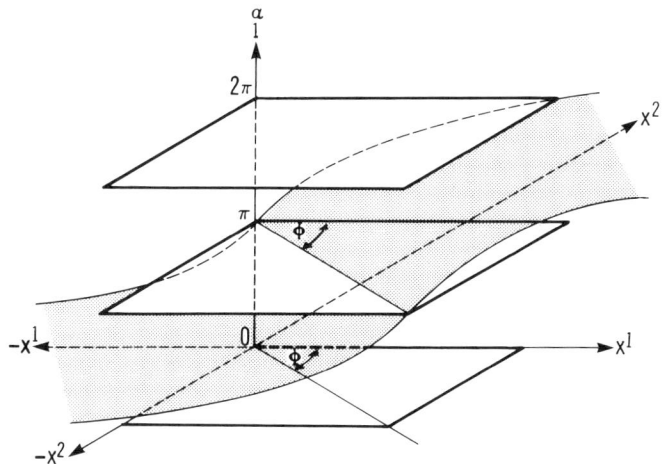

Fig. 1.10 A one-fluxon solution of $\nabla^2 \alpha = \sin \alpha$.

whence

$$\tan\left\{\frac{\underset{1}{\alpha} + v^1}{4}\right\} = c \exp\{x^1 \cos \phi + x^2 \sin \phi\}$$

or,

$$\underset{1}{\alpha} = 4 \tan^{-1}\left\{\frac{c \exp\{x^1 \cos \phi + x^2 \sin \phi\} - \tan(v^1/4)}{1 + c \tan(v^1/4) \exp\{x^1 \cos \phi + x^2 \sin \phi\}}\right\}, \qquad (1.95)$$

where c is a constant of integration. Such a single fluxon solution with $v^1 = 0$ is illustrated in Fig. 1.10.

First-Generation β Solutions

To establish a first-generation β solution $\underset{1}{\beta}$ of (1.89), we employ now a "vacuum" solution $\underset{0}{\alpha} + v^1 = 0$ of (1.88) as a starting point. Substitution of $\underset{0}{\alpha} = -v^1$ into (1.90)$_{1,2}$ yields

$$\frac{\partial}{\partial x^2}\left\{\underset{1}{\beta} + v^2\right\} = -2 \sin \phi \sinh\left\{\frac{\underset{1}{\beta} + v^2}{2}\right\},$$

$$\frac{\partial}{\partial x^1}\left\{\underset{1}{\beta} + v^2\right\} = -2 \cos \phi \sinh\left\{\frac{\underset{1}{\beta} + v^2}{2}\right\},$$

so that (Leibbrandt [26])

$$\beta_1 = \begin{cases} 4\tanh^{-1}\left\{\dfrac{\bar{c}_1 \exp\{x^1\cos\phi + x^2\sin\phi\} - \tanh(v^2/4)}{1 - \bar{c}_1 \tanh(v^2/4)\exp\{x^1\cos\phi + x^2\sin\phi\}}\right\} \\ \qquad \text{if} \quad x^1\cos\phi + x^2\sin\phi \leq 0, \\[1em] 4\coth^{-1}\left\{\dfrac{1 - \bar{c}_2 \coth(v^2/4)\exp\{x^1\cos\phi + x^2\sin\phi\}}{\bar{c}_2 \exp\{x^1\cos\phi + x^2\sin\phi\} - \coth(v^2/4)}\right\} \\ \qquad \text{if} \quad x^1\cos\phi + x^2\sin\phi > 0, \end{cases} \quad (1.96)$$

where $\bar{c}_i$, $i = 1, 2$, are arbitrary constants of integration.

Permutability Theorems

Infinite sequences of additional higher generation α and β solutions may now be constructed by appeal to the following pair of permutability theorems due to Leibbrandt. Since they are established similarly, only the second of the results is demonstrated here.

THEOREM 1.4 If $\underset{0}{\alpha} + v^1$ is a solution of (1.88) and $\underset{1}{\beta^{(j)}} + v^2$, $j = 1, 2$, are two distinct solutions of (1.89) such that

$$i(\underset{1}{\beta^{(1)}} + v^2) = \mathbb{B}_{\phi_1}(\underset{0}{\alpha} + v^1) \quad \text{and} \quad i(\underset{1}{\beta^{(2)}} + v^2) = \mathbb{B}_{\phi_2}(\underset{0}{\alpha} + v^1),$$

then there exists a new α solution $\underset{2}{\alpha} + v^1$ such that

$$\underset{2}{\alpha} + v^1 = \mathbb{B}_{\phi_1}\mathbb{B}_{\phi_3}(\underset{0}{\alpha} + v^1) = \mathbb{B}_{\phi_2}\mathbb{B}_{\phi_1}(\alpha_0 + v^1),$$

given by

$$\tan\left\{\frac{\underset{2}{\alpha} + v^1 - (\underset{0}{\alpha} + v^1)}{4}\right\} = \cot\left(\frac{\phi_1 - \phi_2}{2}\right)\tanh\left\{\frac{\underset{1}{\beta^{(1)}} + v^2 - (\underset{1}{\beta^{(2)}} + v^2)}{4}\right\},$$

$$\phi_2 \neq \phi_1 \pm N\pi, \quad N = 0, 1, 2, \ldots. \quad \blacksquare \quad (1.97)$$

THEOREM 1.5 If $\underset{0}{\beta} + v^2$ is a solution of (1.89) and $\underset{1}{\alpha^{(j)}} + v^1$, $j = 1, 2$, are two distinct solutions of (1.88) such that $\underset{1}{\alpha^{(1)}} + v^1 = i\mathbb{B}_{\sigma_1}(\underset{0}{\beta} + v^2)$ and $\underset{1}{\alpha^{(2)}} + v^1 = i\mathbb{B}_{\sigma_2}(\underset{0}{\beta} + v^2)$, then there exists a new β solution $\underset{2}{\beta} + v^2$ such that

$$\underset{2}{\beta} + v^2 = \mathbb{B}_{\sigma_1}\mathbb{B}_{\sigma_2}(\underset{0}{\beta} + v^2) = \mathbb{B}_{\sigma_2}\mathbb{B}_{\sigma_1}(\underset{0}{\beta} + v^2)$$

1.4 LONG JOSEPHSON JUNCTION: ELLIPTIC SINE-GORDON EQUATION

given by

$$\tanh\left\{\frac{\beta_2 + v^2 - (\beta_0 + v^2)}{4}\right\} = \cot\left(\frac{\sigma_1 - \sigma_2}{2}\right)\tan\left\{\frac{\alpha^{(2)}_1 + v^1 - (\alpha^{(1)}_1 + v^1)}{4}\right\}$$

$$\sigma_2 \neq \sigma_1 \pm N\pi, \quad N = 0, 1, 2, \ldots. \quad \blacksquare \quad (1.98)$$

The derivation of the preceding result is given by Leibbrandt [26]. It is as follows:

The relations $\alpha^{(j)}_1 + v^1 = \mathbb{B}_{\sigma_j}\{i(\beta_0 + v^2)\}, j = 1, 2$, yield

$$\left\{\frac{\partial}{\partial x^1} + i\frac{\partial}{\partial x^2}\right\}\left\{\frac{i(\beta_0 + v^2) - (\alpha^{(1)}_1 + v^1)}{2}\right\}$$

$$= \exp(i\sigma_1)\sin\left\{\frac{i(\beta_0 + v^2) + (\alpha^{(1)}_1 + v^1)}{2}\right\}, \quad (1.99)$$

$$\left\{\frac{\partial}{\partial x^1} + i\frac{\partial}{\partial x^2}\right\}\left\{\frac{i(\beta_0 + v^2) - (\alpha^{(2)}_1 + v^1)}{2}\right\}$$

$$= \exp(i\sigma_2)\sin\left\{\frac{i(\beta_0 + v^2) + (\alpha^{(2)}_1 + v^1)}{2}\right\}, \quad (1.100)$$

while application of $\mathbb{B}_{\sigma_2}$ and $\mathbb{B}_{\sigma_1}$ to $\alpha^{(1)}_1 + v^1$ and $\alpha^{(2)}_1 + v^1$ in turn leads to further solution classes $\mathbb{B}_{\sigma_2}\{\alpha^{(1)}_1 + v^1\}$ and $\mathbb{B}_{\sigma_1}\{\alpha^{(2)}_1 + v^1\}$. If these are assumed to have nonzero intersection, then if $i(\beta_2 + v^2)$ is a common element, it follows that

$$\left\{\frac{\partial}{\partial x^1} + i\frac{\partial}{\partial x^2}\right\}\left\{\frac{\alpha^{(1)}_1 + v^1 - i(\beta_2 + v^2)}{2}\right\}$$

$$= \exp(i\sigma_2)\sin\left\{\frac{\alpha^{(1)}_1 + v^1 + i(\beta_2 + v^2)}{2}\right\}, \quad (1.101)$$

$$\left\{\frac{\partial}{\partial x^1} + i\frac{\partial}{\partial x^2}\right\}\left\{\frac{\alpha^{(2)}_1 + v^1 - i(\beta_2 + v^2)}{2}\right\}$$

$$= \exp(i\sigma_1)\sin\left\{\frac{\alpha^{(2)}_1 + v^1 + i(\beta_2 + v^2)}{2}\right\}. \quad (1.102)$$

The combination (1.99) − (1.100) + (1.101) − (1.102) leads to

$$0 = \exp(i\sigma_1)\left[\sin\left\{\frac{i(\beta_0 + v^2) + \alpha_1^{(1)} + v^1}{2}\right\} - \sin\left\{\frac{\alpha_2^{(2)} + v^1 + i(\beta_2 + v^2)}{2}\right\}\right]$$

$$- \exp(i\sigma_2)\left[\sin\left\{\frac{i(\beta_0 + v^2) + \alpha_1^{(2)} + v^1}{2}\right\} - \sin\left\{\frac{\alpha_1^{(1)} + v^1 + i(\beta_2 + v^2)}{2}\right\}\right]$$

or

$$\exp(i\sigma_1)\sin\left\{\frac{i(\beta_0 - \beta_2) + (\alpha_1^{(1)} - \alpha_2^{(2)})}{4}\right\}$$

$$= \exp(i\sigma_2)\sin\left\{\frac{i(\beta_0 - \beta_2) + (\alpha_1^{(2)} - \alpha_1^{(1)})}{4}\right\}, \quad (1.103)$$

provided that

$$\cos\tfrac{1}{4}\left[\alpha_1^{(1)} + \alpha_1^{(2)} + 2v^1 + i(\beta_0 + \beta_2 + 2v^2)\right] \neq 0. \quad (1.104)$$

Use of the relation

$$\frac{e^{i\sigma_1} + e^{i\sigma_2}}{e^{i\sigma_1} - e^{i\sigma_2}} = -i\cot\left(\frac{\sigma_1 - \sigma_2}{2}\right) \quad (1.105)$$

now produces the required result, namely,

$$\tanh\left\{\frac{\beta_2 + v^2 - (\beta_0 + v^2)}{4}\right\} = \cot\left(\frac{\sigma_1 - \sigma_2}{2}\right)\tan\left\{\frac{\alpha_1^{(2)} + v^1 - (\alpha_1^{(1)} + v^1)}{4}\right\},$$

$$\sigma_2 \neq \sigma_1 \pm N\pi, \quad N = 0, 1, 2, \ldots \quad (1.106)$$

It may now be readily shown that $\beta_2 + v^2$ as given by the permutability relation (1.106) does indeed satisfy the Bäcklund relations (1.101) and (1.102). This completes the demonstration. ∎

As a consequence of Theorems 1.4 and 1.5, the construction of second-generation α and β solutions may be conveniently represented, in turn, by the commutative Bianchi-type diagrams shown in Figs. 1.11 and 1.12.

1.4 LONG JOSEPHSON JUNCTION: ELLIPTIC SINE-GORDON EQUATION

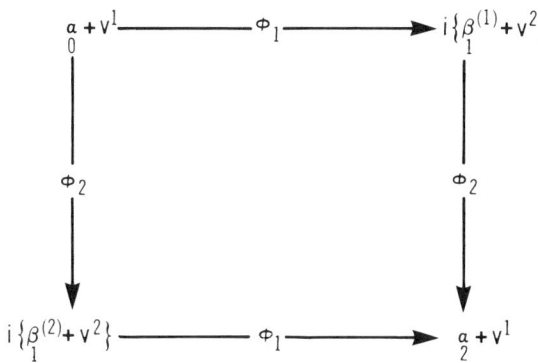

Fig. 1.11 The Bianchi diagram for second-generation α solutions.

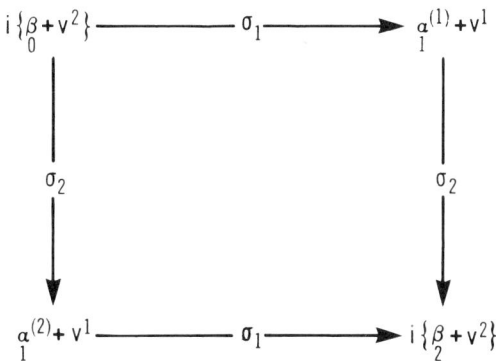

Fig. 1.12 The Bianchi diagram for second-generation β solutions.

The permutability properties embodied in Theorems 1.4 and 1.5 may be used iteratively to generate multiple α and β solutions according to the sequence of relations

$$\tan\left\{\frac{(\alpha_{2n} + v^1) - (\alpha_{2n-2} + v^1)}{4}\right\}$$

$$= \cot\left(\frac{\phi_{2n-1} - \phi_{2n}}{2}\right)\tanh\left\{\frac{\beta^{(1)}_{2n-1} + v^2 - (\beta^{(2)}_{2n-1} + v^2)}{4}\right\},$$

$$n = 1, 2, 3, \ldots, \quad \phi_{2n} \neq \phi_{2n-1} \pm N\pi, \quad N = 0, 1, 2, \ldots, \quad (1.107)$$

and

$$\tanh\left\{\frac{(\beta_{2n} + v^2) - (\beta_{2n-2} + v^2)}{4}\right\}$$

$$= \cot\left(\frac{\sigma_{2n-1} - \sigma_{2n}}{2}\right)\tan\left\{\frac{\alpha^{(2)}_{2n-1} + v^1 - (\alpha^{(1)}_{2n-1} + v^1)}{4}\right\},$$

$$n = 1, 2, 3, \ldots, \quad \sigma_{2n} \neq \sigma_{2n-1} \pm N\pi, \quad N = 0, 1, 2, \quad (1.108)$$

respectively. The relations (1.107) and (1.108) can, in turn, be written symbolically in terms of the Bäcklund operators $\mathbb{B}_\phi$, $\mathbb{B}_\sigma$ as

$$\alpha_{2n} + v^1 = \mathbb{B}_{\phi_{2n-1}}\mathbb{B}_{\phi_{2n}}\{\alpha_{2n-2} + v^1\} = \mathbb{B}_{\phi_{2n}}\mathbb{B}_{\phi_{2n-1}}\{\alpha_{2n-2} + v^1\}, \quad (1.109)$$

$$i\{\beta_{2n} + v^2\} = \mathbb{B}_{\sigma_{2n-1}}\mathbb{B}_{\sigma_{2n}}\{i(\beta_{2n-2} + v^2)\} = \mathbb{B}_{\sigma_{2n}}\mathbb{B}_{\sigma_{2n-1}}\{i(\beta_{2n-2} + v^2)\},$$

$$n = 1, 2, 3, \ldots. \quad (1.110)$$

1.5 THE COLE–HOPF REDUCTION OF BURGERS' EQUATION TO THE HEAT EQUATION. SOLUTION OF INITIAL VALUE PROBLEMS

The particular Monge–Ampère equation

$$u_2 + uu_1 - vu_{11} = 0, \quad (1.111)$$

known as *Burgers' equation*, represents the simplest equation which incorporates both amplitude dispersion and diffusion effects. It seems to have been first set down by Bateman [174] who obtained a particular solution. Subsequently, the equation was rederived by Burgers [175] as a simple model in the theory of turbulence. Lighthill [176] and later Blackstock [177], on the other hand, used Burgers' equation to describe the propagation of one-dimensional acoustic signals of moderate amplitude. A derivation of (1.111) in this context is given by Karpman [178]. A general discussion of the role of (1.111) in the analytic description of small amplitude waves is given by Whitham [29].

It was shown independently by Cole [40] and Hopf [39] that Burgers' equation (1.111) may be mapped to the *linear heat equation*

$$u'_2 - vu'_{11} = 0 \quad (1.112)$$

1.5 REDUCTION OF BURGERS' EQUATION TO HEAT EQUATION

by the Bäcklund transformation

$$u'_1 = \mathbb{B}'_1(u; u') = -\frac{1}{2v} uu',$$

$$u'_2 = \mathbb{B}'_2(u, u_1; u') = \frac{u^2 u'}{4v} - \frac{u' u_1}{2}, \quad (1.113)$$

$$x'^a = x^a, \quad a = 1, 2.$$

This important result is readily demonstrated. Thus, imposition of the integrability requirement

$$\frac{\partial \mathbb{B}'_1}{\partial x^2} - \frac{\partial \mathbb{B}'_2}{\partial x^1} = 0$$

and use of the Bäcklund relations $(1.113)_{1,2}$ give

$$u_2 + uu_1 - vu_{11} = 0,$$

provided that $u' \neq 0$. On the other hand, the relations $(1.113)_{1,2}$ yield

$$u'_2 - vu'_{11} = \frac{u}{4v}[uu' + 2vu'_1] = 0,$$

whence the reduction is established.

Lighthill [176] utilized the Cole–Hopf reduction to analyze the competition between nonlinearity and diffusion in a review of aspects of finite amplitude sound waves involving shock formation, interaction, spreading, and decay. Here, application to the solution of the initial value problem

$$u_2 + uu_1 - vu_{11} = 0, \quad t > 0, \quad u = \Phi(x^1), \quad t = 0 \quad (x^2 = t), \quad (1.114)$$

is summarized. A more detailed account is given by Whitham [29].

Use of the Bäcklund transformation (1.113) shows that the initial value problem (1.114) for Burgers' equation corresponds to the initial value problem

$$u'_2 - vu'_{11} = 0, \quad t > 0,$$

$$u' = \exp\left[-\frac{1}{2v}\int_0^{x^1} \Phi(\sigma)\,d\sigma\right], \quad t = 0, \quad (1.115)$$

for the classical heat equation. The solution of this problem is well known, namely,

$$u' = \frac{1}{\sqrt{4\pi vt}} \int_{-\infty}^{\infty} \exp\left[-\frac{1}{2v}\left\{\int_0^{\tau} \Phi(\sigma)\,d\sigma - \frac{(x^1-\tau)^2}{2t}\right\}\right]d\tau, \quad (1.116)$$

so that the solution of the original initial value problem (1.114) is

$$u = \frac{-2vu'_1}{u'} = \frac{\int_{-\infty}^{\infty} \frac{(x^1 - \tau)}{t} \exp\left[-\frac{1}{2v}\left\{\int_0^\tau \Phi(\sigma)\,d\sigma + \frac{(x^1-\tau)^2}{2t}\right\}\right] d\tau}{\int_{-\infty}^{\infty} \exp\left[-\frac{1}{2v}\left\{\int_0^\tau \Phi(\sigma)\,d\sigma + \frac{(x^1-\tau)^2}{2t}\right\}\right] d\tau}. \quad (1.117)$$

The uniqueness of this solution for suitably restricted $\Phi(x^1)$ was demonstrated by Cole [40]. We now proceed to specific cases.

(i) *Initial Step Data* In this case, the evolution of a shock wave into a steady state is considered. The initial conditions are taken to be

$$\Phi(x^1) = \begin{cases} U_1, & x^1 < 0, \\ -U_1, & x^1 > 0, \end{cases} \quad (1.118)$$

and substitution into (1.117) yields (Cole [40])

$$u(x^1, t) = -U_1 \frac{2\,\text{sh}(U_1 x^1/2v) + \exp[U_1 x^1/2v]\,\text{erf}[(x^1 + U_1 t)/2(vt)^{1/2}]}{+ \exp[-U_1 x^1/2v]\,\text{erf}[(x^1 - U_1 t)/2(vt)^{1/2}]}}{2\,\text{ch}(U_1 x^1/2v) + \exp[U_1 x^1/2v]\,\text{erf}[(x^1 + U_1 t)/2(vt)^{1/2}]} - \exp[-U_1 x^1/2v]\,\text{erf}[(x^1 - U_1 t)/2(vt)^{1/2}]}. \quad (1.119)$$

For large values of t/v ($|x^1| \neq U_1 t$), use of the asymptotic result

$$\text{erf } z \sim 1 - \frac{1}{\pi^{1/2} z}\exp(-z^2) + \exp(-z^2) O\left(\frac{1}{z^3}\right)$$

in (1.119) shows that the solution approaches the steady state

$$u = -U_1 \tanh(U_1 x^1/2v). \quad (1.120)$$

In the more general situation when the initial conditions are

$$\Phi(x^1) = \begin{cases} U_3, & x^1 < 0 \\ U_4, & x^1 > 0 \end{cases} \quad (U_3 > U_4), \quad (1.121)$$

introduction of a coordinate system moving with speed $U = \tfrac{1}{2}(U_3 + U_4)$ allows the use of the solution (1.119) with $U_1 = U_3 - U$ and the changes $x^1 \to x^1 - Ut$, $u \to u - U$.

(ii) *δ Function Initial Data* For the single-hump initial condition with

$$\Phi(x^1) = A\delta(x^1), \quad (1.122)$$

1.5 REDUCTION OF BURGERS' EQUATION TO HEAT EQUATION

(1.117) produces the similarity solution (Whitham [29])

$$u = \sqrt{\frac{v}{t}} f\left\{\frac{x^1}{\sqrt{vt}}; \frac{A}{v}\right\}$$

$$= \sqrt{\frac{v}{t}} \left[\frac{(\exp(A/2v) - 1)\exp[-(x^1)^2/4vt]}{\sqrt{\pi} + (\exp(A/2v) - 1) \int_{x^1/\sqrt{4vt}}^{\infty} \exp[-\zeta^2]\,d\zeta}\right]. \quad (1.123)$$

As $A/2v \to 0$, the diffusion dominates the nonlinearity and (1.123) approaches the source solution of the classical heat equation, namely,

$$u = \frac{A}{\sqrt{4\pi vt}} \exp[-(x^1)^2/4vt]. \quad (1.124)$$

Instead of a direct use of the Cole–Hopf reduction, an inverse approach may be adopted wherein a known solution u' of (1.112) is taken and the corresponding u under the Bäcklund transformation (1.113) interpreted as the solution of an initial value problem for Burgers' equation. Two examples of this procedure are presented below. The reader is recommended to consult the works of Whitham [29] and Burgers [117] for additional details.

(iii) *N Waves* If we select a source solution u' of the heat equation (1,112), namely,

$$u' = 1 + \sqrt{a/t}\exp[-(x^1)^2/4vt], \quad (1.125)$$

then the corresponding solution of Burgers' equation under the Bäcklund transformation (1.113) is

$$u = \frac{x^1}{t} \frac{\sqrt{a/t}\exp[-(x^1)^2/4vt]}{1 + \sqrt{a/t}\exp[-(x^1)^2/4vt]}. \quad (1.126)$$

The latter corresponds to an *N* wave, so-called because of its shape (Fig. 1.13).

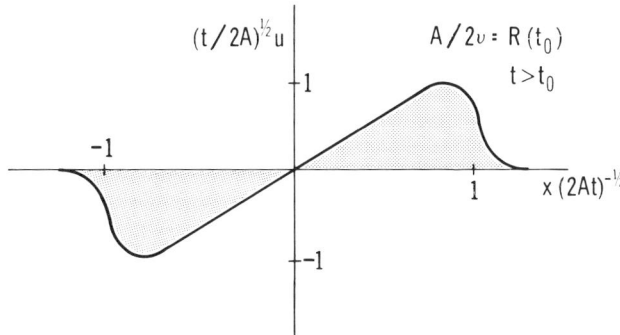

Fig. 1.13 *N*-wave solution of Burgers' equation. (Whitham [29].)

The area under the positive phase of the pulse profile is

$$\int_0^\infty u\, dx^1 = -2\nu \int_0^\infty u'_1 u'^{-1}\, dx^1 = 2\nu \ln\{1 + \sqrt{a/t}\} = R(t),$$

where $R(t)$ is a Reynolds number. As $t \to \infty$, $R(t) \to 0$ and diffusion becomes dominant.

(iv) *Periodic Wave Solutions* If the initial condition on the heat equation has a Fourier series expansion

$$u'|_{t=0} = \frac{1}{\lambda}\left[1 + 2\sum_{n=1}^\infty \cos\left(\frac{2n\pi x^1}{\lambda}\right)\right], \qquad (1.127)$$

then the solution of the initial value problem is

$$u' = \frac{1}{\lambda}\left[1 + 2\sum_{n=1}^\infty \exp\left\{\frac{-4n^2\pi^2\nu t}{\lambda^2}\right\}\cos\left(\frac{2n\pi x^1}{\lambda}\right)\right]. \qquad (1.128)$$

The associated solution of Burgers' equation under the Bäcklund transformation (1.113) is

$$u = \frac{8\pi\nu}{\lambda}\sum_{n=1}^\infty n\exp\left\{\frac{-4n^2\pi^2\nu t}{\lambda^2}\right\}\sin\left(\frac{2n\pi x^1}{\lambda}\right) \bigg/ \left[1 + 2\sum_{n=1}^\infty \exp\left\{\frac{-4n^2\pi^2\nu t}{\lambda^2}\right\}\cos\left(\frac{2n\pi x^1}{\lambda}\right)\right],$$

(1.129)

corresponding to a *periodic wave* (Whitham [29]).

It is remarked that (1.128) is the Fourier series expansion of the solution

$$u' = (4\pi\nu t)^{-1/2} \sum_{n=-\infty}^\infty \exp\left\{\frac{-(x^1 - n\lambda)^2}{4\nu t}\right\} \qquad (1.130)$$

of the heat equation (1.112) due to a distribution of heat sources set at a distance λ apart.

Some progress has been made in direct extensions of the Cole–Hopf reduction, notably by Chu [41], Sachdev [179], and Tasso and Teichmann [180]. On the other hand, a Bäcklund transformation has yet to be constructed for a generalized Burgers' equation recently discussed in the context of nonlinear acoustics by Parker [181].

Despite the undoubted intrinsic significance of the Cole–Hopf reduction, perhaps its greatest importance lies in the impetus it gave to the search for similar Bäcklund transformations for other nonlinear evolution equations, notably, in the first instance, the celebrated Korteweg–deVries equation. This subject is taken up in the following sections.

1.6 FINITE AMPLITUDE DISPERSIVE WAVES AND THE KORTEWEG–deVRIES EQUATION

It was in 1844 that Scott-Russell [182] recorded his dramatic observation of a solitary wave propagating along a canal, "a rounded, smooth and well-defined heap of water, which continued its course along the channel apparently without change of form or diminution of speed."

Korteweg and deVries [30], in a subsequent analytic study of the evolution of long water waves along a canal of rectangular cross section, were led to introduce the model equation that now bears their names, viz.,

$$u_t - \sqrt{g/l}\{\alpha u_x + 3/2 u u_x + (\sigma/2) u_{xxx}\} = 0. \tag{1.131}$$

Here, u represents surface elevation above the equilibrium level l, g is the gravitational constant and $\sigma = \frac{1}{3}l^3 - (Tl/\rho g)$, where T is the surface capillary tension and ρ is the density of the fluid; α is a parameter associated with the underlying motion of the medium.

In 1960, a Korteweg–deVries equation was rederived in an analysis of collision-free hydromagnetic waves by Gardner and Morikawa [31]. Since that time, the Korteweg–deVries equation has been shown to model a diversity of important finite amplitude dispersive wave phenomena in the theory of solids, liquids, gases, and plasmas. Review articles by Jeffrey and Kakutani [183], and Miura [184] have been devoted to the topic, while additional details are given in [185–187]. Kruskal and Zabusky [32, 188] showed that the Fermi–Pasta–Ulam problem [189] associated with the propagation of longitudinal waves in a one-dimensional lattice of equal masses coupled by certain nonlinear springs is modeled by the Korteweg–deVries equation. In plasma physics, its application to the analysis of finite amplitude waves in a low-density plasma was noted by Berezin and Karpman [190, 191] while Washimi and Taniuti [34] used the Korteweg–deVries equation to model the transmission of ion-acoustic waves in a cold plasma. On the other hand, Shen [192] derived the Korteweg–deVries equation in connection with the propagation of three-dimensional waves in channels of arbitrary cross section, while its relevance to the theory of rotating fluids was described by Leibovich [193]. Its occurrence in the analysis of pressure waves in liquid–gas bubble mixtures has been documented by Wijngaarden [194] whereas its application in the nonlinear theory of longitudinal dispersive waves in elastic rods and in the context of thermally excited phonon packets in low-temperature nonlinear crystals has been set out by Nariboli [195] and Tappert and Varma [196], respectively.

In more general discussions, Su and Gardner [38], Taniuti and Wei [197] have demonstrated that a large class of nonlinear evolution equations may be reduced to consideration of the Korteweg–deVries equation. Thus, it may be regarded as an important canonical form.

The rich tapestry of physical applications of the Korteweg–deVries equation is apparent. That this important nonlinear equation admits a Bäcklund transformation is as remarkable as the consequences thereof. Thus, not only can the Bäcklund transformation be used to generate *multi-soliton* solutions of the Korteweg–deVries equation, but it may also be adduced to construct both the associated inverse scattering problem and a multiplicity of conservation laws. Here, the Korteweg–deVries equation is derived in the context of shallow-water wave theory: the presentation is based on that given by Whitham [29]. In the following section, a Bäcklund transformation is constructed by the Clairin procedure as described by Lamb [35]. Multi-soliton solutions of the Korteweg–deVries equation are thereby generated as a consequence of a permutability theorem.

The Navier–Stokes equations governing the motion of an inviscid incompressible fluid under a constant gravitational field are

$$\nabla \cdot \mathbf{q} = 0, \tag{1.132}$$

$$\rho\{\mathbf{q}_t + (\mathbf{q} \cdot \nabla)\} = -\nabla p - \rho \mathbf{g}, \tag{1.133}$$

where $\mathbf{q} = (u, v, w)$ is the velocity vector, ρ is the water density, and $\mathbf{g}$ is the gravitational acceleration vector. The Korteweg–deVries water wave equation emerges out of a boundary value problem for this system in a manner now described.

In water wave theory, problems typically are concerned with situations where there is propagation into regions with no initial vorticity. Thus, any subsequent vorticity will only derive out of diffusion from boundaries where the effects of viscosity are important and there is formation of boundary layers. If the latter are thin in comparison with the depth, then over times that are not too large, the effect of boundary layers on the wave motion may be neglected, whence it may be assumed that the irrotationality condition

$$\nabla \times \mathbf{q} = 0, \tag{1.134}$$

applies in addition to (1.132) and (1.133). However, it should be noted that a long-time analysis should take account of the boundary layer (Phillips [198]). Moreover, the present analysis only has validity up to the occurrence of wave breaking since strong shear layers are then formed and vorticity is transmitted into the main body of the fluid with an accompanying onset of turbulence.

The body of water is assumed to be bounded above by air, with the interface given by

$$\Phi(x, y, z, t) = 0. \tag{1.135}$$

1.6 FINITE AMPLITUDE DISPERSIVE WAVES AND THE K–DV EQUATION

Thus, since there is no transport across the boundary, the velocity of the fluid normal to (1.135) relative to the velocity of the boundary normal to itself must be zero, whence

$$\frac{\nabla \Phi}{|\nabla \Phi|} \cdot \mathbf{q} - \left\{ \frac{-\Phi_t}{|\nabla \Phi|} \right\} = 0,$$

so that,

$$\Phi_t + \mathbf{q} \cdot \nabla \Phi = 0. \tag{1.136}$$

In particular, if the interface is described by

$$z = \zeta(x, y, t), \tag{1.137}$$

where the z axis is taken vertically upward, then we may set

$$\Phi(x, y, z, t) \equiv \zeta(x, y, t) - z, \tag{1.138}$$

whence the kinematic boundary condition becomes

$$\zeta_t + u\zeta_x + v\zeta_y - w = 0 \quad \text{at} \quad z = \zeta(x, y, t). \tag{1.139}$$

In the absence of surface tension, the water and atmospheric pressure must balance on $z = \zeta(x, y, t)$. It is usual to assume that the change in the air pressure due to the motion of the water wave is negligible, whence the air pressure may be approximated by its undisturbed value. Accordingly,

$$p = p_0 \quad \text{on} \quad z = \zeta(x, y, t), \tag{1.140}$$

where p is the water pressure and p_0 is the constant pressure in the undisturbed atmosphere.

If the lower boundary of the water region is taken to be both rigid and impermeable, the appropriate boundary condition requires that the normal velocity of the fluid vanish at the boundary, so that

$$\mathbf{n} \cdot \mathbf{q} = 0 \tag{1.141}$$

there.

The irrotationality condition (1.134) allows the introduction of a potential $\phi^*(x, y, z, t)$ such that

$$\mathbf{q} = \nabla \phi^*, \tag{1.142}$$

whence on substitution in the equation of motion (1.133) and integration, a Bernoulli integral

$$\rho\{\phi_t^* + \tfrac{1}{2}(\nabla \phi^*)^2 + gz\} = p_0 - p + \rho \mathbb{B}(t) \tag{1.143}$$

is obtained. On introduction of a new potential

$$\phi = \phi^* - \int_{t_0}^{t} \mathbb{B}(\sigma)\,d\sigma,$$

the relation (1.143) reduces to

$$\rho\{\phi_t + \tfrac{1}{2}(\nabla\phi)^2 + gz\} = p_0 - p, \qquad (1.144)$$

and the boundary condition (1.140) becomes

$$\phi_t + \tfrac{1}{2}\{\phi_x^2 + \phi_y^2 + \phi_z^2\} + g\zeta = 0 \quad \text{on} \quad z = \zeta(x,y,t), \qquad (1.145)$$

while the continuity requirement (1.132) shows that

$$\nabla^2 \phi = 0. \qquad (1.146)$$

In addition, if the lower boundary is given by

$$z = -h^*(x,y), \qquad (1.147)$$

then condition (1.141) yields

$$\phi_x h_x^* + \phi_y h_y^* + \phi_z = 0 \quad \text{at} \quad z = -h^*(x,y). \qquad (1.148)$$

Thus to summarize, it is required to solve

$$\nabla^2 \phi = 0, \qquad -h^* < z < \zeta, \qquad (1.149)$$

subject to the boundary conditions

$$\left.\begin{array}{l} \zeta_t + \phi_x \zeta_x + \phi_y \zeta_y - \phi_z = 0 \\ \phi_t + \tfrac{1}{2}\{\phi_x^2 + \phi_y^2 + \phi_z^2\} + g\zeta = 0 \end{array}\right\} \quad \text{at} \quad z = \zeta(x,y,t), \qquad (1.150)$$

$$\phi_x h_x^* + \phi_y h_y^* + \phi_z = 0 \quad \text{at} \quad z = -h^*(x,y), \qquad (1.151)$$

augmented by the appropriate initial conditions.

Attention is now restricted to the case of *two-dimensional* gravity waves propagating on the free surface of a layer of water bounded below by a horizontal, plane, impervious, and rigid base. It is convenient to introduce the vertical translation $z \to z + h^*$ so that the lower boundary becomes $z = 0$. Thus, we obtain the moving boundary value problem consisting of Laplace's equation

$$\phi_{xx} + \phi_{zz} = 0, \qquad 0 < z < h(x,t), \qquad (1.152)$$

subject to the boundary conditions

$$\left.\begin{array}{l} h_t + \phi_x h_x - \phi_z = 0 \\ \phi_t + \tfrac{1}{2}\{\phi_x^2 + \phi_y^2\} + g(h - h^*) = 0 \end{array}\right\} \quad \text{at} \quad z = h(x,t), \qquad (1.153)$$

$$\phi_z = 0 \quad \text{at} \quad z = 0 \qquad (1.154)$$

$$(h := \zeta + h^*),$$

1.6 FINITE AMPLITUDE DISPERSIVE WAVES AND THE K–DV EQUATION

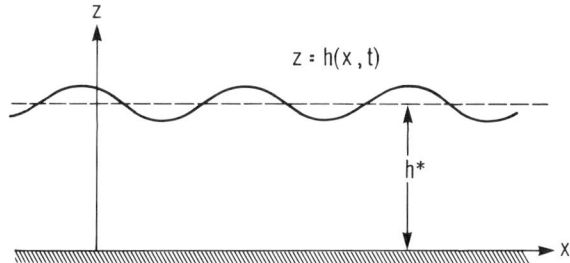

Fig. 1.14 Geometric configuration for two-dimensional gravity waves.

and appropriate initial conditions. The geometric configuration is shown in Fig. 1.14.

In the case of small perturbations on water initially at rest, the linearized version of the boundary value problem (1.152)–(1.154) reduces to consideration of

$$\phi_{xx} + \phi_{zz} = 0, \quad 0 < z < h^*, \tag{1.155}$$

together with

$$\left. \begin{array}{l} h_t - \phi_z = 0 \\ \phi_t + g(h - h^*) = 0 \end{array} \right\} \quad \text{at} \quad z = h^*, \tag{1.156}$$

$$\phi_z = 0 \quad \text{at} \quad z = 0, \tag{1.157}$$

where it is noted that the conditions at the free surface $z = h(x,t)$ are applied at the undistributed surface position $z = h^*$. The system (1.155)–(1.157) is to be augmented by appropriate initial conditions.

On elimination of h in (1.156), we obtain

$$\phi_{tt} + g\phi_z = 0 \quad \text{at} \quad z = h^*; \tag{1.158}$$

whence a linear boundary value problem for ϕ alone is derived. Once ϕ has been determined, the free surface is given by

$$z = h(x,t) = h^* - g^{-1}\phi_t(x, h^*, t). \tag{1.159}$$

Sinusoidal solutions

$$\phi(x, z, t) = Z(z)e^{i(kx - wt)}, \tag{1.160}$$

$$h(x, t) - h^* = He^{i(kx - wt)} \tag{1.161}$$

that are appropriate to horizontal wave propagation are now sought. Substitution of (1.160) into (1.155) yields

$$Z'' - k^2 Z = 0, \tag{1.162}$$

and combination of this requirement with condition (1.157) shows that

$$Z(z) = Z(0)\cosh(kz). \tag{1.163}$$

Moreover, (1.159) and (1.161), together with (1.160) and (1.163), yield

$$H = (iw/g)Z(h^*),$$

whence

$$\phi = \frac{-igH}{w}\frac{\cosh(kz)}{\cosh(kh^*)}e^{i(kx-wt)}, \tag{1.164}$$

$$h = He^{i(kx-wt)} + h^*. \tag{1.165}$$

The remaining condition (1.158) provides the *dispersion relation*

$$w^2 = gk\tanh(kh^*), \tag{1.166}$$

connecting the *wave frequency* $w = 2\pi/T$ and *wave number* $k = w/V_p = 2\pi/\lambda$ (namely, the number of cycles per unit distance). Here, in the usual notation, T is the *wave period*, $V_p = w/k$ is the *wave speed* with which the disturbance propagates to the right, while $\lambda = V_p T$ is the *wave length* (that is, the distance between successive wave crests); H/λ is called the *wave steepness*, while $\theta = kx - wt$ is known as the *phase* of the motion.

In general, associated with a dispersion relation $w = w(k)$ such as (1.166), the notion of *wave group velocity* V_g may be introduced according to

$$V_g = \partial w/\partial k. \tag{1.167}$$

If $V_p \neq V_g$, as in the present case (excluding the limit $k \to 0$), then the wave is said to be *dispersive*. If the dispersion relation $w = w(k)$ has a complex solution w with $\text{im}(w) < 0$, then there is amplitude attenuation with increasing time. This phenomenon is known as *attenuation dissipation*. If $\text{im}(w) > 0$, the wave amplitude increases without bound as time progresses and the wave, accordingly, is said to exhibit *instability*. If $w = w(k)$ is real and $\partial V_p/\partial k \neq 0$, then there is neither dissipation nor stability, but rather there is said to be *pure dispersion*.

In the shallow water wave approximation, the phase speed $V_p \sim \sqrt{gh^*}$ and so is independent of k; hence dispersion effects are not included. If we proceed to the next approximation wherein we consider weak dispersion with the requirement that $k \ll 1$ corresponding to waves of long wavelength, then

$$w^2 = gk(kh^* - (kh^*)^3/3) = V_0^2 k^2 - \tfrac{1}{3}V_0^2 h^{*2}k^4 \quad (V_0 = \sqrt{gh^*}), \tag{1.168}$$

with phase velocity given by

$$V_p = V_0\{1 - \tfrac{1}{6}h^{*2}k^2 + O(k^4)\}. \tag{1.169}$$

1.6 FINITE AMPLITUDE DISPERSIVE WAVES AND THE K–DV EQUATION

In the sequel, it is convenient to normalize the variables according to

$$x \to x/l, \quad z \to z/h^*, \quad t \to V_0 t/l,$$
$$h \to h/a, \quad \phi \to V_0 \phi/gla, \quad \zeta \to \zeta/a, \quad (1.170)$$

where l is a horizontal length scale for the waves and a is a typical amplitude. The stretchings in x and z reduce the problem in normalized variables to (Whitham [29])

$$\beta \phi_{xx} + \phi_{zz} = 0 \quad 0 < z < 1 + \alpha\zeta, \quad (1.171)$$

$$\left.\begin{array}{r}\zeta_t + \alpha\phi_x \zeta_x - \beta^{-1}\phi_z = 0 \\ \zeta + \phi_t + \frac{1}{2}\{\alpha\phi_x^2 + \alpha\beta^{-1}\phi_z^2\} = 0\end{array}\right\} \text{ at } z = 1 + \alpha\zeta, \quad (1.172)$$

$$\phi_z = 0 \quad \text{at} \quad z = 0, \quad (1.173)$$

where $\alpha = a/h^*$, $\beta = h^{*2}/l^2$ are measures of the nonlinearity and dispersion, respectively.

A formal expansion in terms of the small parameter β is now introduced, namely

$$\phi = \sum_0^\infty \{z^m \phi_m(x,t)\} \beta^m, \quad (1.174)$$

whence, on insertion into (1.171) and subsequent use of the boundary condition (1.173), we see that

$$\phi = \sum_0^\infty \frac{(-1)^m \beta^m z^{2m}}{2m!} \frac{\partial^{2m} \phi_0}{\partial x^{2m}}. \quad (1.175)$$

Substitution of (1.175) into the free surface conditions (1.172) now yields

$$\zeta_t + \{(1 + \alpha\zeta)\phi_{0,x}\}_x - \{\tfrac{1}{6}(1+\alpha\zeta)^3 \phi_{0,xxxx} + \tfrac{1}{2}\alpha(1+\alpha\zeta)^2 \zeta_x \phi_{0,xxx}\}\beta + O(\beta^2), \quad (1.176)$$

$$\zeta + \phi_{0,t} + \tfrac{1}{2}\alpha\phi_{0,x}^2 - \tfrac{1}{2}(1+\alpha\zeta)^2\{\phi_{0,xxt} + \alpha\phi_{0,x}\phi_{0,xxx} - \alpha(\phi_{0,xx})^2\}\beta + O(\beta^2). \quad (1.177)$$

If all the terms in β are neglected, the well-known *nonlinear shallow water equations*

$$\zeta_t + \{(1+\alpha\zeta)w\}_x = 0, \quad (1.178)$$

$$w_t + \alpha w w_x + \zeta_x = 0, \quad (1.179)$$

$$(w := \phi_{0,x}),$$

are retrieved.[†] This nondispersive system provides a reasonable approximation for certain bore, breaker, and hydraulic jump phenomena. On the other hand, a characteristic analysis of simple wave propagation as described by the above shallow water model predicts progressive steepening of the disturbance. This feature, which is known as *amplitude dispersion*, ultimately requires that the wave steepens to such an extent that breaking occurs (Peregrine [199]). However, observation shows that breaking does not always occur in shallow water situations. This leads us to investigate how a combination of nonlinear and dispersive effects may act to inhibit breaking.

If, as in Whitham [29], the approximation is introduced wherein terms in β^1 are retained but terms of $O(\alpha\beta)$ are neglected, then we obtain a *Boussinesq*-type system

$$\zeta_t + \{(1+\alpha\zeta)w\}_x - \tfrac{1}{6}\beta w_{xxx} + O(\alpha\beta, \beta^2) = 0, \qquad (1.180)$$

$$w_t + \alpha w w_x + \zeta_x - \tfrac{1}{2}\beta w_{xxt} + O(\alpha\beta, \beta^2) = 0. \qquad (1.181)$$

Following Whitham, a solution of (1.180) and (1.181) is now sought in the form

$$w = \zeta + \alpha\lambda(\zeta, \zeta_x, \zeta_{xx}, \ldots) + \beta\mu(\zeta, \zeta_x, \zeta_{xx}, \ldots) + O(\alpha^2 + \beta^2), \qquad (1.182)$$

whence, on insertion, we obtain

$$\zeta_t + \zeta_x + \alpha(\lambda_x + 2\zeta\zeta_x) + \beta(\mu_x - \tfrac{1}{6}\zeta_{xxx}) + O(\alpha^2 + \beta^2) = 0, \qquad (1.183)$$

$$\zeta_t + \zeta_x + \alpha(\lambda_t + \zeta\zeta_x) + \beta(\mu_t - \tfrac{1}{2}\zeta_{xxt}) + O(\alpha^2 + \beta^2) = 0. \qquad (1.184)$$

Since $\zeta_t = -\zeta_x + O(\alpha, \beta)$, consistency requires

$$\lambda = -\tfrac{1}{4}\zeta^2, \qquad \mu = \tfrac{1}{3}\zeta_{xx},$$

so that

$$w = \zeta - \tfrac{1}{4}\alpha\zeta^2 + \tfrac{1}{3}\beta\zeta_{xx} + O(\alpha^2 + \beta^2), \qquad (1.185)$$

where

$$\zeta_t + \zeta_x + \tfrac{3}{2}\alpha\zeta\zeta_x + \tfrac{1}{6}\beta\zeta_{xxx} + O(\alpha^2 + \beta^2) = 0. \qquad (1.186)$$

The latter is a normalized version of the celebrated Korteweg–deVries equation descriptive of the propagation of a progressive wave in the positive x direction. The change of variables

$$t' = (6/\beta)^{1/2}t, \qquad x' = (6/\beta)^{1/2}x, \qquad \zeta' = \tfrac{1}{4}(\alpha\zeta + \tfrac{2}{3}) \qquad (1.187)$$

leads to the standard form of the Korteweg–deVries equation, namely,

$$\zeta'_{t'} + 6\zeta'\zeta'_{x'} + \zeta'_{x'x'x'} = 0. \qquad (1.188)$$

[†] That similar systems admit Bäcklund transformations will be established in Chapter 3.

1.7 THE WAHLQUIST-ESTABROOK INVARIANT TRANSFORMATION OF THE KORTEWEG-deVRIES EQUATION. THE SOLITON LADDER

In 1973, Wahlquist and Estabrook [36] published an ingenious auto-Bäcklund transformation of the Korteweg-deVries equation. This result was later derived in another manner by Lamb [35] by means of a natural generalization of the Bäcklund transformation (1.9) wherein higher derivatives are included. Thus, transformations of the type

$$\mathbb{B}_i(x^a, u, u_a, u_{ab}; x'^a, u', u'_a, u'_{ab}) = 0, \quad i = 1, \ldots, 4, \quad a, b, = 1, 2 \quad (1.189)$$

were introduced and used to derive permutability theorems for a series of important nonlinear evolution equations.[†] These are discussed in this and subsequent sections of this chapter. Here, a derivation of the auto-Bäcklund transformation of the Korteweg-deVries equation is carried out "in extenso." The method adopted is that of Lamb, and the result is embodied in the following:

THEOREM 1.6 The Korteweg-deVries equation

$$v_2 + 6vv_1 + v_{111} = 0 \quad (1.190)$$

is invariant under the Bäcklund transformation

$$u'_1 = \mathbb{B}'_1(u, u_1; u') = \beta - u_1 - \tfrac{1}{2}[u - u']^2,$$

$$u'_2 = \mathbb{B}'_2(u, u_1, u_{11}, u_2; u', u'_1, u'_{11}) \quad (1.191)$$

$$= -u_2 + [u - u'][u_{11} - u'_{11}] - 2[u_1^2 + u_1 u'_1 + u'^2_1],$$

$$x'^a = x^a, \quad a = 1, 2,$$

where

$$u(x^a) := \int_{-\infty}^{x^1} v(X, x^2) \, dX, \quad (1.192)$$

and β is an arbitrary Bäcklund parameter. ∎

In order to construct the Bäcklund transformation it is convenient to proceed in terms of the integral of the pulse profile. Thus, on introduction of (1.192) into the Korteweg-deVries equation (1.190) and subsequent integration, we obtain

$$u_2 + 3u_1^2 + u_{111} + \Phi(x^2) = 0, \quad (1.193)$$

[†] Systematic extension to include higher order derivatives is introduced in Chapter 2, where a theory of Bäcklund transformations is established based on a jet-bundle formalism.

1. BÄCKLUND TRANSFORMATIONS AND NONLINEAR EQUATIONS

where $\Phi(x^2)$ is arbitrary. The change of variable

$$u \to u - \int_{-\infty}^{x^2} \Phi(X) \, dX \tag{1.194}$$

reduces (1.193) to

$$u_2 + 3u_1^2 + u_{111} = 0. \tag{1.195}$$

A Bäcklund transformation of the type

$$\begin{aligned} u_i &= \mathbb{B}_i(u, u_{jk}; u', u'_j, u'_{jk}), \quad i, j, k = 1, 2, \\ x'^a &= x^a, \quad a = 1, 2, \end{aligned} \tag{1.196}$$

is now sought for (1.195) rather than for (1.190). In order that the integrability condition

$$\Omega = \frac{\partial \mathbb{B}_1}{\partial x^2} - \frac{\partial \mathbb{B}_2}{\partial x^1} = 0 \tag{1.197}$$

should not introduce derivatives that are absent in (1.195), the Bäcklund transformation must be specialized to the form

$$\begin{aligned} u_1 &= \mathbb{B}_1(u; u', u'_1), \\ u_2 &= \mathbb{B}_2(u, u_{11}; u', u'_1, u'_2, u'_{11}), \\ x'^a &= x^a, \quad a = 1, 2. \end{aligned} \tag{1.198}$$

Moreover,

$$u_{11} = \frac{\partial \mathbb{B}_1}{\partial u} \mathbb{B}_1 + \frac{\partial \mathbb{B}_1}{\partial u'} u'_1 + \frac{\partial \mathbb{B}_1}{\partial u'_1} u'_{11},$$

so that u_{11} can be expressed entirely in terms of u, u', u'_1, and u'_{11} in $\mathbb{B}_2$.

The integrability condition (1.197) now requires that

$$\Omega = \left[\frac{\partial \mathbb{B}_1}{\partial u'_1} - \frac{\partial \mathbb{B}_2}{\partial u'_2}\right] u'_{12} + \frac{\partial \mathbb{B}_1}{\partial u} \mathbb{B}_2 + \frac{\partial \mathbb{B}_1}{\partial u'} u'_2$$

$$- \frac{\partial \mathbb{B}_2}{\partial u} \mathbb{B}_1 - \frac{\partial \mathbb{B}_2}{\partial u'} u'_1 - \frac{\partial \mathbb{B}_2}{\partial u'_1} u'_{11} + \frac{\partial \mathbb{B}_2}{\partial u'_{11}} [u'_2 + 3u'^2_1] = 0, \tag{1.199}$$

an immediate consequence of which is that

$$\frac{\partial \mathbb{B}_1}{\partial u'_1} - \frac{\partial \mathbb{B}_2}{\partial u'_2} = 0. \tag{1.200}$$

Hence,

$$\frac{\partial \Omega}{\partial u'_2} = \frac{\partial \mathbb{B}_1}{\partial u} \frac{\partial \mathbb{B}_1}{\partial u'_1} + \frac{\partial \mathbb{B}_1}{\partial u'} - \frac{\partial^2 \mathbb{B}_1}{\partial u \, \partial u'_1} \mathbb{B}_1 - \frac{\partial^2 \mathbb{B}_1}{\partial u' \, \partial u'_1} u'_1 - \frac{\partial^2 \mathbb{B}_1}{\partial u'^2_1} u'_{11} + \frac{\partial \mathbb{B}_2}{\partial u'_{11}},$$

1.7 INVARIANT TRANSFORMATION OF THE K–DV EQUATION

so that

$$\frac{\partial^2 \Omega}{\partial u'_2 \, \partial u'_{11}} = -\frac{\partial^2 \mathbb{B}_1}{\partial u'^2_1} + \frac{\partial^2 \mathbb{B}_2}{\partial u'^2_{11}} = 0. \tag{1.201}$$

The particular solution

$$\begin{aligned}
\mathbb{B}_1 &= \beta_{11} u'_1 + \beta_{12}(u, u'), \\
\mathbb{B}_2 &= \beta_{11} u'_2 + \beta_{21}(u, u') u'_{11} + \beta_{22}(u, u') u'^2_1 \\
&\quad + \beta_{23}(u, u') u'_1 + \beta_{24}(u, u'), \qquad (\beta_{11} = \text{const}),
\end{aligned} \tag{1.202}$$

of (1.201) is now introduced, and substitution back into (1.199) produces the following overdetermined system for the β_{ij}:

$$\begin{aligned}
2\beta_{22} + \beta_{11}\beta_{21,u} + \beta_{21,u'} &= 0, \\
\beta_{21} + \beta_{11}\beta_{12,u} + \beta_{12,u'} &= 0, \\
\beta_{23} + \beta_{12}\beta_{21,u} - \beta_{21}\beta_{12,u} &= 0, \\
\beta_{22}\beta_{12,u} - \beta_{12}\beta_{22,u} + 3\beta_{21} - \beta_{11}\beta_{23,u} - \beta_{23,u'} &= 0, \\
\beta_{23}\beta_{12,u} - \beta_{12}\beta_{23,u} - \beta_{24,u'} - \beta_{11}\beta_{24,u} &= 0, \\
\beta_{11}\beta_{22,u} + \beta_{22,u'} &= 0, \\
\beta_{24}\beta_{12,u} - \beta_{12}\beta_{24,u} &= 0.
\end{aligned} \tag{1.203}$$

We now impose the requirement that u be a solution of (1.195); it is recalled that the corresponding condition for u' has already been used in (1.199). A routine calculation shows that

$$\begin{aligned}
u_{111} &= \beta_{11} u'_{111} - \beta_{21} u'_{11} + 2\beta_{22} u'^2_1 \\
&\quad + [2\beta_{11}\beta_{12}\beta_{12,uu} + 2\beta_{12}\beta_{12,uu'} \\
&\quad + \beta_{12,u}\{\beta_{11}\beta_{12,u} + \beta_{12,u'}\}] u'_1 \\
&\quad + (\beta_{12})^2 \beta_{12,uu} + \beta_{12}(\beta_{12,u})^2,
\end{aligned}$$

whence, on substitution into (1.195), the following additional requirements on the β_{ij} emerge:

$$\begin{aligned}
\beta_{22} - \beta_{11} + (\beta_{11})^2 &= 0, \\
\beta_{11}\beta_{12,uu} + 2\beta_{11} + \beta_{12,uu'} &= 0, \\
(\beta_{12})^2 \beta_{12,uu} + \beta_{12}(\beta_{12,u})^2 + \beta_{24} + 3(\beta_{12})^2 &= 0.
\end{aligned} \tag{1.204}$$

The combined system (1.203)–(1.204) is readily seen to admit the particular solution

$$\begin{aligned}
\beta_{11} &= -1, & \beta_{12} &= \beta - \tfrac{1}{2}[u - u']^2, & \beta_{21} &= -2[u - u'], \\
\beta_{22} &= -2, & \beta_{23} &= 2\beta + [u - u']^2, & \beta_{24} &= \beta[(u - u')^2 - 2\beta],
\end{aligned}$$

54   1. BÄCKLUND TRANSFORMATIONS AND NONLINEAR EQUATIONS

where β is an arbitrary Bäcklund parameter. Substitution into (1.202) now produces the auto-Bäcklund transformation of (1.195) as set out by Wahlquist and Estabrook [36]. Moreover, use of the Bäcklund relation

$$u'_1 = \beta - u_1 - \tfrac{1}{2}[u - u']^2$$

yields

$$u_{111} + u'_{111} = -[u - u'][u_{11} - u'_{11}] - [u_1 - u'_1]^2,$$

which equation, together with (1.195) and its primed counterpart, shows that

$$u'_2 = -u_2 + [u - u'][u_{11} - u'_{11}] - 2[u_1^2 + u_1 u'_1 + u'^2_1];$$

whence the Bäcklund transformation is obtained in the form given by Lamb [35]. ∎

The Wahlquist–Estabrook transformation leads to the following permutability theorem whereby an infinite sequence of solutions of the Korteweg–deVries equation may be generated from a known solution:

THEOREM 1.7 If $u_{\beta_i} = \mathbb{B}_{\beta_i} u_0$, $i = 1, 2$, are solutions of (1.195) that are generated by means of the Bäcklund transformation (1.191) from a known solution u_0 via, in turn, the Bäcklund parameters $\beta = \beta_i$, $i = 1, 2$, then a new solution ϕ is given by

$$\phi = u_0 + 2\{\beta_1 - \beta_2\}/\{u_{\beta_1} - u_{\beta_2}\}, \tag{1.205}$$

where $\phi = \mathbb{B}_{\beta_1}\mathbb{B}_{\beta_2} u_0 = \mathbb{B}_{\beta_2}\mathbb{B}_{\beta_1} u_0$. ∎

The above result is a simple consequence of the Bäcklund transformation (1.191). Thus the latter yields

$$u_{0,1} + u_{\beta_1,1} = \beta_1 - \tfrac{1}{2}\{u_0 - u_{\beta_1}\}^2, \tag{1.206}$$

$$u_{0,1} + u_{\beta_2,1} = \beta_2 - \tfrac{1}{2}\{u_0 - u_{\beta_2}\}^2, \tag{1.207}$$

$$u_{\beta_1,1} + u_{\beta_1\beta_2,1} = \beta_2 - \tfrac{1}{2}\{u_{\beta_1} - u_{\beta_1\beta_2}\}^2, \tag{1.208}$$

$$u_{\beta_2,1} + u_{\beta_2\beta_1,1} = \beta_1 - \tfrac{1}{2}\{u_{\beta_2} - u_{\beta_2\beta_1}\}^2 \tag{1.209}$$

$$(u_{\beta_i,j} := \partial u_{\beta_i}/\partial x_j),$$

where $u_{\beta_1\beta_2} := \mathbb{B}_{\beta_2}\mathbb{B}_{\beta_1} u_0$, $u_{\beta_2\beta_1} := \mathbb{B}_{\beta_1}\mathbb{B}_{\beta_2} u_0$. If it is assumed that ϕ exists such that $\phi = u_{\beta_1\beta_2} = u_{\beta_2\beta_1}$, then the operations (1.206) − (1.207) and (1.208) − (1.209) produce, in turn,

$$u_{\beta_1,1} - u_{\beta_2,1} = \beta_1 - \beta_2 + \tfrac{1}{2}\{u_{\beta_1} - u_{\beta_2}\}\{2u_0 - u_{\beta_1} - u_{\beta_2}\}, \tag{1.210}$$

$$u_{\beta_1,1} - u_{\beta_2,1} = \beta_2 - \beta_1 + \tfrac{1}{2}\{u_{\beta_2} - u_{\beta_1}\}\{u_{\beta_1} + u_{\beta_2} - 2\phi\}. \tag{1.211}$$

1.7 INVARIANT TRANSFORMATION OF THE K–DV EQUATION

Subtraction of (1.210) from (1.211) now produces the required expression

$$\phi = u_0 + 2\{\beta_1 - \beta_2\}/\{u_{\beta_1} - u_{\beta_2}\},$$

and it is readily verified that this is indeed a solution of (1.195). ∎

In Theorem 1.7, u_0 represents an arbitrary solution of (1.195). Accordingly, the more general recursion relation

$$u_{(n)} = u_{(n-2)} + 2\{\beta_n - \beta_{n-1}\}/\{u_{(n-1)'} - u_{(n-1)}\},$$
$$n > 1, \quad u_{(0)} = u_0, \qquad (1.212)$$

is immediate, where the subscript (n) denotes the set of n parameters $\{\beta_1, \ldots, \beta_n\}$, while the subscript $(n)'$ denotes $\{\beta_1, \ldots, \beta_{n-1}, \beta_{n+1}\}$.

With $n = 3$ in (1.212), we obtain

$$u_{\beta_1\beta_2\beta_3} = u_{\beta_1} + 2\{\beta_3 - \beta_2\}/\{u_{\beta_1\beta_3} - u_{\beta_1\beta_2}\},$$

which expression may be written entirely in terms of first-generation solutions as

$$u_{\beta_1\beta_2\beta_3} = \frac{\beta_1 u_{\beta_1}(u_{\beta_2} - u_{\beta_3}) + \beta_2 u_{\beta_2}(u_{\beta_3} - u_{\beta_1}) + \beta_3 u_{\beta_3}(u_{\beta_1} - u_{\beta_2})}{\beta_1(u_{\beta_2} - u_{\beta_3}) + \beta_2(u_{\beta_3} - u_{\beta_1}) + \beta_3(u_{\beta_1} - u_{\beta_2})}. \qquad (1.213)$$

Corresponding to $n = 4$, Wahlquist and Estabrook [36] recorded the relation

$$u_{\beta_1\beta_2\beta_3\beta_4} = u_0 + \frac{\begin{array}{c}N_{\beta_1\beta_2\beta_3\beta_4} + N_{\beta_2\beta_3\beta_1\beta_4} + N_{\beta_1\beta_3\beta_4\beta_2} + N_{\beta_3\beta_4\beta_1\beta_2}\\ + N_{\beta_1\beta_4\beta_2\beta_3} + N_{\beta_4\beta_2\beta_1\beta_3}\end{array}}{D_{\beta_1\beta_2\beta_3\beta_4} + D_{\beta_2\beta_3\beta_1\beta_4} + D_{\beta_1\beta_3\beta_4\beta_2}}, \qquad (1.214)$$

where

$$N_{\beta_p\beta_q\beta_r\beta_s} := \beta_p\beta_q(\beta_p - \beta_q)(u_{\beta_r} - u_{\beta_s}), \qquad (1.215)$$

$$D_{\beta_p\beta_q\beta_r\beta_s} := (\beta_p\beta_q + \beta_r\beta_s)(u_{\beta_p} - u_{\beta_q})(u_{\beta_r} - u_{\beta_s}). \qquad (1.216)$$

It is remarkable that the hierarchy of solutions of the Korteweg–deVries equation, as generated by the theorem of permutability with starting solution $u_0 = 0$, correspond to *multi-soliton pulses*. Thus Zabusky and Kruskal [188], in a computer study of the Korteweg–deVries equation, had earlier predicted the existence of solitary waves with the extraordinary property that they asymptotically preserve their shape and velocity upon collision with other solitary waves, emerging with no more than a phase shift. Such solitary waves are termed *solitons*. A wide variety of nonlinear evolution equations have now been shown to admit soliton solutions and the literature on the subject is now extensive. An account of recent developments in soliton theory is to be found, for example, in Lonngren and Scott [187]. An illustration of soliton typical interaction given in that text is reproduced in Fig. 1.15.

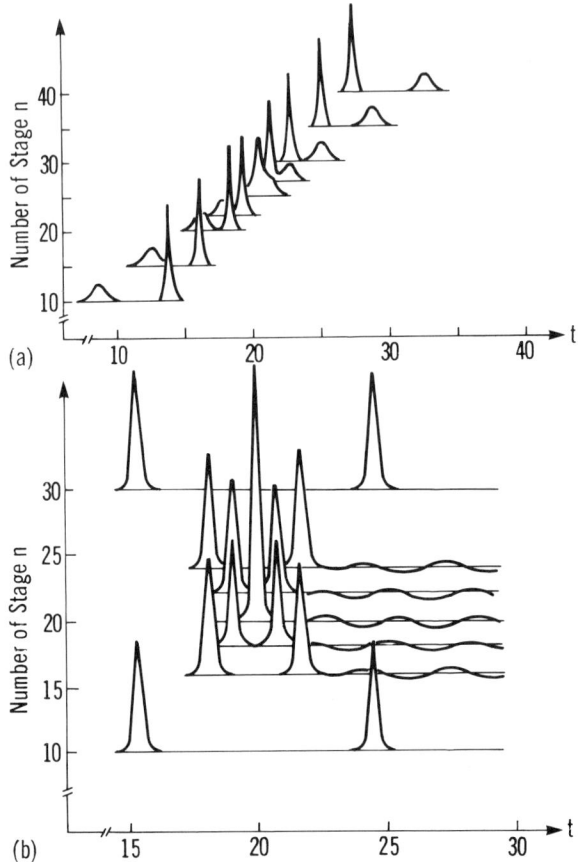

Fig. 1.15 Numerical observation of the collision of two solitons (a) "Overtaking"; (b) "Head-on." (Daikoku et al. [200].)

Hirota [118] introduced a bilinear operator method whereby an *n-soliton* solution of the Korteweg–deVries equation may be constructed. That an identical result may be obtained by means of the permutability theorem associated with the Bäcklund method has been recently demonstrated by Hirota and Satsuma [201]. Indeed this and other similar results indicate a strong connection between the existence of solitons and Bäcklund transformations (Herrmann [202]).

The soliton ladder is initiated by the insertion of the vacuum solution $u = 0$ in the Bäcklund relations (1.191), whence

$$u'_1 = \beta - \tfrac{1}{2}u'^2, \qquad u'_2 = u'u'_{11} - 2u'^2_1, \qquad (1.217)$$

1.8 AUTO-BÄCKLUND TRANSFORMATION OF MODIFIED K–DV EQUATION

the regular solution of which is

$$u' = (2\beta)^{1/2} \tanh[(\beta/2)^{1/2}(x^1 - 2\beta x^2)], \tag{1.218}$$

with associated soliton solution of the Korteweg–deVries equation (1.190) given by

$$v' = \beta \operatorname{sech}^2[(\beta/2)^{1/2}(x^1 - 2\beta x^2)]. \tag{1.219}$$

Hence, the Bäcklund parameter β corresponds to the amplitude of the single soliton while it also determines its speed 2β.

In view of the invariance of (1.217) under the transformation $u' \to 2\beta/u'$, the system also admits the singular solution

$$u^* = (2\beta)^{1/2} \coth[(\beta/2)^{1/2}(x^1 - 2\beta x^2)], \tag{1.220}$$

with corresponding solution of the Korteweg–deVries equation

$$v^* = -\beta \operatorname{csch}^2[(\beta/2)^{1/2}(x^1 - 2\beta x^2)]. \tag{1.221}$$

If the regular solution (1.218) is used as the starting point in the soliton ladder generated by the recursion relation (1.212), then the only regular second-generation solution obtained is given by

$$u_{\beta_1 \beta_2} = 2\{\beta_1 - \beta_2\}/\{u_{\beta_1} - u^*_{\beta_2}\}, \qquad \beta_1 < \beta_2, \tag{1.222}$$

where $u^*_{\beta_2}$ is the singular first-generation solution corresponding to the parameter β_2. Similarly, the regular three-soliton solution is generated via the theorem of permutability and use of $\{u_{\beta_1}, u^*_{\beta_2}, u_{\beta_3}\}$, where $\beta_1 < \beta_2 < \beta_3$ and u_{β_3} is the regular β_3 soliton. In general, it may be asserted that in order to maintain regularity in the multi-soliton solutions as constructed by this iterative procedure, it is required that Bäcklund parameters increase monotonically so that

$$\beta_1 < \beta_2 < \cdots < \beta_n,$$

(Wahlquist and Estabrook [36]). An expression for the multi-soliton of the Korteweg–deVries equation is subsequently presented in Section 1.12.

1.8 AN AUTO-BÄCKLUND TRANSFORMATION OF THE MODIFIED KORTEWEG–deVRIES EQUATION. THE PERMUTABILITY THEOREM AND GENERATION OF SOLUTIONS

The modified Korteweg–deVries equation arises out of the analysis of nonlinear Alfvén waves in a collisionless plasma (Kakutani and Ono [203]). Further, its connection with acoustic wave propagation in anharmonic lattices has been described by Zabusky [33]. In Section 1.17 a Bäcklund transformation is constructed for a nonlinear lattice equation which, in its continuum approximation, includes the sine-Gordon and modified Korteweg–deVries equation as but special cases.[†] Here, a procedure analogous to that adopted in the preceding section for the Korteweg–deVries equation leads to a Bäcklund transformation which leaves the modified Korteweg–deVries equation invariant. The result, which is due to Lamb [35], is as follows:

THEOREM 1.8   The modified Korteweg–deVries equation

$$v_2 + 6v^2 v_1 + v_{111} = 0 \qquad (1.223)$$

is invariant under the Bäcklund transformation

$$u_1 = \mathbb{B}_1(u; u', u'_1) = \alpha u'_1 + \beta \sin\{u + \alpha u'\},$$

$$u_2 = \mathbb{B}_2(u, u_1, u_{11}; u'_1, u'_{11})$$
$$= \alpha u'_2 - \beta[2\alpha u'_{11} \cos\{u + \alpha u'\} + 2u'^2_1 \sin\{u + \alpha u'\} + \beta(u_1 + \alpha u'_1)] \qquad (1.224)$$

$$x'^a = x^a \quad a = 1, 2, \quad \alpha = \pm 1,$$

where

$$u(x^a) := \int_{-\infty}^{x^1} v(X, x^2)\, dX \qquad (1.225)$$

and β is an arbitrary Bäcklund parameter. ■

The above result is readily established. Thus, under the transformation (1.225), subsequent integration, and incorporation of a translation of the type (1.194), the equation for the integral of the pulse profile reduces to

$$u_2 + 2u_1^3 + u_{111} = 0. \qquad (1.226)$$

If a Bäcklund transformation of the form (1.196) is sought for (1.226), considerations similar to those set out in Theorem 1.6 allow a specialization of the type (1.202). The restrictions on the β_{ij} in this case consist of the system (1.203), but with (1.203)$_{4,6}$ replaced by

$$\beta_{22}\beta_{12,u} - \beta_{12}\beta_{22,u} - \beta_{11}\beta_{23,u} - \beta_{23,u'} = 0,$$
$$\beta_{11}\beta_{22,u} + \beta_{22,u'} - 2\beta_{21} = 0, \qquad (1.227)$$

together with the conditions imposed by the requirement that u satisfy (1.226), namely,

$$\beta_{11}(\beta_{11}^2 - 1) = 0, \qquad \beta_{22} + 2\beta_{11}^2 \beta_{12} = 0,$$
$$2\beta_{11}\beta_{12} + \beta_{12,uu'} + \beta_{11}\beta_{12,uu} = 0, \qquad (1.228)$$
$$\beta_{12}[\beta_{12}\beta_{12,uu} + \beta_{12,u}^2 + 2\beta_{12}^2] + \beta_{24} = 0.$$

[†] A Bäcklund transformation and associated permutability theorem have also been developed by Wadati [294] for a nonlinear lattice described, in the long wave length continuum approximation, by a combined Korteweg–deVries–modified-Korteweg–deVries equation.

1.8 AUTO-BÄCKLUND TRANSFORMATION OF MODIFIED K–DV EQUATION

A routine calculation reveals that the system (1.227)–(1.228) admits the particular solution (Lamb [35])[†]

$$\beta_{11}^2 = +1, \quad \beta_{12} = \beta \sin\{u + \beta_{11} u'\},$$
$$\beta_{21} = -2\beta\beta_{11} \cos\{u + \beta_{11} u'\}, \quad \beta_{22} = -2\beta \sin\{u + \beta_{11} u'\}, \quad (1.229)$$
$$\beta_{23} = -2\beta^2 \beta_{11}, \quad \beta_{24} = -\beta^3 \sin\{u + \beta_{11} u'\},$$

which, on insertion in (1.202), produces the required result.[‡] ∎

An immediate consequence of the Bäcklund transformation (1.224) is the following:

THEOREM 1.9 If u_{β_i}, $i = 1, 2$, represent solutions of (1.226) generated via the Bäcklund transformation (1.225) with a starting solution u_0 and, in turn, the Bäcklund parameters $\beta = \beta_i$, $i = 1, 2$, then a new solution ϕ is given by

$$\tan\left\{\frac{\phi - u_0}{2}\right\} = \alpha \left(\frac{\beta_1 + \beta_2}{\beta_1 - \beta_2}\right) \tan\left\{\frac{u_{\beta_1} - u_{\beta_2}}{2}\right\}, \quad (1.230)$$

where $\phi = \mathbb{B}_{\beta_1} \mathbb{B}_{\beta_2} u_0 = \mathbb{B}_{\beta_2} \mathbb{B}_{\beta_1} u_0$. ∎

Application of the Bäcklund transformation (1.225) with parameters $\beta = \beta_i$, $i = 1, 2$, and starting vacuum solution $u_0 = 0$ produces the first-generation solutions

$$u_{\beta_i} = 2 \tan^{-1} e^{\mu_i}, \quad (1.231)$$

where

$$u_i = \beta_i x^1 - \beta_i^3 x^2 + \gamma_i, \quad (1.232)$$

and γ_i, $i = 1, 2$, are integration constants. The permutability relation (1.230) now yields

$$\phi = \pm 2 \tan^{-1}\left[\left(\frac{\beta_1 + \beta_2}{\beta_1 - \beta_2}\right) \frac{\sinh\{1/2(\mu_1 - \mu_2)\}}{\cosh\{1/2(\mu_1 + \mu_2)\}}\right], \quad (1.233)$$

which accords with a solution obtained by Wadati [119] by means of an inverse scattering formalism.

Multiple soliton solutions of the modified Korteweg–deVries equation may be constructed either by iteration of the permutability relation (1.230) or by use of the Hirota method [201, 204].

[†] Here $\alpha = \beta_{11} = \pm 1$, the ambiguity of sign corresponding to the invariance of (1.226) under the transformation $u \to -u$.
[‡] A detailed account of a more general result is presented in Section 1.17.

1.9 THE MIURA TRANSFORMATION

Whereas the Bäcklund transformations set out in the preceding two sections were invariant transformations of the Korteweg–deVries and modified Korteweg–deVries equations, Miura's celebrated transformation links these nonlinear equations. The result, here derived by Clairin's classical procedure as described by Lamb [35], is as follows:

THEOREM 1.10 The Korteweg–deVries equation

$$u'_2 + 6u'u'_1 + u'_{111} = 0 \qquad (1.234)$$

and modified Korteweg–deVries equation

$$u_2 - 6u^2 u_1 + u_{111} = 0 \qquad (1.235)$$

are linked by the Bäcklund transformation

$$\begin{aligned} u_1 &= \mathbb{B}_1(u; u') = \pm\{u' + u^2\}, \\ u_2 &= \mathbb{B}_2(u, u_1; u', u'_1, u'_{11}) = \mp u'_{11} - 2\{uu'_1 + u'u_1\}, \\ x'^a &= x^a, \quad a = 1, 2. \end{aligned} \qquad (1.236)$$

Thus, introduction of the Ansatz

$$u_1 = \mathbb{B}_1(u; u'), \qquad u_2 = \mathbb{B}_2(u; u', u'_1, u'_{11}) \qquad (1.237)$$

and application of the integrability condition (1.197) yield

$$\Omega = \frac{\partial \mathbb{B}_1}{\partial x^2} - \frac{\partial \mathbb{B}_2}{\partial x^1} = \frac{\partial \mathbb{B}_1}{\partial u}\mathbb{B}_2 + \frac{\partial \mathbb{B}_1}{\partial u'}u'_2 - \frac{\partial \mathbb{B}_2}{\partial u}\mathbb{B}_1 - \frac{\partial \mathbb{B}_2}{\partial u'}u'_1$$

$$-\frac{\partial \mathbb{B}_2}{\partial u'_1}u'_{11} + \frac{\partial \mathbb{B}_2}{\partial u'_{11}}\{u'_2 + 6u'u'_1\} = 0, \qquad (1.238)$$

so that

$$\Omega_{u'_2} = \frac{\partial \mathbb{B}_1}{\partial u'} + \frac{\partial \mathbb{B}_2}{\partial u'_{11}} = 0, \qquad (1.239)$$

$$\Omega_{u'_2 u'_1} = \frac{\partial^2 \mathbb{B}_2}{\partial u'_{11}\,\partial u'_1} = 0, \qquad (1.240)$$

$$\Omega_{u'_1 u'_{11}} = -\frac{\partial^2 \mathbb{B}_2}{\partial u'\,\partial u'_{11}} - \frac{\partial^2 \mathbb{B}_2}{\partial u'_1{}^2} = \frac{\partial^2 \mathbb{B}_1}{\partial u'^2} - \frac{\partial^2 \mathbb{B}_2}{\partial u'_1{}^2} = 0, \qquad (1.241)$$

$$\Omega_{u'_1 u'_1} = \frac{\partial \mathbb{B}_1}{\partial u}\frac{\partial^2 \mathbb{B}_1}{\partial u'^2} - \frac{\partial^3 \mathbb{B}_1}{\partial u\,\partial u'^2}\mathbb{B}_1 - 2\frac{\partial^2 \mathbb{B}_2}{\partial u'\,\partial u'_1} - u'_1\frac{\partial^3 \mathbb{B}_1}{\partial u'^3} = 0, \qquad (1.242)$$

$$\Omega_{u'_1 u'_{11} u'_1} = -\frac{\partial^3 \mathbb{B}_2}{\partial u'_1{}^3} = 0, \qquad (1.243)$$

1.9 THE MIURA TRANSFORMATION

$$\Omega_{u'_1 u'_1 u'_1} = -3 \frac{\partial^3 \mathbb{B}_1}{\partial u'^3} = 0. \tag{1.244}$$

Integration of (1.244) and use of (1.241) give

$$\frac{\partial^2 \mathbb{B}_1}{\partial u'^2} = \frac{\partial^2 \mathbb{B}_2}{\partial u'^2_1} = \bar{\alpha}(u),$$

whence, if we set $\alpha = 0$,

$$\frac{\partial \mathbb{B}_1}{\partial u'} = -\frac{\partial \mathbb{B}_2}{\partial u'_{11}} = \beta_1(u) \tag{1.245}$$

and

$$\frac{\partial \mathbb{B}_2}{\partial u'_1} = \beta_2(u; u', u'_{11}). \tag{1.246}$$

Moreover, (1.240)–(1.242) immediately show that β_2 is dependent on u alone. Accordingly, integration of (1.246) and subsequent appeal to the relations (1.239) and (1.245) show that

$$\mathbb{B}_2 = \beta_2(u)u'_1 - \beta_1(u)u'_{11} + \alpha(u, u'), \tag{1.247}$$

where $\alpha(u, u')$ is determined by the requirements

$$\Omega_{u'u'u'_1} = -\frac{\partial^3 \mathbb{B}_2}{\partial u'^3} = 0, \tag{1.248}$$

$$\Omega_{u'u'u'} = 3\left[\frac{\partial^2 \mathbb{B}_1}{\partial u \partial u'}\frac{\partial^2 \mathbb{B}_2}{\partial u'^2} - \frac{\partial^3 \mathbb{B}_2}{\partial u \partial u'^2}\frac{\partial \mathbb{B}_1}{\partial u'}\right] = 0. \tag{1.249}$$

Thus, as a consequence of (1.247)–(1.249), the Bäcklund transformation adopts the reduced form

$$\begin{aligned} u_1 &= \beta_1 u' + \beta_3, \\ u_2 &= \beta_2 u'_1 - \beta_1 u'_{11} + \gamma \beta_1 u'^2 + \beta_4 u' + \beta_5, \end{aligned} \tag{1.250}$$

where $\beta_i = \beta_i(u)$, $i = 1, \ldots, 5$, remain to be determined; γ is an arbitrary constant of integration.

Substitution of the Bäcklund relations (1.250) back into (1.238) now produces the following system of equations for the β_i:

$$\begin{aligned} \beta_1 \beta'_2 - \beta_2 \beta'_1 + 2\beta_1(3 + \gamma) &= 0, \\ \beta_1 \beta'_4 - \beta_4 \beta'_1 + \gamma(\beta_3 \beta'_1 - \beta_1 \beta'_3) &= 0, \\ \beta_1 \beta'_5 - \beta_5 \beta'_1 + \beta_3 \beta'_4 - \beta_4 \beta'_3 &= 0, \\ \beta_4 = \beta_2 \beta'_3 - \beta_3 \beta'_2, \quad \beta_5 \beta'_3 - \beta_3 \beta'_5 &= 0, \quad \beta_2 = \beta_3 \beta'_1 - \beta_1 \beta'_3, \end{aligned} \tag{1.251}$$

while the requirement that the Bäcklund relations (1.250) produce the modified Korteweg–deVries equation (1.235) leads to the further set of conditions

$$\beta_2 + \beta_1\beta_3' = 0, \quad \gamma + \beta_1\beta_3'' = 0,$$
$$2\beta_1\beta_3\beta_3'' + \beta_1\beta_3'^2 - 6\beta_1 u^2 + \beta_4 = 0, \quad (1.252)$$
$$\beta_3''\beta_3^2 + \beta_3'^2\beta_3 - 6u^2\beta_3 + \beta_5 = 0,$$

where, in the derivation of the system (1.252), β_1 has been assumed to be a nonzero constant.

The combined system (1.251)–(1.252) is readily seen to admit the solution

$$\beta_1 = \pm 1, \quad \beta_2 = -2u, \quad \beta_3 = \pm u^2, \quad \beta_4 = \mp 2u^2, \quad \beta_5 = 0, \quad \gamma = -2,$$

which, on insertion in (1.250), leads to the required Bäcklund transformation (1.236).

Relation (1.236) is known as the Miura transformation which accordingly represents part of a Bäcklund transformation.

1.10 A NONLINEAR SCHRÖDINGER EQUATION

The nonlinear cubic Schrödinger equation

$$iu_2 + u_{11} + vu^2\bar{u} = 0 \qquad (1.253)$$

is an important wave evolution model. Thus, it arises in the description of self-focusing of optical beams in nonlinear media [205–207], modulation of monochromatic waves [208–211], and the propagation of Langmuir waves in plasmas [212–214]. It is also related to the Ginzburg–Landau equation in superconductivity [215], while its occurrence in low temperature physics has been documented by Tsuzuki [216], and in vortex motion by Hasimoto [217] and Yuen [218]. Recent results associating (1.253) with the acceleration of wave packets in weakly inhomogeneous plasmas are to be found in [219, 220].

Equation (1.253) was originally derived in the context of deep water waves via a spectral method by Zakharov [221] and later, independently, by Hasimoto and Ono [222] and Davey [223] using multiple scale techniques. Subsequently, Yuen and Lake [226] rederived the same equation using Whitham's averaged variational principle. Extensions to two dimensions and finite depth were made by Benney and Roskes [224] and Davey and Stewartson [225]. Derivative nonlinear Schrödinger equations have been investigated in [268–271].

Here, the relevance of the cubic Schrödinger equation (1.253) to the evolution of weakly nonlinear deep water gravity wave trains is summarized. Further, its importance as a canonical form is highlighted by two recent

1.10 THE NONLINEAR SCHRÖDINGER EQUATION

results. In the next section, an inverse scattering formalism is introduced whereby Bäcklund transformations for a wide class of nonlinear evolution equations, including (1.253), may be readily constructed.

Deep Water Gravity Waves

Whitham's averaged variational principle shows that the evolution of a slowly varying weakly nonlinear wave system characterized by wave vector $\mathbf{k}$, wave amplitude a, and frequency w is governed by the equations

$$\frac{\partial k^i}{\partial t} + \frac{\partial w}{\partial x^i} = 0, \tag{1.254}$$

$$\frac{\partial a^2}{\partial t} + \frac{\partial}{\partial x^j}\left\{\frac{\partial w}{\partial k^j} a^2\right\} = 0, \tag{1.255}$$

where x^i are the spatial coordinates and t is the time. Equations (1.254) and (1.255) must be augmented by an appropriate dispersion relation for weakly nonlinear deep-water gravity waves, namely, the *Stokes relation*

$$w^2 = gk(1 + k^2 a^2), \tag{1.256}$$

where $k = |\mathbf{k}|$ is the wave number. Following Yuen and Lake [226], attention is confined to a wave train wherein

$$\mathbf{k} = \mathbf{k}^0 + \mathbf{k}' = (k^0, 0) + (k'^1, k'^2), \tag{1.257}$$

with $|\mathbf{k}'|$ small. Accordingly, expansion of (1.256) shows that

$$\begin{aligned}w(\mathbf{k}) &= w(\mathbf{k}^0) + \frac{\partial w}{\partial k^i}(\mathbf{k}^0)k'^i + \frac{1}{2}\frac{\partial^2 w}{\partial k^i \partial k^j}(\mathbf{k}^0)k'^i k'^j + \cdots \\ &= w^0 + \frac{w_0}{2k^0} k'^1 - \frac{w^0}{8(k^0)^2}(k'^1)^2 + \frac{w^0}{4(k^0)^2}(k'^2)^2 \\ &\quad - \frac{1}{2}w^0(k^0)^2 a^2 + O(|\mathbf{k}'|^2, a^2),\end{aligned} \tag{1.258}$$

where $w_0 = \sqrt{gk^0}$.

On introduction of the complex wave envelope

$$A(x^1, x^2, t) = a(x^1, x^2, t) \exp i\theta(x^1, x^2, t), \tag{1.259}$$

where

$$\frac{\partial \theta}{\partial t} = w^0 - w, \quad \frac{\partial \theta}{\partial x^1} = k'^1, \quad \frac{\partial \theta}{\partial x^2} = k'^2, \tag{1.260}$$

the governing system (1.254)–(1.255), together with (1.258), lead to the *Davey–Stewartson equation*

$$i\left\{\frac{\partial A}{\partial t} + \frac{w^0}{2k^0}\frac{\partial A}{\partial x^1}\right\} - \frac{w^0}{8(k^0)^2}\frac{\partial^2 A}{\partial x^{1\,2}} + \frac{w^0}{4(k^0)^2}\frac{\partial^2 A}{\partial x^{2\,2}} - \frac{1}{2}w^0(k^0)^2|A|^2 A = 0. \quad (1.261)$$

In the one-dimensional case, the change of variable $x'^1 = x^1 - (w^0/2k^0)t$, $t' = t$ and appropriate rescaling reduces (1.261) to the cubic Schrödinger equation (1.253).

Slowly Varying Solitary Waves

In a recent paper by Grimshaw [227], a slowly varying solitary wave solution of the variable coefficient cubic Schrödinger equation

$$iv_2 + \lambda(x^2)v_{11} + v(x^2)|v|^2 v = 0 \quad (1.262)$$

was constructed. In particular, it was shown that under the transformation

$$v = |v/\lambda|^{1/2} u \exp\{-\tfrac{1}{4}i\mu|v/\lambda|(x^1)^2\},$$
$$\xi = |v/\lambda|x^1, \qquad \sigma = (1/\mu)\{|v/\lambda| - |v/\lambda|_{x^2=0}\}, \quad (1.263)$$

and subject only to the constraint

$$(\lambda/v)_{x^2} = -\lambda\mu, \quad (1.264)$$

Eq. (1.262) is transformed to the canonical form

$$iu_\sigma + u_{\xi\xi} \pm |u|^2 u = 0, \quad (1.265)$$

where the sign $\pm$ corresponds to the sign of the quantity λv; μ is an arbitrary nonzero constant.

Nonlinear Waves in a Weakly Inhomogeneous Plasma

Zakharov [228] has shown that the analysis of self-modulation of plasma oscillations in one dimension leads to an equation of the type

$$i\frac{\partial E}{\partial \tau} + \frac{\partial^2 E}{\partial z^2} + 2[|E|^2 - \alpha z]E = 0 \quad (1.266)$$

for an appropriately nondimensionalized electric field E. That (1.266) is reducible to the canonical form (1.253) was demonstrated by Chen and Liu [219]. Thus, it was shown that under the transformation

$$E = u\exp[-2i\alpha\xi\sigma + \tfrac{8}{3}i\alpha^2\sigma^3], \qquad z = \xi - 2\alpha\sigma^2, \qquad \tau = \sigma, \quad (1.267)$$

Eq. (1.266) reduces to the cubic Schrödinger equation

$$iu_\sigma + u_{\xi\xi} + 2|u|^2 u = 0. \quad (1.268)$$

This result has recently been used by Motz et al. [441] in the study of the acceleration and slowing down of nonlinear wave packets in a weakly non-uniform plasma.

Thus, it is seen that the cubic Schrödinger equation has importance not only as a basic model for the evolution of the wave envelope of a weakly nonlinear system with a carrier wave, but also as a canonical form for associated inhomogeneous systems. In the next section, an auto-Bäcklund transformation for the cubic Schrödinger equation is set forth. A permutability theorem is available, albeit of an implicit nature, whereby solutions of this and other associated nonlinear equations may, in principle, be generated.

1.11 BÄCKLUND TRANSFORMATIONS AND THE AKNS SYSTEM. AN INVARIANT TRANSFORMATION OF THE CUBIC SCHRÖDINGER EQUATION

As in Ablowitz et al. [121], the inverse scattering formalism

$$\frac{\partial \psi_1}{\partial x} - \lambda \psi_1 = q(x,t)\psi_2, \qquad \frac{\partial \psi_2}{\partial x} + \lambda \psi_2 = r(x,t)\psi_1 \qquad (1.269)$$

is introduced wherein the eigenfunctions ψ_1, ψ_2 evolve in time according to

$$\begin{aligned}\partial \psi_1/\partial t &= A(x,t;\lambda)\psi_1 + B(x,t;\lambda)\psi_2, \\ \partial \psi_2/\partial t &= C(x,t;\lambda)\psi_1 - A(x,t;\lambda)\psi_2.\end{aligned} \qquad (1.270)$$

If, as in [121], time invariance of the eigenvalues λ is required, then

$$\frac{\partial A}{\partial x} = qC - rB,$$

$$\frac{\partial B}{\partial x} - 2\lambda B = \frac{\partial q}{\partial t} - 2Aq, \qquad (1.271)$$

$$\frac{\partial C}{\partial x} + 2\lambda C = \frac{\partial r}{\partial t} + 2Ar.$$

Specializations of A, B, and C lead to members of the AKNS system of nonlinear evolution equations amenable to the inverse scattering method [120, 121, 260, 308].[†] The link between the latter technique and the construction of Bäcklund transformations is well documented[‡]: it will be alluded

---

[†] The formalism has recently been extended to embrace further important nonlinear evolution equations [264, 272, 273].
[‡] Recent accounts have been given by Konopelchenko [231] and Calogero [309].

to again both in Section 1.14 in connection with Chen's method and in the next chapter in a jet-bundle context. Here, our concern is the construction of certain well-known Bäcklund transformations by means of the formulation. The following members of the AKNS system generated by the indicated specializations of A, B, C, wherein λ is taken as a constant, are noteworthy:

The Korteweg–deVries Equation

$$q_t + 6qq_x + q_{xxx} = 0$$

$$A = -4\lambda^3 - 2\lambda q - q_x, \quad B = -q_{xx} - 2\lambda q_x - 4\lambda^2 q - 2q^2,$$
$$C = 4\lambda^2 + 2q, \quad r = -1. \quad (1.272)$$

The Sine-Gordon Equation

$$u_{xt} = \sin u$$

$$A = \frac{1}{4\lambda}\cos u, \quad B = C = \frac{1}{4\lambda}\sin u, \quad r = -q = \frac{u_x}{2}. \quad (1.273)$$

The Modified Korteweg–deVries Equation

$$q_t + 6q^2 q_x + q_{xxx} = 0$$

$$A = -4\lambda^3 - 2\lambda q^2, \quad B = -q_{xx} - 2\lambda q_x - 4\lambda^2 q - 2q^3,$$
$$C = q_{xx} - 2\lambda q_x + 4\lambda^2 q + 2q^3, \quad r = -q. \quad (1.274)$$

The Cubic Schrödinger Equation

$$iq_t + q_{xx} + 2|q|^2 q = 0$$

$$A = 2i\lambda^2 + i|q|^2, \quad B = iq_x + 2i\lambda q,$$
$$C = i\bar{q}_x - 2i\lambda\bar{q}, \quad r = -\bar{q}. \quad (1.275)$$

Wadati et al. [310] have shown that Bäcklund transformations for subclasses of the AKNS system that include the above nonlinear evolution equations may be readily generated by appeal to an invariance property contained in a paper by Crum [311] on associated *Sturm–Liouville* systems.[†] This states that the Sturm–Liouville equation

$$-\frac{\partial^2 \psi_2}{\partial x^2} + \lambda^2 \psi_2 = q\psi_2 \quad (1.276)$$

is mapped to the associated Sturm–Liouville equation

$$-\frac{\partial^2 \psi'_2}{\partial x^2} + \lambda^2 \psi'_2 = q'\psi'_2, \quad (1.277)$$

[†] A Wronskian representation of N-soliton solutions of the Korteweg–deVries and modified Korteweg–deVries equations based on Crum's work has been given by Satsuma [312].

1.11 BÄCKLUND TRANSFORMATIONS AND THE AKNS SYSTEM

under the transformations

$$q' = q - 2(\log \psi'_2)_{xx}, \qquad \psi'_2 = 1/\psi_2. \tag{1.278}$$

In the application of this result, it will prove convenient to introduce $\Gamma(x, t)$ according to

$$\Gamma = \psi_1/\psi_2, \tag{1.279}$$

so that the scattering equations (1.269) and (1.270) combine to produce the *Riccati* forms

$$\frac{\partial \Gamma}{\partial x} = 2\lambda\Gamma + q - r\Gamma^2, \tag{1.280}$$

$$\frac{\partial \Gamma}{\partial t} = 2A\Gamma + B - C\Gamma^2. \tag{1.281}$$

The following three subclasses of the AKNS system are now considered:

Class I $(r = -1)$ In this case, the scattering equations (1.269) lead to the linear Schrödinger equation (1.276), and under the above invariance property,

$$w' - w = 2\psi_{2,x}/\psi_2 = -2(\lambda + \Gamma), \tag{1.282}$$

where $q = w_x$, $q' = w'_x$. On substitution of Γ as given by (1.282) into the Riccati equations (1.280) and (1.281), we obtain the Bäcklund transformation

$$\begin{aligned} w'_x + w_x &= 2\lambda^2 - \tfrac{1}{2}(w - w')^2, \\ w'_t - w_t &= 4A[\tfrac{1}{2}(w' - w) + \lambda] - 2B + 2C[\tfrac{1}{2}(w' - w) - \lambda]^2 \end{aligned} \tag{1.283}$$

for this subclass of the AKNS system. In particular, with the specializations incorporated in (1.272), the relations (1.283) give the Wahlquist–Estabrook auto-Bäcklund transformation (1.191) of the Korteweg–deVries equation.

Class II $(r = -q)$ In this case, the scattering equations (1.269) yield

$$\phi_{1,xx} - \lambda^2 \phi_1 = U\phi_1 \quad \text{and} \quad \bar{\phi}_{1,xx} - \lambda^2 \bar{\phi}_1 = \bar{U}\bar{\phi}_1, \tag{1.284}$$

where

$$\phi_1 = \psi_1 + i\psi_2, \qquad U = -iq_x - q^2.$$

The system (1.284) is invariant under the transformations

$$\begin{aligned} \phi'_1 &= 1/\phi_1, & U' &= U + 2(\log \phi'_1)_{xx}, \\ \bar{\phi}'_1 &= 1/\bar{\phi}_1, & \bar{U}' &= \bar{U} + 2(\log \bar{\phi}'_1)_{xx}, \end{aligned} \tag{1.285}$$

whence

$$w' = w - i\log(\phi_1/\bar{\phi}_1) = w + 2\tan^{-1}(\psi_2/\psi_1),$$

where $q = w_x$, $q' = w'_x$. Thus, under the transformations (1.285),

$$\Gamma = \cot\{\tfrac{1}{2}(w' - w)\},$$

and insertion of this expression into the Riccati equations (1.280) and (1.281) produces the Bäcklund transformation in the form

$$\begin{aligned} w_x + w'_x &= 2\lambda \sin\{w - w'\}, \\ w_t - w'_t &= 2A\sin\{w' - w\} - (B + C)\cos\{w' - w\} + B - C. \end{aligned} \quad (1.286)$$

The specialization

$$w = -u/2, \qquad \beta = 1/2\lambda,$$

together with (1.273), gives the Bäcklund transformation (1.19) for the sine-Gordon equation. On the other hand, the choice

$$w = -u, \qquad \alpha = -1, \qquad \beta = 2\lambda,$$

together with (1.274) and use of (1.226), produces the Bäcklund transformation (1.224) for the modified Korteweg–deVries equation.

Class III $(r = -\bar{q})$ Konno and Wadati [229] constructed a Bäcklund transformation for this subclass of the AKNS system via Γ' and q', which leave the Riccati equations (1.280) and (1.281) invariant. In particular, the specializations contained in (1.275) were shown to lead to an auto-Bäcklund transformation for the cubic Schrödinger equation, namely,

$$\begin{aligned} q_x + q'_x &= \{q - q'\}\sqrt{4\lambda^2 - |q + q'|^2}, \\ q_t + q'_t &= i\{q_x - q'_x\}\sqrt{4\lambda^2 - |q + q'|^2} \\ &\quad + \frac{i}{2}\{q + q'\}[|q + q'|^2 + |q - q'|^2]. \end{aligned} \quad (1.287)$$

Combination of this result with the invariance of the cubic Schrödinger equation, under the transformations

$$x^* = x - 2kt, \qquad t^* = t, \qquad q^* = qe^{-ikx + ik^2 t}, \quad (1.288)$$

leads to Lamb's auto-Bäcklund transformation as generated by Clairin's procedure.[†]

If q_1 and q_2 are solutions of the cubic Schrödinger equation generated by the Bäcklund transformation (1.287) from the solution q_0 and with Bäcklund parameters λ_1 and λ_2, respectively, and if, in accordance with the usual Bianchi diagram, $q_{12} = \mathbb{B}_{\lambda_2}\mathbb{B}_{\lambda_1}q_0 = \mathbb{B}_{\lambda_1}\mathbb{B}_{\lambda_2}q_0$, then the Bäcklund transfor-

---

[†] A Bäcklund transformation for the cubic Schrödinger equation in $2 + 1$ dimensions, namely, the Davey–Stewartson equation (1.261), has recently been obtained by Levi, Pilloni and Santini [267].

1.12 THE HIROTA BILINEAR OPERATOR FORMULATION 69

mation (1.287) produces the algebraic permutability relation

$$(q_0 - q_1)\sqrt{4\lambda_1^2 - |q_0 + q_1|^2} - (q_0 - q_2)\sqrt{4\lambda_2^2 - |q_0 + q_2|^2}$$
$$+ (q_2 - q_{12})\sqrt{4\lambda_1^2 - |q_2 + q_{12}|^2} \qquad (1.289)$$
$$- (q_1 - q_{12})\sqrt{4\lambda_2^2 - |q_1 + q_{12}|^2} = 0,$$

which determines, albeit implicitly, the further solution q_{12} of the cubic Schrödinger equation. It is noted that an alternative implicit nonlinear superposition principle has been given by Gerdzhikov and Kulish [230] in a recent paper on the derivation of the Bäcklund transformation for the cubic nonlinear Schrödinger equation via the Gel'fand–Levitan–Marchenko integral equation.

In conclusion, we remark that recently Boiti and Pempinelli [433] have used an auto-Bäcklund transformation of the cubic Schrödinger equation to reveal, via the language of Painlevé transcendents, hidden symmetries of its similarity solutions. Further, if the general stationary solution of the cubic Schrödinger equation is expressed in terms of the Weierstrass elliptic ℘-functions, then an application of the Bäcklund transformation may be used to retrieve Bianchi's celebrated formula for the zeros of a fourth order polynomial in terms of elliptic ℘-forms (Bianchi [434]). These and other related results indicate the growing interest in Bäcklund transformations in the context of the theory of ordinary differential equations [435–440].

1.12 THE HIROTA BILINEAR OPERATOR FORMULATION OF BÄCKLUND TRANSFORMATIONS. THE BOUSSINESQ EQUATION. KAUP'S HIGHER ORDER WATER WAVE EQUATION

Hirota and Satsuma [201] have recently shown that the permutability theorems for a wide class of nonlinear evolution equations adopt the same simple structure when written in terms of certain bilinear operators. The latter formulation was originally adopted by Hirota [118] in connection with a direct expansion technique for the generation of multi-soliton solutions.[†] Here, Hirota's bilinear operator formalism is introduced and used to derive Bäcklund transformations associated with both Boussinesq's equation and Kaup's higher order water wave equation. In subsequent sections, the same method is used to construct Bäcklund transformations for higher order Korteweg–deVries equations, the Kadomtsev–Petviashvili equation, and the Benjamin–Ono equation.

The Boussinesq equation [232, 237]

$$u_{tt} - u_{xx} - 3(u^2)_{xx} - u_{xxxx} = 0, \qquad (1.290)$$

[†] A review of the bilinear operator method is given by Hirota in [234].

like the Korteweg–deVries equation, arises in the analysis of plane gravity waves but is not restricted to unidirectional wave propagation.
If we set

$$u = 2(\log f)_{xx}, \quad f > 0, \quad (1.291)$$

then, on introduction of the *Hirota bilinear operators* D_x and D_t according to

$$D_t^n D_x^m a \circ b := \left(\frac{\partial}{\partial t} - \frac{\partial}{\partial t'}\right)^n \left(\frac{\partial}{\partial x} - \frac{\partial}{\partial x'}\right)^m a(t,x) b(t',x')\bigg|_{x'=x,\,t'=t}, \quad (1.292)$$

we appeal to properties set out in Appendix I in order to associate (1.290) with a bilinear form. Thus, on use of (I.2) and (I.3), respectively, we see that

$$u = f^{-2} D_x^2 f \circ f, \quad (1.293)$$

$$u_{xx} = f^{-2} D_x^4 f \circ f - 12[f^{-2} D_x^2 f \circ f/2]^2,$$

$$= f^{-2} D_x^4 f \circ f - 3u^2. \quad (1.294)$$

Again, from (I.2),

$$2(\log f)_{tt} = f^{-2} D_t^2 f \circ f, \quad (1.295)$$

whence, on use of (1.293) and (1.294),

$$2(\log f)_{tt} - u - 3u^2 - u_{xx} = f^{-2}[(D_t^2 - D_x^2 - D_x^4) f \circ f]. \quad (1.296)$$

Accordingly,

$$2(\log f)_{ttxx} - u_{xx} - 3(u^2)_{xx} - u_{xxxx} = \frac{\partial^2}{\partial x^2}\left[\frac{1}{f^2}(D_t^2 - D_x^2 - D_x^4) f \circ f\right],$$

that is,

$$u_{tt} - u_{xx} - 3(u^2)_{xx} - u_{xxxx} = \frac{\partial^2}{\partial x^2}\left[\frac{1}{f^2}(D_t^2 - D_x^2 - D_x^4) f \circ f\right]. \quad (1.297)$$

Hence, if f is a solution of

$$(D_t^2 - D_x^2 - D_x^4) f \circ f = 0, \quad (1.298)$$

then u as given by (1.291) is a solution of the Boussinesq equation.

A Bäcklund transformation is now established for Eq. (1.298) rather than (1.290). Thus, relations

$$(D_t + \alpha D_x^2) f \circ f' - \alpha \beta f f' = 0, \quad (1.299)$$

$$\{(1 + 3\beta) D_x + D_x^3 + \alpha D_t D_x\} f \circ f' = 0 \quad (1.300)$$

are introduced between f and f' and it is shown that, subject to an appropriate condition on α, if f is a solution of (1.298), then so also is f', and conversely. Specifically, it is demonstrated that relations (1.299) and (1.300)

1.12 THE HIROTA BILINEAR OPERATOR FORMULATION

imply that

$$\mathbb{P} := [(D_t^2 - D_x^2 - D_x^4)f \circ f]f'f' - ff[(D_t^2 - D_x^2 - D_x^4)f' \circ f']$$

vanishes, provided that $\alpha^2 = -3$.

Thus, use of the properties (I.4) and (I.5) yields

$$(D_t^2 f \circ f)f'f' - ff(D_t^2 f' \circ f') = 2D_t(D_t f \circ f') \circ ff',$$

$$(D_x^2 f \circ f)f'f' - ff(D_x^2 f' \circ f') = 2D_x(D_x f \circ f') \circ ff',$$

$$(D_x^4 f \circ f)f'f' - ff(D_x^4 f' \circ f') = 2D_x(D_x^3 f \circ f') \circ ff'$$
$$- 6D_x(D_x^2 f \circ f') \circ (D_x f \circ f'),$$

whence

$$\mathbb{P} = 2[D_t(D_t f \circ f') \circ ff' - D_x[(D_x + D_x^3)f \circ f'] \circ ff'$$
$$+ 3D_x(D_x^2 f \circ f') \circ (D_x f \circ f')]. \quad (1.301)$$

Now, from (1.299) and (1.300),

$$D_t f \circ f' = \alpha \beta ff' - \alpha D_x^2 f \circ f', \quad (D_x + D_x^3)f \circ f' = -[\alpha D_t D_x + 3\beta D_x]f \circ f',$$

and on insertion into (1.301), it is seen that

$$\mathbb{P} = 2[D_t(\alpha \beta ff' - \alpha D_x^2 f \circ f') \circ ff' + D_x[(\alpha D_t D_x + 3\beta D_x)f \circ f'] \circ ff'$$
$$+ 3D_x(D_x^2 f \circ f') \circ (D_x f \circ f')]$$
$$= 2[\alpha[D_x(D_t D_x f \circ f') \circ ff' - D_t(D_x^2 f \circ f') \circ ff']$$
$$+ 3[D_x(D_x^2 f \circ f' - \beta ff') \circ (D_x f \circ f')]],$$

on use of (1.4).

Finally, from (I.6),

$$D_x(D_t D_x f \circ f') \circ ff' - D_t(D_x^2 f \circ f') \circ ff' = -D_x(D_t f \circ f') \circ (D_x f \circ f'),$$

whence

$$\mathbb{P} = -2D_x[(\alpha D_t - 3D_x^2)f \circ f' + 3\beta ff'] \circ (D_x f \circ f'). \quad (1.302)$$

If we now set $\alpha^2 = -3$, then (1.302) yields

$$\mathbb{P} = -2\alpha D_x[(D_t + \alpha D_x^2)f \circ f' - \alpha \beta ff'] \circ (D_x f \circ f') = 0$$

by virtue of relation (1.299). This completes the demonstration. ■

Let f_0 be a solution of (1.298) and suppose that f_1 and $\hat{f}_1$ are solutions of (1.298) given by the Bäcklund relations (1.299) and (1.300) with starting solution $f = f_0$ and Bäcklund parameters $\beta = \beta_1$ and $\beta = \beta_2$, respectively. Let us assume that f_2 exists such that $f_2 = \mathbb{B}_{\beta_1}\mathbb{B}_{\beta_2}f_0 = \mathbb{B}_{\beta_2}\mathbb{B}_{\beta_1}f_0$ so that a commutative Bianchi diagram may be constructed (Fig. 1.16). This embodies

1. BÄCKLUND TRANSFORMATIONS AND NONLINEAR EQUATIONS

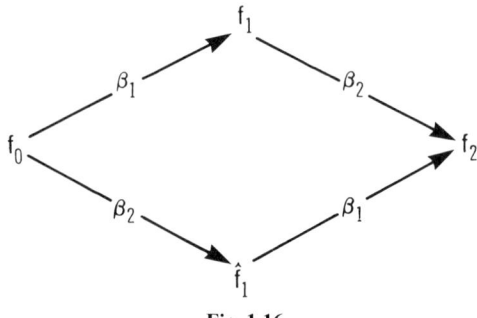

Fig. 1.16

the relations

$$(D_t + \alpha D_x^2)f_0 \circ f_1 = \alpha\beta_1 f_0 f_1, \tag{1.303}$$

$$(D_t + \alpha D_x^2)f_0 \circ \hat{f}_1 = \alpha\beta_2 f_0 \hat{f}_1, \tag{1.304}$$

$$(D_t + \alpha D_x^2)\hat{f}_1 \circ f_2 = \alpha\beta_1 \hat{f}_1 f_2, \tag{1.305}$$

$$(D_t + \alpha D_x^2)f_1 \circ f_2 = \alpha\beta_2 f_1 f_2, \tag{1.306}$$

where $\alpha^2 = -3$. Multiplication of (1.303) by $\hat{f}_1 f_2$ and of (1.305) by $f_0 f_1$ and subsequent subtraction yield

$$(D_t f_0 \circ f_1)\hat{f}_1 f_2 - f_0 f_1(D_t \hat{f}_1 \circ f_2)$$
$$+ \alpha[(D_x^2 f_0 \circ f_1)\hat{f}_1 f_2 - f_0 f_1(D_x^2 \hat{f}_1 \circ f_2)] = 0. \tag{1.307}$$

Similarly, (1.304) and (1.306) give

$$(D_t f_0 \circ \hat{f}_1)f_1 f_2 - f_0 \hat{f}_1(D_t f_1 \circ f_2)$$
$$+ \alpha[(D_x^2 f_0 \circ \hat{f}_1)f_1 f_2 - f_0 \hat{f}_1(D_x^2 f_1 \circ f_2)] = 0. \tag{1.308}$$

On use of (I.4) and (I.12), Eqs. (1.307) and (1.308) become

$$(D_t f_0 \circ \hat{f}_1)f_1 f_2 - f_0 \hat{f}_1(D_t f_1 \circ f_2)$$
$$+ \alpha D_x[(D_x f_0 \circ f_2) \circ f_1 \hat{f}_1 + f_0 f_2 \circ (D_x \hat{f}_1 \circ f_1)] = 0,$$

$$(D_t f_0 \circ \hat{f}_1)f_1 f_2 - f_0 \hat{f}_1(D_t f_1 \circ f_2)$$
$$+ \alpha D_x[(D_x f_0 \circ f_2) \circ f_1 \hat{f}_1 + f_0 f_2 \circ (D_x f_1 \circ \hat{f}_1)] = 0,$$

respectively, so that on subtraction and use of (I.1),

$$D_x[(f_0 f_2) \circ (D_x f_1 \circ \hat{f}_1)] = 0.$$

Accordingly,

$$f_0 f_2 = \text{const } D_x f_1 \circ \hat{f}_1, \tag{1.309}$$

and it is readily demonstrated that f_2 as given by (1.309) is indeed a solution of (1.298).

1.12 THE HIROTA BILINEAR OPERATOR FORMULATION

More generally,
$$f_{N-1}f_{N+1} = \text{const } D_x f_N \circ \hat{f}_N, \qquad (1.310)$$
where
$$f_{N-1} = f_{N-1}(\beta_1, \beta_2, \ldots, \beta_{N-1}), \quad f_N = f_N(\beta_1, \beta_2, \ldots, \beta_N),$$
$$\hat{f}_N = \hat{f}_N(\beta_1, \beta_2, \ldots, \beta_{N-1}, \beta_{N+1}),$$
$$f_{N+1} = f_{N+1}(\beta_1, \beta_2, \ldots, \beta_{N-1}, \beta_N, \beta_{N+1})$$
(see Fig. 1.17).

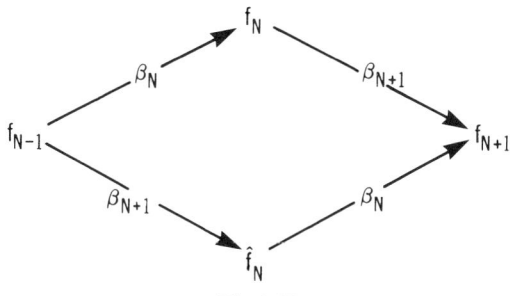

Fig. 1.17

Relation (1.310) represents the permutability theorem for the Boussinesq equation in bilinear form. In fact, it may be shown that it is generic for a wide class of nonlinear evolution equations (Hirota and Satsuma [201]). Here we show that indeed the same superposition theorem obtains for the Korteweg–deVries equation. It is then used to generate N-soliton solutions.

If, again, u is introduced as in (1.291), then (I.2) shows that
$$u_t = \frac{\partial}{\partial x}\left[\frac{1}{f^2} D_x D_t f \circ f\right], \qquad (1.311)$$
while from (1.294),
$$6uu_x + u_{xxx} = \frac{\partial}{\partial x}\left[\frac{1}{f^2} D_x^4 f \circ f\right]. \qquad (1.312)$$

Combination of (1.311) and (1.312) yields
$$u_t + 6uu_x + u_{xxx} = \frac{\partial}{\partial x}\left[\frac{1}{f^2} D_x(D_t + D_x^3) f \circ f\right].$$

Accordingly, u is a solution of the Korteweg–deVries equation
$$u_t + 6uu_x + u_{xxx} = 0 \qquad (1.313)$$
if f is a solution of the associated bilinear equation
$$D_x(D_t + D_x^3) f \circ f = 0. \qquad (1.314)$$

74   1. BÄCKLUND TRANSFORMATIONS AND NONLINEAR EQUATIONS

Hirota [233] has constructed the auto-Bäcklund transformation

$$(D_t + 3\beta D_x + D_x^3)f \circ f' = 0, \tag{1.315}$$

$$(D_x^2 - \beta)f \circ f' = 0 \tag{1.316}$$

for (1.314). The invariance of the latter bilinear equation under the Bäcklund relations (1.315) and (1.316) is readily established. Thus, (I.5) and (I.8) show in turn that

$$(D_x^4 f \circ f)f'f' - ff(D_x^4 f' \circ f') = 2D_x(D_x^3 f \circ f') \circ f'f$$
$$+ 6D_x(D_x^2 f \circ f') \circ (D_x f' \circ f),$$

$$(D_x D_t f \circ f)f'f' - ff(D_x D_t f' \circ f') = 2D_x[(D_t f \circ f') \circ f'f],$$

and, on addition of these two relations we obtain

$$[D_x(D_t + D_x^3)f \circ f]f'f' - ff[D_x(D_t + D_x^3)f' \circ f']$$
$$= 2D_x[(D_t + D_x^3)f \circ f'] \circ f'f + 6D_x(D_x^2 f \circ f') \circ (D_x f' \circ f)$$
$$= 2D_x[(D_t + D_x^3)f \circ f'] \circ f'f + 6\beta D_x(D_x f \circ f') \circ f'f$$
$$= 2D_x[(D_t + 3\beta D_x + D_x^3)f \circ f'] \circ f'f = 0.$$

Consequently,

$$ff[D_x(D_t + D_x^3)f' \circ f'] = f'f'[D_x(D_t + D_x^3)f \circ f],$$

so that if f is a solution of (1.314), so is f', and conversely. Thus, the relations (1.315) and (1.316) do indeed determine an auto-Bäcklund transformation of (1.314). The Estabrook–Wahlquist transformation (1.191) for the Korteweg–deVries equation is readily shown to be a consequence of this result for the bilinear representation.

Let f_0 be a solution of (1.314) and suppose that f_1 and $\hat{f}_1$ are new solutions generated by the Bäcklund transformation (1.315) and (1.316) with starting solution $f = f_0$ and Bäcklund parameters $\beta = \beta_1$ and $\beta = \beta_2$, respectively. Let us assume that f_2 exists such that $f_2 = \mathbb{B}_{\beta_1}\mathbb{B}_{\beta_2}f_0 = \mathbb{B}_{\beta_2}\mathbb{B}_{\beta_1}f_0$, whence a commutative Bianchi diagram may be constructed (Fig. 1.16).

Now (1.316) shows that

$$D_x^2 f_0 \circ f_1 = \beta_1 f_0 f_1, \tag{1.317}$$

$$D_x^2 f_0 \circ \hat{f}_1 = \beta_2 f_0 \hat{f}_1, \tag{1.318}$$

$$D_x^2 \hat{f}_1 \circ f_2 = \beta_1 \hat{f}_1 f_2, \tag{1.319}$$

$$D_x^2 f_1 \circ f_2 = \beta_2 f_1 f_2. \tag{1.320}$$

Multiplication of (1.317) by $\hat{f}_1 f_2$ and of (1.319) by $f_0 f_1$ and subsequent subtraction yield

$$\hat{f}_1 f_2 (D_x^2 f_0 \circ f_1) - f_0 f_1 (D_x^2 \hat{f}_1 \circ f_2) = 0$$

1.12 THE HIROTA BILINEAR OPERATOR FORMULATION

or, on use of (I.4),
$$D_x[(D_x f_0 \circ f_2) \circ f_1 \hat{f}_1 + f_0 f_2 \circ (D_x \hat{f}_1 \circ f_1)] = 0. \tag{1.321}$$

Similarly, (1.318) and (1.320) may be combined to give
$$D_x[(D_x f_1 \circ \hat{f}_1) \circ f_0 f_2 + f_1 \hat{f}_1 \circ (D_x f_0 \circ f_2)] = 0. \tag{1.322}$$

Addition of (1.321) and (1.322) shows that
$$D_x[(f_0 f_2) \circ (D_x f_1 \circ \hat{f}_1)] = 0, \tag{1.323}$$

while subtraction yields
$$D_x[(f_1 \hat{f}_1) \circ (D_x f_0 \circ f_2)] = 0. \tag{1.324}$$

The relations (1.323) and (1.324) imply, in turn, the alternative superposability laws
$$f_0 f_2 = \text{const } D_x f_1 \circ \hat{f}_1 \tag{1.325}$$
and
$$f_1 \hat{f}_1 = \text{const } D_x f_0 \circ f_2. \tag{1.326}$$

In particular, the permutability theorem (1.205) for the Korteweg–deVries equation (1.313) may be shown to be a consequence of (1.325). Moreover, the latter explicit law leads to the same permutability relation
$$f_{N-1} f_{N+1} = \text{const } D_x f_N \circ \hat{f}_N \tag{1.327}$$

as was obtained for the Boussinesq equation. Furthermore, Hirota and Satsuma [201] have demonstrated that the same law applies for both the modified Korteweg–deVries equation and Kadomtsev–Petviashvili equation. A Bäcklund transformation and its associated permutability theorem are constructed for the Kadomtsev–Petviashvili equation in Section 1.14.

N-Soliton Solutions

We now turn to the application of the bilinear Bäcklund transformation (1.315)–(1.316) of (1.314) to the generation of N-soliton solutions of the Korteweg–deVries equation.

Thus, insertion of the vacuum solution $f_0 = 1$ of (1.314) into the Bäcklund relations (1.315) and (1.316) yields
$$f'_t + 3\beta f'_x + f'_{xxx} = 0, \qquad f'_{xx} - \beta f' = 0 \tag{1.328}$$

on use of the result
$$D_x^m a \circ 1 = \frac{\partial^m}{\partial x^m} a.$$

It is readily seen that the *linear* system (1.328) leads to a nontrivial solution of (1.314) in the form

$$f' = \exp[k_1(x - 4k_1^2 t) + \gamma_1] + \exp[-k_1(x - 4k_1^2 t) - \gamma_1] \quad (\beta = k_1^2), \quad (1.329)$$

where γ_1 is an arbitrary phase constant.

A routine calculation shows that if

$$f(x, t) = e^{ax + b(t)} g(x, t), \quad (1.330)$$

then

$$D_x(D_t + D_x^3) f \circ f = e^{2(ax + b(t))} D_x(D_t + D_x^3) g \circ g,$$

so that, if g is a solution of (1.314), then so also is f. Moreover,

$$u = 2(\log f)_{xx} = 2(\log g)_{xx},$$

so that f and g generate the *same* solution of the Korteweg–deVries equation (1.313). Accordingly,

$$f_1 = \exp[k_1(x - 4k_1^2 t) + \gamma_1] f'(x, t) = 1 + e^{\eta_1}, \quad (1.331)$$

where

$$\eta_1 = 2k_1(x - 4k_1^2 t) + 2\gamma_1, \quad (1.332)$$

leads to the same solution of (1.313) as does (1.329), namely,

$$u_1 = 2 \frac{\partial^2}{\partial x^2} \log(1 + e^{\eta_1}) = 2k_1^2 \operatorname{sech}^2\left(\frac{\eta_1}{2}\right). \quad (1.333)$$

Thus, the first-generation Bäcklund solution represents a one soliton.

Application of the permutability law (1.325) with $f_0 = 1$ and

$$\begin{aligned} f' &= e^{\eta_1/2} + e^{-\eta_1/2}, \\ \hat{f}' &= e^{\eta_2/2} + e^{-\eta_2/2}, \end{aligned} \qquad \eta_i = 2k_i(x - 4k_i^2 t) + 2\gamma_i, \quad i = 1, 2, \quad (1.334)$$

now leads to a new solution of (1.314), namely,

$$f_2' = \operatorname{const} D_x f' \circ \hat{f}' = \operatorname{const} \{f_x' \hat{f}' - f' \hat{f}_x'\},$$

that is,

$$f_2' = A \exp\left[-\frac{1}{2}(\eta_1 + \eta_2)\right]$$
$$\times \left\{1 - \frac{k_1 + k_2}{k_1 - k_2} \exp(\eta_1) + \frac{k_1 + k_2}{k_1 - k_2} \exp(\eta_2) - \exp(\eta_1 + \eta_2)\right\}, \quad (1.335)$$

where A is an arbitrary constant.

1.12 THE HIROTA BILINEAR OPERATOR FORMULATION

If we now define

$$\gamma_1^{(0)} = \log\left(\frac{k_1 + k_2}{k_2 - k_1}\right), \qquad \gamma_2^{(0)} = \log\left(\frac{k_1 + k_2}{k_1 - k_2}\right),$$

$$\bar{\eta}_i = \eta_i + \gamma_i^{(0)}, \qquad A_{ij} = \log\left(\frac{k_i - k_j}{k_i + k_j}\right)^2, \tag{1.336}$$

then f'_2 may be written as

$$f'_2 = A \exp[-\tfrac{1}{2}(\bar{\eta}_1 + \bar{\eta}_2 - \gamma_1^{(0)} - \gamma_2^{(0)})]$$
$$\times \{1 + \exp(\bar{\eta}_1) + \exp(\bar{\eta}_2) + \exp(\bar{\eta}_1 + \bar{\eta}_2 + A_{12})\}.$$

Hence,

$$f_2 = 1 + \exp(\bar{\eta}_1) + \exp(\bar{\eta}_2) + \exp(\bar{\eta}_1 + \bar{\eta}_2 + A_{12}) \tag{1.337}$$

may be taken as the second-generation Bäcklund solution of (1.314). It may be shown to correspond to a two-soliton solution of the Korteweg–deVries equation (Hirota [118]).

A second application of the permutability law leads to a third-generation Bäcklund solution in the form

$$f_3 = 1 + \exp(\bar{\eta}_1) + \exp(\bar{\eta}_2) + \exp(\bar{\eta}_3)$$
$$+ \exp(\bar{\eta}_1 + \bar{\eta}_2 + A_{12}) + \exp(\bar{\eta}_2 + \bar{\eta}_3 + A_{23}) + \exp(\bar{\eta}_3 + \bar{\eta}_1 + A_{31})$$
$$+ \exp(\bar{\eta}_1 + \bar{\eta}_2 + \bar{\eta}_3 + A_{12} + A_{13} + A_{23}),$$

$$= \sum_{\mu=0,1} \exp\left\{\sum_{i<j}^{(3)} \mu_i \mu_j A_{ij} + \sum_{i=1}^{3} \mu_i \bar{\eta}_i\right\},$$

where $\sum_{\mu=0,1}$ here indicates summation over all possible combinations of $\mu_1 = 0, 1$, $\mu_2 = 0, 1$, $\mu_3 = 0, 1$, while $\sum_{i<j}^{(3)}$ denotes summation over all possible combinations of i and j from 1 to 3 subject to the requirement $i < j$.

In general, an Nth-generation Bäcklund solution may be constructed in the form

$$f_N = \sum_{\mu=0,1} \exp\left\{\sum_{i<j}^{(N)} \mu_i \mu_j A_{ij} + \sum_{i=1}^{N} \mu_i \bar{\eta}_i\right\}, \tag{1.338}$$

where $\sum_{\mu=0,1}$ now indicates summation over all possible combinations of $\mu_1 = 0, 1$, $\mu_2 = 0, 1, \ldots, \mu_N = 0, 1$, while $\sum_{i<j}^{(N)}$ designates summation over all possible combinations of i and j from 1 to N subject to the requirement $i < j$. That the expression (1.338) corresponds to an N-soliton solution of the Korteweg–deVries equation was demonstrated by Hirota [118]. Extensions to slowly varying solitary waves and cylindrical solitons have been made possible by recent work of Grimshaw [238] and Hirota [239, 240], respectively, wherein certain variable coefficient Korteweg–deVries equations have been shown to be reducible to the canonical form (1.313). Indeed, multi-soliton solutions for a wide class of nonlinear evolution equations adopt the functional form (1.338) (see Hirota [233]).

In fact, solutions of nonlinear evolution equations other than the Boussinesq equation may be generated based on multi-soliton solutions of the Bäcklund relations (1.299) and (1.300). In particular, solutions of a higher order water wave equation due to Kaup [236] may be so derived in the manner indicated below.

Thus, consider the propagation of water waves in an infinite narrow channel of constant mean depth h. The free surface conditions are

$$\eta_T + U\eta_X - V = 0$$
$$U_T + UU_X + VV_X + \eta_X g - \rho^{-1}\tau\eta_{XXX} = 0 \quad \text{at} \quad Y = h + \eta, \quad (1.339)$$

where X is the coordinate parallel to the channel, Y the vertical coordinate, η the amplitude of the wave, (U, V) are the velocity components, τ is the surface tension, and ρ the mean density of the inviscid, incompressible fluid. The motion is assumed to be irrotational so that a velocity potential Φ may be introduced such that

$$\nabla\Phi = (U, V). \quad (1.340)$$

On appropriate scaling and expansion in terms of the quantities δ and ε, where δ is the ratio of depth/wave length and ε is the ratio of the wave amplitude/depth, it emerges that Φ (evaluated at $Y = 0$) satisfies the equations (Kaup [236])

$$\pi_t = \Phi_{xx} + \delta^2(\tfrac{1}{3} - \sigma)\Phi_{xxxx} - \varepsilon(\Phi_x\pi)_x + O(\delta^4, \varepsilon\delta^2), \quad (1.341)$$

$$\pi = \Phi_t + \tfrac{1}{2}\varepsilon\Phi_x^2, \quad (1.342)$$

where $\sigma = \tau/(\rho g h^2)$ and (x, t) are scaled, unitless (X, T) coordinates.

We now return to the Bäcklund relations (1.299) and (1.300) originally introduced in association with the Boussinesq equation corresponding to $\alpha^2 = -3$. It is natural to inquire as to whether or not any other value of α leads to results of interest. In fact, Hirota and Satsuma [235] have shown that the specialization $\alpha = 1$ has importance in connection both with a nonlinear Schrödinger equation and with Kaup's higher order water wave equation. The method is illustrated for the latter case.

Thus, introduction of the change of variables

$$\phi = \log(f'/f), \quad (1.343)$$

$$\psi = \log(ff'), \quad (1.344)$$

1.13 HIGHER ORDER KORTEWEG–deVRIES EQUATIONS

into the Bäcklund relations (1.299) and (1.300) yields

$$\begin{aligned}\phi_t - \alpha[\psi_{xx} + (\phi_x)^2] + \alpha\beta &= 0, \\ \alpha(\psi_{xt} + \phi_x\phi_t) - (1 + 3\beta)\phi_x - \phi_{xxx} - 3\phi_x\psi_{xx} - (\phi_x)^3 &= 0.\end{aligned} \quad (1.345)$$

The transformations

$$\begin{aligned}\Phi &= (-2\gamma/\varepsilon)\phi, & \pi &= (-2/\varepsilon)\psi_{xx} \\ x' &= \gamma x, & t' &= \gamma t\end{aligned} \quad (\gamma^2 = \delta^2(\tfrac{1}{3} - \sigma)), \quad (1.346)$$

together with specializations $\alpha = 1$, $\beta = 0$ now take (1.345) to the Kaup system

$$\begin{aligned}\pi &= \Phi_{t'} + \tfrac{1}{2}\varepsilon(\Phi_{x'})^2, \\ \pi_{t'} &= \Phi_{x'x'} + \delta^2(\tfrac{1}{3} - \sigma)\Phi_{x'x'x'x'} - \varepsilon(\Phi_{x'}\pi)_{x'}.\end{aligned} \quad (1.347)$$

Hirota and Satsuma [235] demonstrated that the Bäcklund relations (1.299) and (1.300) admit N-soliton solutions, wherein f and f' adopt the usual forms

$$\begin{aligned}f &= \sum_{\mu=0,1} \exp\left\{\sum_{i<j}^{(N)} A_{ij}\mu_i\mu_j + \sum_{i=1}^N \mu_i\eta_i\right\}, \\ f' &= \sum_{\mu=0,1} \exp\left\{\sum_{i<j}^{(N)} A_{ij}\mu_i\mu_j + \sum_{i=1}^N \mu_i(\eta_i + \theta_i)\right\},\end{aligned} \quad (1.348)$$

both in the case $\alpha^2 = -3$ corresponding to the Boussinesq equation and in the case $\alpha = 1$ for Kaup's equation. In the latter instance, the relations (1.343), (1.346), and (1.348) combine to give a multiparameter solution

$$\begin{aligned}\Phi = -\frac{2\gamma}{\varepsilon}\Bigg\{&\log\left[\sum_{\mu=0,1} \exp\left\{\sum_{i<j}^{(N)} A_{ij}\mu_i\mu_j + \sum_{i=1}^N \mu_i(\eta_i + \theta_i)\right\}\right] \\ &- \log\left[\sum_{\mu=0,1} \exp\left\{\sum_{i<j}^{(N)} A_{ij}\mu_i\mu_j + \sum_{i=1}^N \mu_i\eta_i\right\}\right]\Bigg\}\end{aligned} \quad (1.349)$$

of Kaup's nonlinear system (1.347).

1.13 BÄCKLUND TRANSFORMATIONS FOR HIGHER ORDER KORTEWEG–deVRIES EQUATIONS

Hierarchies of Korteweg–deVries equations have been investigated both by Lax [241] and Caudrey *et al.* [242]. Some attention in the latter paper was concentrated on the higher order Korteweg–deVries equation (Sawada and Kotera [245])

$$u_t + 180u^2 u_x + 30(uu_{xxx} + u_x u_{xx}) + u_{xxxxx} = 0, \quad (1.350)$$

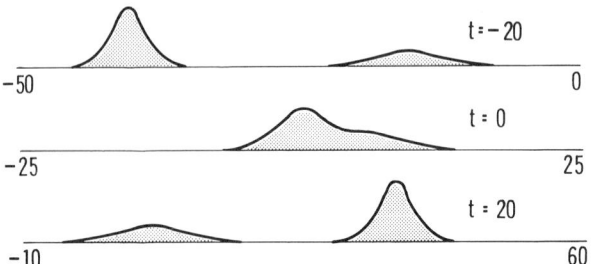

Fig. 1.18 Three plots of $u(x, t)$ against x for $t = -20, 0$, and 20 which describe a two-soliton collision for $u_t + 180u^2 u_x + 30(uu_{xxx} + u_x u_{xx}) + u_{xxxxx} = 0$. (Caudrey et al. [242].)

which was shown to have N-soliton solutions. A two-soliton collision is shown in Fig. 1.18. That (1.350) admits a Bäcklund transformation was subsequently demonstrated by Satsuma and Kaup [243], and it is their result that we now discuss.[†]

Thus, a natural higher order generalization of the bilinear representation (1.314) of the Korteweg–deVries equation (1.313) is the bilinear equation

$$D_x(D_t + D_x^5)f \circ f = 0, \tag{1.351}$$

leading to (1.350). Satsuma and Kaup constructed the auto-Bäcklund transformation

$$(D_t - \tfrac{15}{2}\beta D_x^2 - \tfrac{3}{2}D_x^5)f' \circ f = 0, \tag{1.352}$$

$$(D_x^3 - \beta)f' \circ f = 0 \tag{1.353}$$

of (1.351), where β is a Bäcklund parameter. The method of derivation in [243] was to show that

$$\mathbb{P} := f'f' D_x(D_t + D_x^5)f \circ f - ff D_x(D_t + D_x^5)f' \circ f' \tag{1.354}$$

vanishes if (1.352) and (1.353) hold. Accordingly, on use of (I.8)–(I.10), we see that

$$\mathbb{P} = D_x[2(f'f) \circ (D_t f' \circ f) + \tfrac{3}{4}(f'f) \circ (D_x^5 f' \circ f)$$
$$- \tfrac{15}{4}(D_x f' \circ f) \circ (D_x^4 f' \circ f) + \tfrac{15}{2}(D_x^2 f' \circ f) \circ (D_x^3 f' \circ f)]$$
$$+ \tfrac{5}{4}D_x^3[(f'f) \circ (D_x^3 f' \circ f) - 3(D_x f' \circ f) \circ (D_x^2 f' \circ f)],$$

whence, on substitution of (I.11),

$$\mathbb{P} = D_x[2(f'f) \circ (D_t f'f) - 3(f'f) \circ (D_x^5 f' \circ f)$$
$$+ 15(D_x^2 f' \circ f) \circ (D_x^3 f' \circ f)] + 5D_x^3(f'f) \circ (D_x^3 f' \circ f). \tag{1.355}$$

[†] Bäcklund transformations for the Lax hierarchy of higher order Korteweg–deVries equations have recently been constructed by Matsuno [413].

1.13 HIGHER ORDER KORTEWEG–deVRIES EQUATIONS

Insertion of the Bäcklund relations (1.352) and (1.353) into (1.355) yields

$$\mathbb{P} = D_x[15\beta(f'f) \circ (D_x^2 f' \circ f) + 15\beta(D_x^2 f' \circ f) \circ (f'f)] = 0, \quad (1.356)$$

and the required result is established.

The Bäcklund transformation (1.352)–(1.353) may be reformulated more conventionally in terms of the potential

$$w = \int_{-\infty}^{x} u \, dx \quad (1.357)$$

and its derivatives. Thus, introduction of ϕ and ψ, as defined by (1.343) and (1.344), into (1.352)–(1.353) yields

$$\phi_t = \tfrac{3}{2}\phi_{xxxxx} + \tfrac{15}{2}\phi_x\psi_{xxxx} + 15\phi_{xxx}\psi_{xx} + 15(\phi_x)^2\phi_{xxx}$$
$$+ \tfrac{45}{2}\phi_x(\psi_{xx})^2 + 15(\phi_x)^3\psi_{xx} + \tfrac{3}{2}(\phi_x)^5 + \tfrac{15}{2}\beta\psi_{xx} + \tfrac{15}{2}\beta(\phi_x)^2, \quad (1.358)$$

$$\phi_{xxx} + 3\phi_x\psi_{xx} + (\phi_x)^3 = \beta. \quad (1.359)$$

Introduction of the relations

$$\phi_x = w' - w, \quad (1.360)$$

$$\psi_x = w' + w \quad (1.361)$$

into (1.358) and (1.359) now reduces the Bäcklund transformation to the form obtained independently by Dodd and Gibbon [244] via the inverse scattering formulation, namely,

$$(w' - w)_t = \tfrac{3}{2}[(w' - w)_{xxxx} + 5(w' - w)(w' + w)_{xxx}$$
$$+ 15(w' - w)_{xx}(w' + w)_x$$
$$+ 15(w' - w)^2(w' - w)_{xx} + 30(w' - w)(w' + w)_x^2$$
$$+ 30(w' - w)^3(w' + w)_x + 6(w' - w)^5]_x, \quad (1.362)$$

$$(w' - w)_{xx} + 3(w' - w)(w' + w)_x + (w' - w)^3 = \beta. \quad (1.363)$$

In an interesting recent development, Hirota and Ramani [246] have constructed a Bäcklund transformation which links another higher order Korteweg–deVries equation due to Kaup [247], namely,

$$u_t + 180u^2 u_x + 30(uu_{xxx} + \tfrac{5}{2}u_x u_{xx}) + u_{xxxxx} = 0, \quad (1.364)$$

with a scaled version of the Sawada–Kotera equation (1.350), viz.,

$$\hat{u}_t + 45\hat{u}^2 \hat{u}_x + 15(\hat{u}\hat{u}_{xxx} + \hat{u}_x\hat{u}_{xx}) + \hat{u}_{xxxxx} = 0. \quad (1.365)$$

Thus, integration of (1.364) and (1.365) with respect to x, under the vanishing boundary conditions $u = \hat{u} = 0$ at $|x| = \infty$, leads to the equations

$$\sigma_t + 15(\sigma_x^3 + \sigma_x\sigma_{xxx} + \tfrac{3}{4}\sigma_{xx}^2) + \sigma_{xxxxx} = 0, \quad (1.366)$$

$$\tau_t + 15(\tau_x^3 + \tau_x\tau_{xxx}) + \tau_{xxxxx} = 0 \quad (1.367)$$

for the potentials σ, τ such that

$$\sigma_x = 2u, \quad \tau_x = \hat{u}.$$

Introduction of the Miura-type one-sided Bäcklund transformation

$$2\tau_x + \sigma_x - (\tau - \sigma)^2 = 0 \tag{1.368}$$

leads to the relation

$$\frac{\partial}{\partial x}\left[2\{\tau_t + 15(\tau_x^3 + \tau_x\tau_{xxx}) + \tau_{xxxxx}\}\right.$$

$$+ \sigma_t + 15\left(\sigma_x^3 + \sigma_x\sigma_{xxx} + \frac{3}{4}\sigma_{xx}^2\right) + \sigma_{xxxxx}\bigg]$$

$$+ 2(\tau - \sigma)[\tau_t + 15(\tau_x^3 + \tau_x\tau_{xxx}) + \tau_{xxxxx}$$

$$- \left\{\sigma_t + 15\left(\sigma_x^3 + \sigma_x\sigma_{xxx} + \frac{3}{4}\sigma_{xx}^2\right) + \sigma_{xxxxx}\right\}\bigg] = 0. \tag{1.369}$$

It follows that if τ is a solution of equation (1.367) under the vanishing boundary condition, then σ satisfying (1.368) is a solution of equation (1.366) under the vanishing boundary condition. Since N-soliton solutions of (1.365) have already been obtained in the literature (Sawada and Kotera [245], Caudrey et al. [242]), multiparameter solutions of equation (1.364) are now readily constructed.

In conclusion, we remark that a Miura-type transformation linking scaled versions of (1.350) and (1.364) has also been derived by Fordy and Gibbons [248].

1.14 BÄCKLUND TRANSFORMATIONS IN HIGHER DIMENSIONS. THE SINE-GORDON EQUATION IN 3 + 1 DIMENSIONS. THE KADOMTSEV–PETVIASHVILI EQUATION. THE YANG EQUATIONS

The extension of Bäcklund transformation theory to higher dimensional nonlinear evolution equations remains a subject of current research. Recent work by Leibbrandt [249], Leibbrandt et al. [392], Christiansen [250, 251], Anderson et al. [265], Tenenblat and Terng [253], Case [393] and Wilson and Swamy [394] has concerned Bäcklund transformations for the higher dimensional sine-Gordon equation

$$\left[\nabla^2 - \frac{\partial^2}{\partial t^2}\right]u = \sin u, \tag{1.370}$$

1.14 BÄCKLUND TRANSFORMATIONS IN HIGHER DIMENSIONS

associated with the propagation of magnetic flux through Josephson tunnel junctions. Permutability theorems have been established in both the (2 + 1)- and the (3 + 1)-dimensional cases and multi-soliton solutions of (1.370) thereby constructed.† Moreover, this work has recently been extended by Popowicz [254], who derived Bäcklund transformations and associated permutability theorems for the $O(3)$ nonlinear σ model in both 2 + 1 and 3 + 1 dimensions (Pohlmeyer [255]).

In an independent development, Chen [257] constructed an auto-Bäcklund transformation of a two-dimensional Korteweg–deVries equation

$$q_{xt} + \alpha q_{yy} + q_{xx} + (3q^2)_{xx} + q_{xxxx} = 0, \qquad (1.371)$$

which describes the propagation of disturbances in a weakly dispersive, weakly nonlinear medium. Chen's method represents a variant of the procedure outlined in Section 1.11 for the construction of Bäcklund transformations from the inverse scattering formalism. The procedure is based directly on the *Lax equations*

$$L\psi = \lambda\psi, \qquad (1.372)$$

$$M\psi = \psi_t, \qquad (1.373)$$

wherein, L and M are linear operators subject to the constraint

$$\partial L/\partial t = -[L, M] = -(LM - ML). \qquad (1.374)$$

The link between the Lax system and the inverse scattering method has been discussed by Chen [258]. In particular, it was noted that the AKNS system corresponds to the Lax system with the specializations

$$L = \begin{bmatrix} \dfrac{\partial}{\partial x} & -q \\ r & -\dfrac{\partial}{\partial x} \end{bmatrix}, \quad M = \begin{bmatrix} A & B \\ C & -A \end{bmatrix}, \quad \psi = \begin{bmatrix} \psi_1 \\ \psi_2 \end{bmatrix}. \qquad (1.375)$$

In an adaption of a result due to Dryuma [261], a Lax system was constructed in [257] for the two-dimensional Korteweg–deVries equation (1.371) and, following a simple extension of a procedure outlined in [258, 259] for the one-dimensional case, an auto-Bäcklund transformation was thereby obtained. An alternative method based on the Hirota bilinear operator formalism was subsequently presented by Hirota and Satsuma [201].

Higher dimensional Bäcklund transformations may also be constructed for the Yang equations. Thus in recent years there has been much interest

---

† Ring-shaped quasi-soliton solutions to (1.370) have recently been investigated by Christiansen and Olsen [252].

in the study of solutions of the SU(2) Yang–Mills equations in four-dimensional Euclidean space in the case when the field strengths are self-dual or anti-self-dual [396–402].[†] It has been established by Atiyah and Ward [398] that the problem of constructing the so-called "instanton" solutions may be converted to a problem in algebraic geometry via a series of Ansätze. Corrigan et al. [399] showed that these Ansätze are linked by Bäcklund transformations for the Yang equations in the R-gauge, while Belavin and Zakharov [402] used a linear scattering problem to construct instanton solutions. In a recent paper by Pohlmeyer [401], a formulation was presented which leads to both the inverse scattering problem of Belavin and Zakharov and to Bäcklund transformations for the Yang equations.

Here, an auto-Bäcklund transformation for the sine-Gordon equation is presented along with its associated permutability theorem. Two- and four-soliton solutions of (1.370) are thereby generated.

Next, the derivation of an auto-Bäcklund transformation for a two-dimensional Korteweg–deVries equation via the Hirota bilinear formalism is given. This has the advantage over the Chen method that the resultant permutability theorem adopts a particularly concise form.

Finally, Pohlmeyer's Bäcklund transformation for the four-dimensional Yang equations is recorded:

The sine-Gordon equation in 3 + 1 dimensions

In Section 1.4, it was shown that the sine-Gordon equation in $2 + 0$ dimensions admits a Bäcklund transformation

$$\left\{\frac{\partial}{\partial x^1} + i \frac{\partial}{\partial x^2}\right\}\left\{\frac{\alpha - i\beta}{2}\right\} = \exp[i\theta]\sin\left\{\frac{\alpha + i\beta}{2}\right\} \quad (1.376)$$

that maps a known solution α to a new solution $i\beta = \mathbb{B}(\theta)\alpha$, where θ is a real Bäcklund parameter. Subsequently, this result was extended by both Leibbrandt [249] and Christiansen [250, 251] to the sine-Gordon equation in 3 + 1 dimensions. Thus, it was shown that

$$\left\{\sum_{i=1}^{3} \frac{\partial^2}{\partial x^{i2}} - \frac{\partial^2}{\partial t^2}\right\}u = \sin u \quad (1.377)$$

possesses a Bäcklund transformation

$$\left\{I\frac{\partial}{\partial x^1} + i\sigma_1\frac{\partial}{\partial x^2} + i\sigma_3\frac{\partial}{\partial x^3} + \sigma_2\frac{\partial}{\partial t}\right\}\left\{\frac{\alpha - i\beta}{2}\right\}$$

$$= \exp[i\theta\sigma_1\exp[(-i\phi\sigma_2)\exp(-\tau\sigma_1)]]\sin\left\{\frac{\alpha + i\beta}{2}\right\}, \quad (1.378)$$

[†] A review of the mathematics of the Yang–Mills theory is given in [403].

1.14 BÄCKLUND TRANSFORMATIONS IN HIGHER DIMENSIONS

where $\sigma_1, \sigma_2, \sigma_3$ are the Pauli spin matrices and I is the 2×2 identity matrix. The real Bäcklund parameters (θ, ϕ, τ) are restricted to the domains $0 \leq \theta \leq 2\pi$, $0 \leq \phi \leq 2\pi$, $-\infty < \tau < \infty$, while the real functions α, β are readily shown to satisfy

$$\left\{\sum_{i=1}^{3} \frac{\partial^2}{\partial x^{i^2}} - \frac{\partial^2}{\partial t^2}\right\} \begin{matrix} \alpha(x^1, x^2, x^3, t) = \sin \alpha(x^1, x^2, x^3, t), & (1.379) \\ \beta(x^1, x^2, x^3, t) = \sinh \beta(x^1, x^2, x^3, t), & (1.380) \end{matrix}$$

respectively. The matrix equation (1.378) describes, in analogy with (1.376), a transformation from α to $i\beta = \mathbb{B}(\theta, \phi, \tau)\alpha$.

Single Soliton Solutions

The matrix Bäcklund equation (1.378) may be written more succinctly as

$$\left\{I \frac{\partial}{\partial x^1} + iP\right\}\left\{\frac{\alpha - i\beta}{2}\right\} = [A_1 + iA_2] \sin\left\{\frac{\alpha + i\beta}{2}\right\}, \quad (1.381)$$

where

$$P = \sigma_1 \frac{\partial}{\partial x^2} + \sigma_3 \frac{\partial}{\partial x^3} - i\sigma_2 \frac{\partial}{\partial t},$$

$$\sigma_1 = \begin{bmatrix} 0 & 1 \\ 1 & 0 \end{bmatrix}, \quad \sigma_2 = \begin{bmatrix} 0 & -i \\ i & 0 \end{bmatrix}, \quad \sigma_3 = \begin{bmatrix} 1 & 0 \\ 0 & -1 \end{bmatrix}, \quad (1.382)$$

and

$$A_1 = I \cos \theta, \quad (1.383)$$

$$A_2 = \begin{bmatrix} \sin \theta \sin \phi \cosh \tau & (\cos \phi - \sin \phi \sinh \tau) \sin \theta \\ (\cos \phi + \sin \phi \sinh \tau) \sin \theta & -\sin \theta \sin \phi \cosh \tau \end{bmatrix}. \quad (1.384)$$

Hence, on separation of the real and imaginary components in (1.381), we obtain

$$I \frac{\partial}{\partial x^1} \left\{\frac{\alpha}{2}\right\} + P\left\{\frac{\beta}{2}\right\} = A_1 \sin\left(\frac{\alpha}{2}\right) \cosh\left(\frac{\beta}{2}\right) - A_2 \cos\left(\frac{\alpha}{2}\right) \sinh\left(\frac{\beta}{2}\right), \quad (1.385)$$

$$P\left\{\frac{\alpha}{2}\right\} - I \frac{\partial}{\partial x^1}\left\{\frac{\beta}{2}\right\} = A_1 \cos\left(\frac{\alpha}{2}\right) \sinh\left(\frac{\beta}{2}\right) + A_2 \sin\left(\frac{\alpha}{2}\right) \cosh\left(\frac{\beta}{2}\right). \quad (1.386)$$

The simplest, nontrivial solutions α, β of (1.379) and (1.380), respectively, now emerge by setting first $\beta = \beta_0 = 0$ and then $\alpha = \alpha_0 = 0$ in the Bäcklund relations (1.385) and (1.386). These are given by

$$\alpha_1(x^1, x^2, x^3; \theta, \phi, \tau) = 4 \tan^{-1}\{a_0 \exp R\}, \quad (1.387)$$

$$\beta_1(x^1, x^2, x^3; \theta, \phi, \tau) = \begin{cases} 4 \tan^{-1}\{a_1 \exp R\} & \text{if } R \leq 0, \\ 4 \coth^{-1}\{a_1 \exp R\} & \text{if } R > 0, \end{cases} \quad (1.388)$$

where

$$R = x^1 \cos\theta + x^2 \sin\theta \cos\phi + \sin\theta \sin\phi \, (x^3 \cosh\tau + t \sinh\tau), \quad (1.389)$$

and a_i, $i = 0, 1$, are integration constants. The solution (1.387) corresponds to a single-soliton solution of (1.379); the β solutions, by contrast, lack soliton-like character.

The Permutability Theorem

Multi-soliton solutions of the $3 + 1$ sine-Gordon equation may now be generated by appeal to the following permutability theorem (Leibbrandt [249]):

THEOREM 1.11 If α_0 is a solution of the $3 + 1$ sine-Gordon equation (1.379) and $\beta_1^{(j)}$, $j = 1, 2$, are two distinct solutions of (1.380) generated via the Bäcklund transformation (1.378) according to $i\beta_1^{(j)} = \mathbb{B}(\theta_j, \phi_j, \tau_j)\alpha_0$, then a new solution $\alpha_2(x^1, x^2, x^3, t; \theta_1, \theta_2, \phi_1, \phi_2, \tau_1, \tau_2)$ of (1.379) is given algebraically by

$$\tan\left\{\frac{\alpha_2 - \alpha_0}{4}\right\} = D_{12} \tanh\left\{\frac{\beta_1^{(1)} - \beta_1^{(2)}}{2}\right\}, \quad (1.390)$$

where

$$D_{12} = \pm\sqrt{(1 + L_{12})/(1 - L_{12})} \quad (1.391)$$

and

$$L_{12} = \cos\theta_1 \cos\theta_2 + \sin\theta_1 \sin\theta_2$$
$$\times [\cos\phi_1 \cos\phi_2 + \sin\phi_1 \sin\phi_2 \cosh(\tau_1 - \tau_2)]. \quad ■ \quad (1.392)$$

The above result has an associated Bianchi diagram as illustrated in Fig. 1.19. Iteration generates, without additional quadrature, an infinite sequence of exact α_{2n} solutions of the $3 + 1$ sine-Gordon equation. However, in contrast with the $2 + 0$ case, there are constraints on the $L_{2n-1, 2n}$ $n \geq 2$ that represent restrictions on the Bäcklund parameters and hence on the solutions that may be constructed by the Bäcklund transformation method. The geometrical implications of these constraints have been described recently by Christiansen [406], while computational aspects have been investigated in [392] for the $(2 + 1)$-dimensional analogs of (1.379)–(1.381), namely,

$$\left\{\sum_{i=1}^{2} \frac{\partial}{\partial x^{i2}} - \frac{\partial}{\partial t^2}\right\} \alpha(x^1, x^2, t) = \sin\alpha(x^1, x^2, t), \quad (1.393)$$

$$\left\{\sum_{i=1}^{2} \frac{\partial}{\partial x^{i2}} - \frac{\partial}{\partial t^2}\right\} \beta(x^1, x^2, t) = \sinh\beta(x^1, x^2, t), \quad (1.394)$$

1.14 BÄCKLUND TRANSFORMATIONS IN HIGHER DIMENSIONS

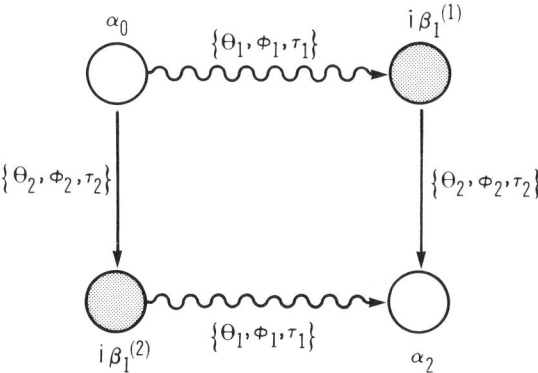

Fig. 1.19 The Bianchi diagram for second-generation α solutions of the 3 + 1 sine-Gordon equation. (Leibbrandt [27].)

with associated Bäcklund transformation

$$\left\{I\frac{\partial}{\partial x^1}+i\sigma_1\frac{\partial}{\partial x^2}+\sigma_2\frac{\partial}{\partial t}\right\}\left\{\frac{\alpha-i\beta}{2}\right\}=\exp[i\theta\sigma_1\exp(\lambda\sigma_3)]\sin\left\{\frac{\alpha+i\beta}{2}\right\} \quad (1.395)$$

and permutability theorem similar to Theorem 1.12 but with the replacement $L_{12}\to\bar{L}_{12}=\cos\theta_1\cos\theta_2+\sin\theta_1\sin\theta_2\cosh(\lambda_1-\lambda_2)$. Leibbrandt et al. [392] used the permutability theorem to construct a four-soliton solution α_4 of (1.393) given by

$$\tan\left\{\frac{\alpha_4-\alpha_2^{(2)}}{4}\right\}=D_{14}\tanh\left\{\frac{\beta_3^{(1)'}-\beta_3^{(2)'}}{4}\right\}, \quad (1.396)$$

where

$$\tanh\left\{\frac{\beta_3^{(s)}-\beta_1^{(s+1)}}{4}\right\}=D_{s,s+2}\tan\left\{\frac{\alpha_2^{(s+1)}-\alpha_2^{(s)}}{4}\right\}, \quad s=1,2,$$

while the two-soliton solutions $\alpha_2^{(p)}$ are given by

$$\tan\left\{\frac{\alpha_2^{(p)}-\alpha_0}{4}\right\}=D_{p,p+1}\tanh\left\{\frac{\beta_1^{(p)}-\beta_1^{(p+1)}}{4}\right\}, \quad p=1,2,3, \quad (1.397)$$

where

$$\tanh\left\{\frac{\beta_1^{(j)}}{4}\right\}=a_j\exp[x^1\cos\theta_j+\sin\theta_j(x^2\cosh\lambda_j+t\sinh\lambda_j)],$$

$$j=1,2,3,4.$$

The structure and time evolution of typical α_2- and α_4-soliton solutions is depicted in Figs. 1.20 and 1.21 (Leibbrandt et al. [392]).

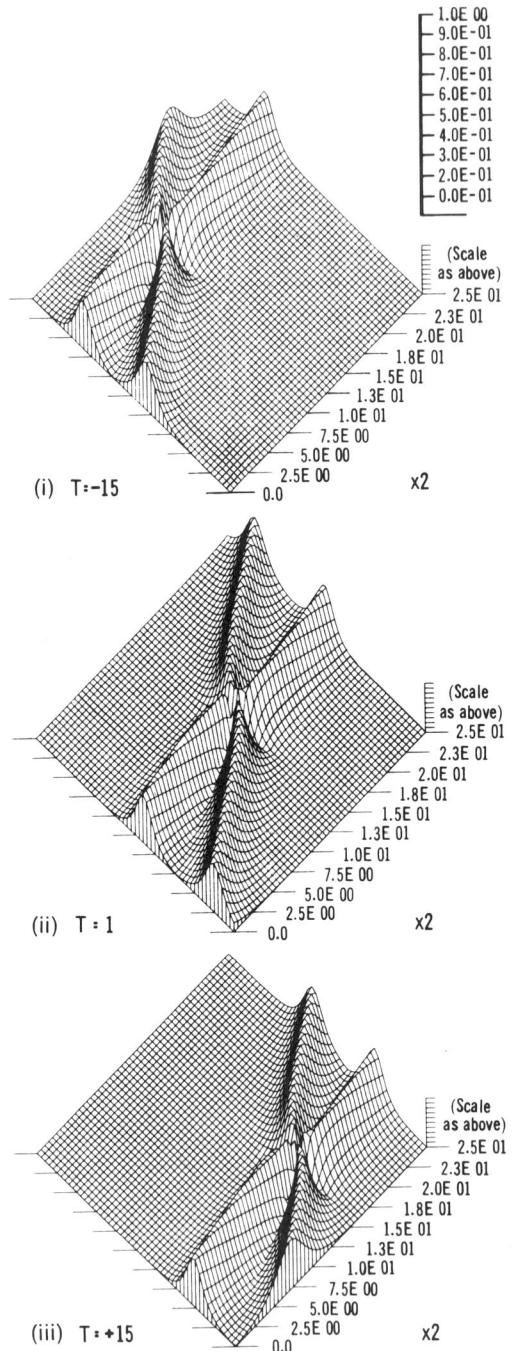

Fig. 1.20 Time evolution of a typical two-soliton solution of the 2 + 1 sine-Gordon equation. (Leibbrandt et al. [392].)

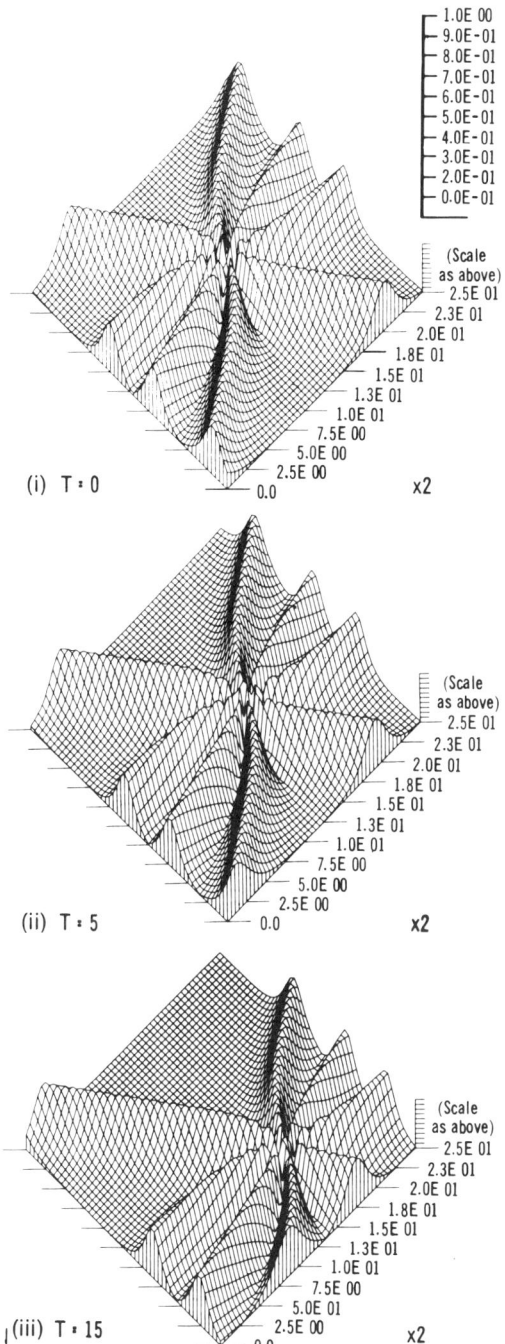

Fig. 1.21 Time evolution of a typical four-soliton solution of the 2 + 1 sine-Gordon equation. (Leibbrandt *et al.* [392].)

The Kadomtsev–Petviashvili Equation

We now present a Bäcklund transformation and associated permutability theorem for the two-dimensional Korteweg–deVries equation in the form originally adopted by Kadomtsev and Petviashvili [256], namely,

$$q_{xt} + \alpha q_{yy} + (3q^2)_{xx} + q_{xxxx} = 0. \tag{1.398}$$

This is obtained from (1.371) by the simple change of variables $x' = x - t$, $t' = t$.

It proves convenient to adopt the Hirota formalism. Thus, the substitution

$$q = 2(\log f)_{xx}, \quad f > 0, \tag{1.399}$$

leads, under the boundary condition $q \to 0$ as $|x| \to \infty$, to the bilinear representation

$$(D_x D_t + D_x^4 + \alpha D_y^2)f \circ f = 0, \tag{1.400}$$

of the Kadomtsev–Petviashvili equation (1.398). Hirota and Satsuma [201] recently presented an auto-Bäcklund transformation for (1.400) in the form

$$(D_x^2 - \gamma D_y)f' \circ f = \beta f'f, \tag{1.401}$$

$$(D_t + 3\beta D_x + D_x^3 + 3\gamma D_x D_y)f' \circ f = 0, \tag{1.402}$$

where $\gamma = \pm(\alpha/3)^{1/2}$ and β is a Bäcklund parameter. That relations (1.401) and (1.402) leave (1.400) invariant is readily demonstrated by means of standard bilinear operator identities.

Thus, if $\mathbb{P}$ is introduced according to

$$\mathbb{P} := \{(D_x D_t + D_x^4 + \alpha D_y^2)f \circ f\}f'f' \\ - ff\{(D_x D_t + D_x^4 + \alpha D_y^2)f' \circ f'\}, \tag{1.403}$$

then (I.4), (I.5), and (I.8) together show that

$$\begin{aligned}\mathbb{P} &= 2[D_x(D_t f \circ f') \circ ff' + D_x(D_x^3 f \circ f') \circ ff' \\ &\quad + 3D_x(D_x^2 f \circ f') \circ (D_x f' \circ f) + \alpha D_y(D_y f \circ f') \circ ff'] \\ &= 2[3\gamma D_x\{(D_x D_y f \circ f') \circ f'f + (D_y f' \circ f) \circ (D_x f' \circ f)\} \\ &\quad + \alpha D_y(D_y f \circ f') \circ ff'] \end{aligned} \tag{1.404}$$

on use of the Bäcklund relations (1.401) and (1.402). Substitution of (1.401) into (1.404) and appeal to the identity (I.6) now yield

$$\mathbb{P} = 2\gamma^{-1}(3\gamma^2 - \alpha)D_y(D_x^2 f \circ f') \circ f'f = 0, \tag{1.405}$$

and the result is established. ∎

The Bäcklund relation (1.401) delivers a convenient permutability theorem. Thus, if f_0 is a solution of (1.400) and if $f_1, \hat{f}_1$, and f_2 are new solutions

1.14 BÄCKLUND TRANSFORMATIONS IN HIGHER DIMENSIONS

introduced after the manner of Fig. 1.17, then

$$(D_x^2 - \gamma D_y)f_1 \circ f_0 = \beta_1 f_1 f_0, \quad (1.406)$$

$$(D_x^2 - \gamma D_y)\hat{f}_1 \circ f_0 = \beta_2 \hat{f}_1 f_0, \quad (1.407)$$

$$(D_x^2 - \gamma D_y)f_2 \circ \hat{f}_1 = \beta_1 f_2 \hat{f}_1, \quad (1.408)$$

$$(D_x^2 - \gamma D_y)f_2 \circ f_1 = \beta_2 f_2 f_1. \quad (1.409)$$

Multiplication of (1.406) by $\hat{f}_1 f_2$ and of (1.408) by $f_0 f_1$ and subtraction yield

$$D_x[(D_x f_0 \circ f_2) \circ f_1 \hat{f}_1 - f_0 f_2 \circ (D_x f_1 \circ \hat{f}_1)] \\ + \gamma(D_y f_0 \circ f_1)\hat{f}_1 f_2 - \gamma f_0 f_1 (D_y \hat{f}_1 \circ f_2) = 0, \quad (1.410)$$

on use of (I.4). Similarly, from (1.407) and (1.409)

$$D_x[(D_x f_0 \circ f_2) \circ f_1 \hat{f}_1 - f_0 f_2 \circ (D_x \hat{f}_1 \circ f_1)] \\ + \gamma(D_y f_0 \circ \hat{f}_1)f_1 f_2 - \gamma f_0 \hat{f}_1 (D_y f_1 \circ f_2) = 0, \quad (1.411)$$

and subtraction of (1.411) from (1.410) shows that

$$D_x(f_0 f_2) \circ (D_x f_1 \circ \hat{f}_1) = 0, \quad (1.412)$$

whence we obtain the permutability theorem in the familiar form

$$f_0 f_2 = \text{const}\, D_x f_1 \circ \hat{f}_1. \quad (1.413)$$

The N-soliton solution of the Kadomtsev–Petviashvili equation (1.398) may now be generated in the usual functional form (1.337) with appropriate phase and interaction terms (Satsuma [262], Hirota and Satsuma [201]). Moreover, Satsuma and Ablowitz [263] have shown that suitable specialization leads to lump solutions which decay to a uniform state in all directions (Fig. 1.22).

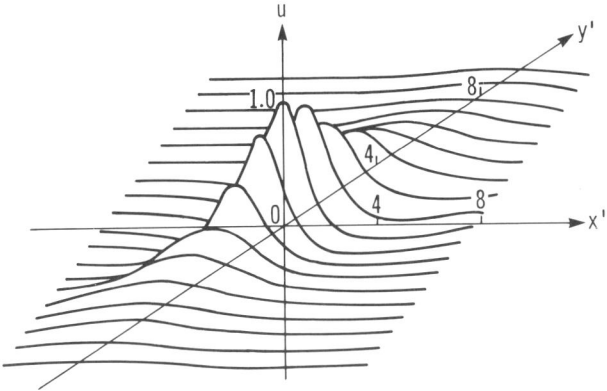

Fig. 1.22 A lump solution of the Kadomtsev–Petviashvili equation as seen at fixed time. (Satsuma and Ablowitz [263].)

The Yang Equations

Bäcklund transformations for the four-dimensional Yang equations have recently been established by Corrigan *et al.* [399], Brihaye and Nuyts [404], and Pohlmeyer [401]. The Pohlmeyer result is recorded below.

Thus, let $x^1, \ldots, x^4$ be coordinates in $\mathbb{R}^4$ and set $y = x^1 + ix^2, z = x^3 + ix^4$. Let M be a four-dimensional Minkowski space with coordinates q^0, q^1, q^2, q^3, and metric $g_{ab} = \text{diag}(1, -1, -1, -1)$. Then Yang's equations may be written in the form

$$\mathbf{q} \cdot \mathbf{q} = 1, \tag{1.414}$$

$$\mathbf{q}_{y\bar{y}} + \mathbf{q}_{z\bar{z}} + (\mathbf{q}_y \cdot \mathbf{q}_{\bar{y}} + \mathbf{q}_z \cdot \mathbf{q}_{\bar{z}})\mathbf{q} + i[\mathbf{q}\,;\mathbf{q}_{\bar{y}}\,;\mathbf{q}_y] + i[\mathbf{q}\,;\mathbf{q}_{\bar{z}}\,;\mathbf{q}_z] = 0, \tag{1.415}$$

where $\mathbf{q} \in M$, $\cdot$ denotes the Minkowski inner product $\mathbf{q} \cdot \mathbf{q} = q^a q_a$, $q_a = g_{ab} q^b$ and $[\,;\,;\,]$ denotes the triple vector product. Thus, $[\mathbf{u}\,;\mathbf{v}\,;\mathbf{w}]$ is the vector in M with coordinates $[\mathbf{u}\,;\mathbf{v}\,;\mathbf{w}]^a = \varepsilon^{abcd} u_b v_c w_d$, where ε^{abcd} is the Levi–Civita symbol.

Pohlmeyer's result states that if

$$\mathbf{q}_y - \mathbf{q}'_{\bar{z}} = -(\mathbf{q} \cdot \mathbf{q}'_{\bar{z}})\mathbf{q} + (\mathbf{q} \cdot \mathbf{q}'_y)\mathbf{q}' - i[\mathbf{q}'\,;\mathbf{q}_{\bar{z}} + \mathbf{q}'_y\,;\mathbf{q}], \tag{1.416}$$

where

$$\mathbf{q} \cdot \mathbf{q} = 1, \qquad \mathbf{q}' \cdot \mathbf{q}' = -1, \qquad \mathbf{q} \cdot \mathbf{q}' = 0, \tag{1.417}$$

then $\mathbf{q}$ is a solution of (1.415) whenever $\mathbf{q}'$ is a solution of

$$\mathbf{q}'_{y\bar{y}} + \mathbf{q}'_{z\bar{z}} - (\mathbf{q}'_y \cdot \mathbf{q}'_{\bar{y}} + \mathbf{q}'_z \cdot \mathbf{q}'_{\bar{z}})\mathbf{q}'$$
$$- i[\mathbf{q}'\,;\mathbf{q}'_{\bar{y}}\,;\mathbf{q}'_y] - i[\mathbf{q}'\,;\mathbf{q}'_{\bar{z}}\,;\mathbf{q}'_z] = \mathbf{0}. \tag{1.418}$$

It was shown in [401] that the preceding Bäcklund transformation may be used to derive an infinite collection of conservation laws. In the next chapter it will be viewed in terms of a jet-bundle framework.

1.15 THE BENJAMIN–ONO EQUATION FOR INTERNAL DEEP WATER WAVES. THE NAKAMURA TRANSFORMATION

The Benjamin–Ono equation

$$u_t + 2uu_x + H[u_{xx}] = 0, \tag{1.419}$$

where H is the Hilbert transform operator defined by the Cauchy principal value integral

$$Hf(x) := \frac{1}{\pi} P \int_{-\infty}^{\infty} \frac{f(z)}{z - x} \, dz, \tag{1.420}$$

1.15 THE BENJAMIN–ONO EQUATION

was introduced by Benjamin [274] and independently by Ono [276] to model internal water wave propagation in a *deep* stratified fluid. Nakamura [277] has recently constructed a Bäcklund transformation for (1.419) in terms of Hirota's bilinear representation.[†] Moreover, this Bäcklund transformation was subsequently used, in an ingenious manner, to derive N-periodic wave and N-soliton solutions of an associated "modified" Benjamin–Ono equation (Nakamura [278]). Here, Nakamura's results are described, while in the next section, extension to Joseph's equation for internal wave propagation in a fluid of *finite depth* is discussed.

In order to obtain a bilinear representation associated with the Benjamin–Ono equation (1.419), we set

$$u(x, t) = i \frac{\partial}{\partial x} (\log[f'/f]), \qquad (1.421)$$

where

$$f \propto \prod_{n=1}^{N} (x - z_n(t)), \qquad (1.422)$$

$$f' \propto \prod_{n=1}^{N} (x - z'_n(t)), \qquad (1.423)$$

and z_n, z'_n are complex with im $z_n > 0$, im $z'_n < 0$ $\forall n$, while $N \in \mathbb{Z}^+$. Accordingly,

$$u = i \left\{ \frac{f'_x}{f'} - \frac{f_x}{f} \right\} = i \sum_{n=1}^{N} \left\{ \frac{1}{x - z'_n} - \frac{1}{x - z_n} \right\}, \qquad (1.424)$$

whence

$$Hu = \frac{i}{\pi} P \int_{-\infty}^{\infty} \frac{1}{(z - x)} \sum_{n=1}^{N} \left\{ \frac{1}{z - z'_n} - \frac{1}{z - z_n} \right\} dz. \qquad (1.425)$$

In order to evaluate (1.425), a contour C is introduced as in Fig 1.23. The residue theorem yields

$$\frac{1}{2\pi i} \oint_C \frac{1}{(z - x)} \left[\frac{1}{z - z'_n} - \frac{1}{z - z_n} \right] dz = \text{res}(z = z_n),$$

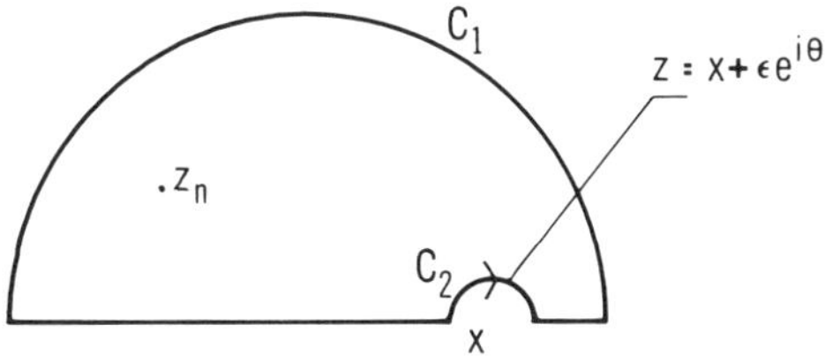

Fig. 1.23

[†] A bilinearization of higher order Benjamin–Ono equations has recently been presented by Matsuno [275].

whence

$$\lim_{\varepsilon \to 0} \frac{1}{2\pi i} \int_{-\infty}^{x-\varepsilon} \frac{1}{z-x} \left[\frac{1}{z-z'_n} - \frac{1}{z-z_n} \right] dz$$

$$+ \lim_{\varepsilon \to 0} \frac{1}{2\pi i} \int_{C_2} \frac{1}{z-x} \left[\frac{1}{z-z'_n} - \frac{1}{z-z_n} \right] dz$$

$$+ \lim_{\varepsilon \to 0} \frac{1}{2\pi i} \int_{x+\varepsilon}^{\infty} \frac{1}{z-x} \left[\frac{1}{z-z'_n} - \frac{1}{z-z_n} \right] dz = \frac{1}{x-z_n}.$$

Thus,

$$\frac{1}{2\pi i} P \int_{-\infty}^{\infty} \frac{1}{z-x} \left[\frac{1}{z-z'_n} - \frac{1}{z-z_n} \right] dz$$

$$= \frac{1}{x-z_n} - \lim_{\varepsilon \to 0} \frac{1}{2\pi i} \int_{C_2} \frac{1}{z-x} \left[\frac{1}{z-z'_n} - \frac{1}{z-z_n} \right] dz$$

$$= \frac{1}{x-z_n} - \lim_{\varepsilon \to 0} \left\{ \frac{1}{2\pi i} \int_{\pi}^{0} \varepsilon^{-1} e^{-i\theta} \left[\frac{1}{x+\varepsilon e^{i\theta} - z'_n} - \frac{1}{x+\varepsilon e^{i\theta} - z_n} \right] \varepsilon i e^{i\theta}\, d\theta \right\}$$

$$= \frac{1}{x-z_n} + \frac{1}{2} \left[\frac{1}{x-z'_n} - \frac{1}{x-z_n} \right] = \frac{1}{2} \left[\frac{1}{x-z'_n} + \frac{1}{x-z_n} \right], \qquad (1.426)$$

and we see that

$$Hu = \frac{i}{\pi} \sum_{n=1}^{N} \pi i \left(\frac{1}{x-z'_n} + \frac{1}{x-z_n} \right) = -\sum_{n=1}^{N} \left(\frac{1}{x-z'_n} + \frac{1}{x-z_n} \right) \qquad (1.427)$$

$$= -\left[\frac{f'_x}{f'} + \frac{f_x}{f} \right] = -\frac{\partial}{\partial x} (\log[f'f]). \qquad (1.428)$$

Insertion of (1.421) and (1.428) into the Benjamin–Ono equation (1.419) shows that

$$\frac{\partial}{\partial x} \left[i \frac{\partial}{\partial t} (\log[f'/f]) - \frac{\partial}{\partial x} (\log[f'/f])^2 - \frac{\partial^2}{\partial x^2} (\log[f'f]) \right] = 0,$$

whence on integration with respect to x and choice of the arbitrary function of time so introduced to be zero, it follows that

$$i \frac{\partial}{\partial t} (\log[f'/f]) - \left[\frac{\partial}{\partial x} (\log[f'/f]) \right]^2 - \frac{\partial^2}{\partial x^2} (\log[f'f]) = 0,$$

so that

$$i(ff'_t - f'f_t) - ff'_{xx} + 2f'_x f_x - f'f_{xx} = 0. \qquad (1.429)$$

1.15 THE BENJAMIN–ONO EQUATION

The latter equation may be rewritten as

$$[i(f'_t f - f' f_t) - (f'_{xx}f - 2f'_x f_x + f' f_{x'x'})]_{x'=x,\, t'=t} = 0$$

or

$$\left[i(f'_t f - f' f_t) - \left(\frac{\partial}{\partial x} - \frac{\partial}{\partial x'}\right)(f'_x f - f' f_x)\right]_{x'=x,\, t'=t} = 0.$$

Consequently,

$$\left[\left[i\left(\frac{\partial}{\partial t} - \frac{\partial}{\partial t'}\right) - \left(\frac{\partial}{\partial x} - \frac{\partial}{\partial x'}\right)^2\right] f'(x,t)f(x',t')\right]_{x'=x,\, t'=t} = 0,$$

that is,

$$(iD_t - D_x^2) f' \circ f = 0. \qquad (1.430)$$

Thus, we conclude that if f and f' are introduced according to relations (1.422) and (1.423), then if the bilinear condition (1.430) holds, it follows that u as given by (1.421) is a solution of the Benjamin–Ono equation (1.419).

Suppose that (f, f') is a solution pair of (1.430) and (g, g') is a pair in similar product form introduced via the Bäcklund relations

$$(iD_t - 2i\lambda D_x - D_x^2 - \mu)f \circ g = 0, \qquad (1.431)$$

$$(iD_t - 2i\lambda D_x - D_x^2 - \mu)f' \circ g' = 0, \qquad (1.432)$$

$$(D_x + i\lambda)f \circ g' = i\nu f' g, \qquad (1.433)$$

where λ, μ, and ν are arbitrary parameters. It may be demonstrated that g and g' are such that

$$(iD_t - D_x^2) g' \circ g = 0, \qquad (1.434)$$

and accordingly,

$$v = i \frac{\partial}{\partial x} (\log[g'/g]) \qquad (1.435)$$

gives a new solution of the Benjamin–Ono equation (1.419). To establish this result, it is shown that Eqs. (1.430)–(1.433) together imply that

$$\mathbb{P} := g'g(iD_t - D_x^2)f' \circ f - f'f(iD_t - D_x^2)g' \circ g = 0, \qquad (1.436)$$

so that condition (1.434) automatically follows.

Thus, (I.12) (see Appendix I) shows that

$$g'g(D_t f' \circ f) - f'f(D_t g' \circ g) = fg(D_t f' \circ g') - f'g'(D_t f \circ g),$$

whence

$$\begin{aligned}\mathbb{P} &= fg(iD_tf' \circ g') - f'g'(iD_tf \circ g) - g'g(D_x^2f' \circ f) + f'f(D_x^2g' \circ g) \\ &= 2i\lambda[fgD_xf' \circ g' - f'g'D_xf \circ g] + [fgD_x^2f' \circ g' - f'g'D_x^2f \circ g] \\ &\quad - g'gD_x^2f' \circ f + f'f(D_x^2g' \circ g)\end{aligned}$$ (1.437)

on use of (1.431) and (1.432). Now, identity (I.13) yields

$$fgD_xf' \circ g' - f'g'D_xf \circ g = D_xf'g \circ fg',$$ (1.438)

while (I.4) shows that

$$fgD_x^2f' \circ g' - f'g'D_x^2f \circ g = D_x[(D_xf' \circ g) \circ fg' + f'g \circ (D_xf \circ g')]$$ (1.439)

and

$$f'fD_x^2g' \circ g - g'gD_x^2f' \circ f = D_x[(D_xg' \circ f) \circ f'g + fg' \circ (D_xf' \circ g)].$$ (1.440)

Addition of (1.439) and (1.440), together with use of the antisymmetric property (I.1), now yields

$$\begin{aligned} fgD_x^2f' \circ g' &- f'g'D_x^2f \circ g + f'fD_x^2g' \circ g - g'gD_x^2f' \circ f \\ &= 2D_x[f'g \circ (D_xf \circ g')],\end{aligned}$$ (1.441)

whence, from (I.13),

$$\begin{aligned}\mathbb{P} &= 2i\lambda D_xf'g \circ fg' + 2D_x[f'g \circ (D_xf \circ g')] \\ &= 2D_x[f'g \circ (D_x + i\lambda)f \circ g'].\end{aligned}$$ (1.442)

Thus, on use of the Bäcklund relation (1.433), it follows that

$$\mathbb{P} = 2D_x[f'g \circ (ivf'g)] = 2ivD_x[f'g \circ f'g] = 0$$

as required.

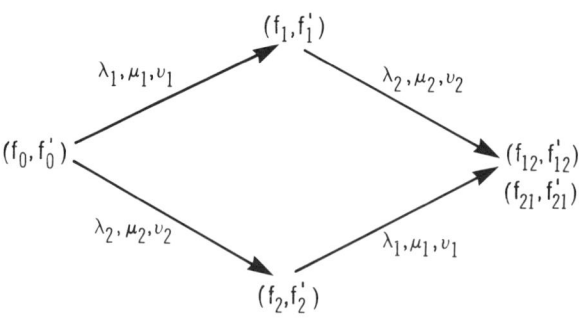

Fig. 1.24

1.15 THE BENJAMIN–ONO EQUATION

We now turn to the superposition theorem associated with the above auto-Bäcklund transformation. The result is embodied in the following:

THEOREM 1.12 If (f_1, f'_1), (f_2, f'_2) are solution pairs of the Benjamin–Ono equation (1.419) generated by the Bäcklund transformation (1.431)–(1.433) with starting solution pair (f_0, f'_0) and parameters $(\lambda_1, \mu_1, \nu_1)$ and $(\lambda_2, \mu_2, \nu_2)$, respectively, then there is a solution pair (f_{12}, f'_{12}) given by the relations

$$f_0 f_{12} = k[D_x + i(\lambda_2 - \lambda_1)] f_1 \circ f_2, \tag{1.443}$$

$$f'_0 f'_{12} = k[D_x + i(\lambda_2 - \lambda_1)] f'_1 \circ f'_2, \tag{1.444}$$

where (f_{12}, f'_{12}) is generated via the Bäcklund transformation (1.431)–(1.433) from $\begin{array}{c}(f_1, f'_1)\\(f_2, f'_2)\end{array}$ with parameters $\begin{array}{c}(\lambda_2, \mu_2, \nu_2)\\(\lambda_1, \mu_1, \nu_1)\end{array}$

(see Fig. 1.24). ∎

The above result is readily established. Thus, the Bäcklund relation (1.433) yields

$$(D_x + i\lambda_1) f_0 \circ f'_1 = i\nu_1 f'_0 f_1, \tag{1.445}$$

$$(D_x + i\lambda_1) f_2 \circ f'_{21} = i\nu_1 f'_2 f_{21}, \tag{1.446}$$

$$(D_x + i\lambda_2) f_0 \circ f'_2 = i\nu_2 f'_0 f_2, \tag{1.447}$$

$$(D_x + i\lambda_2) f_1 \circ f'_{12} = i\nu_2 f'_1 f_{12}. \tag{1.448}$$

We proceed on the assumption that a solution pair $(f_{12}, f'_{12}) = (f_{21}, f'_{21})$ exists, whence on elimination of ν_1 and ν_2 in (1.445)–(1.448), we obtain, in turn,

$$\{(D_x + i\lambda_1) f_0 \circ f'_1\} f'_2 f_{12} - f'_0 f_1 \{(D_x + i\lambda_1) f_2 \circ f'_{12}\} = 0, \tag{1.449}$$

$$\{(D_x + i\lambda_2) f_1 \circ f'_{12}\} f'_0 f_2 - f'_1 f_{12} \{(D_x + i\lambda_2) f_0 \circ f'_2\} = 0. \tag{1.450}$$

Addition of (1.449) and (1.450) and use of the identity (I.14) now show that

$$f'_0 f'_{12} \{D_x + i(\lambda_2 - \lambda_1)\} f_1 \circ f_2 - f_0 f_{12} \{D_x + i(\lambda_2 - \lambda_1)\} f'_1 \circ f'_2 = 0,$$

whence we obtain the *necessary* conditions

$$f_0 f_{12} = k\{D_x + i(\lambda_2 - \lambda_1)\} f_1 \circ f_2,$$
$$f'_0 f'_{12} = k\{D_x + i(\lambda_2 - \lambda_1)\} f'_1 \circ f'_2,$$

where k is an arbitrary function. Here, only the case of constant k is considered. It is interesting to note that when $\lambda_1 = \lambda_2$, the same superposition structure as that obtained previously in Section 1.12 for other Bäcklund transformations is retrieved.

It is now a routine calculation to show that the pair $(f_{12}, f'_{12}) = (f_{21}, f'_{21})$ as given by (1.443) and (1.444) does indeed satisfy the Bäcklund relations (1.446) and (1.448). Moreover, a lengthy but straightforward procedure shows that the superposition relations (1.443) and (1.444) also hold for the other two Bäcklund relations (1.431) and (1.432). These calculations are set out in detail by Nakamura [277] and are not repeated here.

The permutability theorem embodied in the relations (1.443) and (1.444) may now be used to generate solutions of the Benjamin–Ono equation starting with the "vacuum" solution $g = g' = 1$. Thus, with such a solution pair (g, g'), the Bäcklund relations (1.431)–(1.433) adopt the linear form

$$\left(i\frac{\partial}{\partial t} - 2i\lambda \frac{\partial}{\partial x} - \frac{\partial^2}{\partial x^2} - \mu\right)f = 0,$$

$$\left(i\frac{\partial}{\partial t} - 2i\lambda \frac{\partial}{\partial x} - \frac{\partial^2}{\partial x^2} - \mu\right)f' = 0, \quad (1.451)$$

$$\left(\frac{\partial}{\partial x} + i\lambda\right)f = ivf'.$$

With the choice $\lambda = v$, $\mu = 0$, the equations (1.451) admit the solutions

$$f = \theta + \phi, \quad f' = \theta - \phi, \quad (1.452)$$

where

$$\theta = i\{x + \omega t - (x_0 + ix'_0)\}, \quad (1.453)$$

$$\phi = -1/\omega, \quad (1.454)$$

and $\omega = 2\lambda$, x_0, and x'_0 are arbitrary real parameters. The pair (f, f') provides the bilinear representation of the single-soliton solution of the Benjamin–Ono equation as presented by Matsuno [279] and Satsuma and Ishimori [280]. On the other hand, with $\lambda \neq v$ and $\mu \neq 0$, Eqs. (1.451) admit the one-periodic wave solution with

$$f = 2\cosh\tfrac{1}{2}(\theta + \phi), \quad f' = 2\cosh\tfrac{1}{2}(\theta - \phi), \quad (1.455)$$

where θ is as given by (1.453), but now (1.454) is of the form

$$\omega = -\coth\phi. \quad (1.456)$$

The permutability theorem may now be used iteratively to generate either N-soliton, N-periodic, or N-soliton-periodic wave solutions of the Benjamin–Ono equation.[†] Furthermore, in a subsequent paper, Nakamura

---

[†] Computer plots depicting the interaction of Benjamin–Ono solitons have been published by Matsuno [414].

1.15 THE BENJAMIN–ONO EQUATION

[278] showed that the results obtained for the Benjamin–Ono equation may be used in a novel manner to derive associated solutions of a "modified" Benjamin–Ono equation. A description of their procedure follows:

Nakamura's modified Benjamin–Ono equation adopts the form

$$u_t - 2\lambda u_x + 2ve^u u_x + Hu_{xx} + u_x Hu_x = 0, \qquad (1.457)$$

where H is the Hilbert transform operator given by (1.420) and λ, v are constants. In order to construct a bilinear representation associated with (1.457), we set

$$u(x,t) = u_0 + \log\left[\frac{f'g}{fg'}\right], \qquad (1.458)$$

where f, f' are in the product forms (1.422) and (1.423), while

$$g \propto \prod_{n=1}^{N}(x - w_n(t)), \qquad (1.459)$$

$$g' \propto \prod_{n=1}^{N}(x - w'_n(t)), \qquad (1.460)$$

and w_n, w'_n are complex with im $w_n > 0$, im $w'_n < 0$. Thus,

$$\begin{aligned}u_x &= \frac{\partial}{\partial x}\log\left(\frac{f'g}{fg'}\right) = \left\{\frac{f'_x}{f'} + \frac{g_x}{g} - \frac{f_x}{f} - \frac{g'_x}{g'}\right\} \\ &= \sum_{n=1}^{N}\left\{\frac{1}{x - z'_n} + \frac{1}{x - w_n} - \frac{1}{x - z_n} - \frac{1}{x - w'_n}\right\},\end{aligned} \qquad (1.461)$$

so that

$$Hu_x = \frac{1}{\pi}P\int_{-\infty}^{\infty}\frac{1}{(z-x)}\sum_{n=1}^{N}\left\{\frac{1}{z-z'_n} + \frac{1}{z-w_n} - \frac{1}{z-z_n} - \frac{1}{z-w'_n}\right\}dz. \qquad (1.462)$$

Consider the contour integral

$$\oint_C \frac{1}{(z-x)}\left[\frac{1}{z-z'_n} + \frac{1}{z-w_n} - \frac{1}{z-z_n} - \frac{1}{z-w'_n}\right]dz,$$

with C as shown in Fig. 1.23. The residue theorem shows that

$$\frac{1}{2\pi i}P\int_{-\infty}^{\infty}\frac{1}{(z-x)}\left[\frac{1}{z-z'_n} + \frac{1}{z-w_n} - \frac{1}{z-z_n} - \frac{1}{z-w'_n}\right]dz$$

$$+ \lim_{\varepsilon \to 0}\frac{1}{2\pi i}\int_{C_2}\frac{1}{(z-x)}\left[\frac{1}{z-z'_n} + \frac{1}{z-w_n} - \frac{1}{z-z_n} - \frac{1}{z-w'_n}\right]dz$$

$$= \text{res}(z = z_n) + \text{res}(z = w_n) = \frac{1}{x - z_n} - \frac{1}{x - w_n}.$$

Accordingly,

$$\frac{1}{2\pi i} P \int_{-\infty}^{\infty} \frac{1}{(z-x)} \left[\frac{1}{z-z'_n} + \frac{1}{z-w_n} - \frac{1}{z-z_n} - \frac{1}{z-w'_n} \right] dz$$

$$= \frac{1}{2}\left[\frac{1}{x-z'_n} + \frac{1}{x-z_n} - \frac{1}{x-w'_n} - \frac{1}{x-w_n} \right],$$

whence

$$Hu_x = i \sum_{n=1}^{N} \left\{ \frac{1}{x-z'_n} + \frac{1}{x-z_n} - \frac{1}{x-w_n} - \frac{1}{x-w'_n} \right\}$$

$$= i \left\{ \frac{f'_x}{f'} + \frac{f_x}{f} - \frac{g'_x}{g'} - \frac{g_x}{g} \right\} = i \frac{\partial}{\partial x} \log\left(\frac{ff'}{gg'}\right). \tag{1.463}$$

The modified Benjamin–Ono equation (1.457) now yields

$$i\left(\frac{f'_t}{f'} + \frac{g_t}{g} - \frac{f_t}{f} - \frac{g'_t}{g'}\right) - 2i\lambda\left(\frac{f'_x}{f'} + \frac{g_x}{g} - \frac{f_x}{f} - \frac{g'_x}{g'}\right)$$

$$+ 2ive^{u_0}\left(\frac{f'g}{g'f}\right)\left(\frac{f'_x}{f'} + \frac{g_x}{g} - \frac{f_x}{f} - \frac{g'_x}{g'}\right)$$

$$- \left(\frac{f'_{xx}}{f'} + \frac{f_{xx}}{f} - \frac{g'_{xx}}{g'} - \frac{g_{xx}}{g}\right.$$

$$\left. - 2\frac{f_x^2}{f^2} + 2\frac{g_x'^2}{g'^2} + 2\frac{f_x g_x}{fg} - 2\frac{f'_x g'_x}{f'g'}\right) = 0 \tag{1.464}$$

on appropriate substitution. Consider the system (Nakamura [278])

$$(iD_t - 2i\lambda D_x - D_x^2 - \mu)f \circ g = 0, \tag{1.465}$$

$$(iD_t - 2i\lambda D_x - D_x^2 - \mu)f' \circ g' = 0, \tag{1.466}$$

$$(D_x + i\lambda')f \circ g' = iv'f'g, \quad (v' = ve^{u_0}) \tag{1.467}$$

suggested by the form of the Bäcklund transformation (1.431)–(1.433) for the Benjamin–Ono equation. On expansion, the system (1.465)–(1.467) yields

$$i\left(\frac{f_t}{f} - \frac{g_t}{g}\right) - 2i\lambda\left(\frac{f_x}{f} - \frac{g_x}{g}\right) - \frac{f_{xx}}{f} + \frac{2f_x g_x}{fg} - \frac{g_{xx}}{g} - \mu = 0, \tag{1.468}$$

$$i\left(\frac{f'_t}{f'} - \frac{g'_t}{g'}\right) - 2i\lambda\left(\frac{f'_x}{f'} - \frac{g'_x}{g'}\right) - \frac{f'_{xx}}{f'} + \frac{2f'_x g'_x}{f'g'} - \frac{g'_{xx}}{g'} - \mu = 0, \tag{1.469}$$

$$\frac{f_x}{f} - \frac{g'_x}{g'} + i\lambda' - iv'\frac{f'g}{fg'} = 0, \tag{1.470}$$

1.16 THE BÄCKLUND TRANSFORMATION FOR JOSEPH'S EQUATION

and the operations (1.469) − (1.468) − $2(\partial/\partial x)$(1.470) lead to Eq. (1.460). Consequently, *if the product forms f, f', g, g' are subject to conditions* (1.465)–(1.467), *then u as given by* (1.458) *is a solution of the modified Benjamin–Ono equation* (1.457).

In view of the form of the system (1.465)–(1.467), Hirota's procedure for the generation of N-periodic wave and N-soliton solutions of the modified Benjamin–Ono equation follows that of the conventional Benjamin–Ono equation (Satsuma and Ishimori [280]). Thus, in particular, N-periodic wave solutions of the system (1.465)–(1.467) may be constructed in the form (Nakamura [278])

$$f = \sum_{n_j=0,1; j=1,\ldots,N} \exp\left\{\sum_{j=1}^{N} n_j(n_j - \psi_j + \phi_j) + \sum_{i,j=1(i<j)}^{N} n_i n_j \tau_{ij}\right\},$$

$$f' = \sum_{n_j=0,1; j=1,\ldots,N} \exp\left\{\sum_{j=1}^{N} n_j(n_j - \psi_j - \phi_j) + \sum_{i,j=1(i<j)}^{N} n_i n_j \tau_{ij}\right\},$$

$$n_j := ik_j(x - x_{0j} - c_j t), \qquad c_j := k_j \coth \phi_j - 2\lambda,$$

$$\exp(\tau_{ij}) := \frac{(c_i - c_j)^2 - (k_i - k_j)^2}{(c_i - c_j)^2 - (k_i + k_j)^2},$$

$$\exp(2\phi_j) := \frac{c_j + 2\lambda - 2v' + k_j}{c_j + 2\lambda - 2v' - k_j},$$

with expressions for g and g' similar to those for f and f', respectively, but with $\phi_j \to -\phi_j$. The N-soliton solution may be generated by a limiting process out of the N-periodic wave solution [278, 280].

In conclusion, it is noted that recently Bock and Kruskal [281] have obtained a Bäcklund transformation for the Benjamin–Ono equation without recourse to the bilinear formalism.

1.16 A BÄCKLUND TRANSFORMATION FOR JOSEPH'S EQUATION FOR INTERNAL WAVES IN A STRATIFIED FLUID WITH FINITE DEPTH

Joseph's equation, descriptive of wave propagation in a stratified fluid of finite depth, may be written in dimensionless variables as [282, 283]

$$u_t + 2uu_x + G[u_{xx}] = 0, \qquad (1.471)$$

where G is the integral operator defined by

$$G[u(x,t)] = \frac{1}{2}\lambda \int_{-\infty}^{\infty} \left[\coth\{\frac{\pi}{2}\lambda(x' - x)\} - \operatorname{sgn}(x' - x)\right] u(x', t) \, dx' \quad (1.472)$$

and λ^{-1} is a parameter characterizing the depth of the fluid. In the shallow water limit, $\lambda \to \infty$, and (1.471) reduces to the Korteweg–deVries equation while in the deep water limit, $\lambda = 0$, it reduces to the Benjamin–Ono equation.

In order to hold a bilinear representation associated with Joseph's equation for u which are real and finite for all x and t and which are such that the boundary conditions $u \to 0$ as $|x| \to \infty$ hold, we set (Matsuno [284])

$$u(x,t) = i\frac{\partial}{\partial x}\left(\log\left[\frac{\bar{f}}{f}\right]\right), \qquad (1.473)$$

where

$$f(x,t) = \prod_{n=1}^{N} [1 + \exp\{\lambda[\lambda(\text{im } z_n)(x - \lambda t) - \bar{z}_n]\}] \qquad (1.474)$$

and z_n, $n = 1, 2, \ldots, N$, are complex with $0 < \lambda \text{ im } z_n < \pi$.

On insertion of f and its conjugate $\bar{f}$ into (1.473) and use of the result

$$\int_{-\infty}^{\infty} \frac{\coth\{\frac{1}{2}\pi\lambda(x' - x)\}\,dx'}{\cosh(\lambda\gamma x') + \cos(\gamma)} = \frac{-2\lambda^{-1}\csc(\gamma)\sinh(\lambda\gamma x)}{\cosh(\lambda\gamma x) + \cos(\gamma)}$$

$$(0 < \gamma < \pi),$$

we obtain

$$\frac{\partial}{\partial x}G[u] = -\frac{\partial}{\partial x}\left\{\frac{f_x}{f} + \frac{\bar{f}_x}{\bar{f}}\right\} + \lambda u, \qquad (1.475)$$

whence, on appropriate substitution of (1.473) and (1.475) into Joseph's equation, integration, and use of the boundary conditions, it is seen that

$$(iD_t + i\lambda D_x - D_x^2)\bar{f} \circ f = 0. \qquad (1.476)$$

Suppose that $(f, \bar{f})$ is a solution pair of Joseph's equation and $(g, \bar{g})$ is a pair in similar product form introduced via the Bäcklund relations

$$(iD_t + i(\lambda - 2\lambda')D_x - D_x^2 - \mu')f \circ g = 0, \qquad (1.477)$$

$$(iD_t + i(\lambda - 2\lambda')D_x - D_x^2 - \mu')\bar{f} \circ \bar{g} = 0, \qquad (1.478)$$

$$(D_x + i\lambda')f \circ \bar{g} = iv'\bar{f}g, \qquad (1.479)$$

where λ', μ', v' are arbitrary parameters. It is readily established that $(g, \bar{g})$ as determined by (1.477)–(1.479) is also a solution pair associated with Joseph's equation. Thus, if $\mathbb{P}$ is introduced according to

$$\mathbb{P} := [(iD_t + i\lambda D_x - D_x^2)\bar{f} \circ f]\bar{g}g - \bar{f}f[(iD_t + i\lambda D_x - D_x^2)\bar{g} \circ g], \qquad (1.480)$$

1.16 THE BÄCKLUND TRANSFORMATION FOR JOSEPH'S EQUATION

then use, in turn, of the identity (I.12) and of the Bäcklund relations (1.477) and (1.478) shows that

$$\begin{aligned}\mathbb{P} &= fg[(iD_t + i\lambda D_x)\bar{f} \circ \bar{g}] - \bar{f}\bar{g}[i(D_t + i\lambda D_x)f \circ g] \\ &\quad - \bar{g}gD_x^2(\bar{f} \circ f) + \bar{f}fD_x^2(\bar{g} \circ g) \\ &= fg[(2i\lambda' D_x + D_x^2 + \mu')\bar{f} \circ \bar{g}] - \bar{f}\bar{g}[(2i\lambda' D_x + D_x^2 + \mu')f \circ g] \\ &\quad - \bar{g}gD_x^2(\bar{f} \circ f) + \bar{f}fD_x^2(\bar{g} \circ g). \end{aligned} \quad (1.481)$$

Now, from identities (I.13) and (I.14),

$$fgD_x(\bar{f} \circ \bar{g}) - \bar{f}\bar{g}D_x(f \circ g) = D_x(\bar{f}g \circ \bar{g}f), \quad (1.482)$$

$$fgD_x^2(\bar{f} \circ \bar{g}) - \bar{f}\bar{g}D_x^2(f \circ g) = D_x\{[D_x\bar{f} \circ g] \circ f\bar{g} + \bar{f}g \circ [D_xf \circ \bar{g}]\}, \quad (1.483)$$

and

$$\bar{f}fD_x^2(\bar{g} \circ g) - \bar{g}gD_x^2(\bar{f} \circ f) = D_x\{[D_x\bar{g} \circ f] \circ \bar{f}g + f\bar{g} \circ [D_x\bar{f} \circ g]\}. \quad (1.484)$$

On substitution of (1.482)–(1.484) into (1.481), it is seen that

$$\begin{aligned}\mathbb{P} &= 2i\lambda' D_x(\bar{f}g \circ \bar{g}f) + D_x\{[D_x\bar{f} \circ g] \circ f\bar{g} + \bar{f}g \circ [D_xf \circ \bar{g}]\} \\ &\quad + D_x\{[D_x\bar{g} \circ f] \circ \bar{f}g + f\bar{g} \circ [D_x\bar{f} \circ g]\} \\ &= 2D_x\{\bar{f}g \circ (D_x + i\lambda')f \circ \bar{g}\} = 0 \end{aligned}$$

by virtue of the third Bäcklund relation (1.479).

Thus, if $(f, \bar{f})$ is a solution pair of (1.476), then so also is $(g, \bar{g})$ given by the Bäcklund transformation (1.477)–(1.479). Moreover, since the latter differs from that of the Benjamin–Ono equation only by the replacement $\lambda - 2\lambda' \to 2\lambda$, the superposition property of the Bäcklund transformation may be established in a similar manner. The N-soliton and one- and two-periodic solutions of Joseph's equation as set out by Matsuno [284] and Nakamura and Matsuno [285] may then be generated as for the Benjamin–Ono equation.

Just as the Bäcklund transformation for the Benjamin–Ono equation leads to solutions of the "modified" Benjamin–Ono equation, so the Backlund transformation for Joseph's equation is connected to solutions of a "modified" Joseph equation (Nakamura [286). Thus, if we set

$$\begin{aligned}\phi_+ &:= \log(g/f), & \phi_- &:= \log(\bar{g}/\bar{f}), \\ \rho_+ &:= \log(gf), & \rho_- &:= \log(\bar{g}\bar{f}),\end{aligned} \quad (1.485)$$

substitution into the expanded versions of (1.477)–(1.479), namely, the analogs of (1.468)–(1.470), yields

$$i\phi_{+t} + i(\lambda - 2\lambda')\phi_{+x} + \rho_{+xx} + \phi_{+x}^2 + \mu = 0, \quad (1.486)$$

$$i\phi_{-t} + i(\lambda - 2\lambda')\phi_{-x} + \rho_{-xx} + \phi_{-x}^2 + \mu = 0, \quad (1.487)$$

104   1. BÄCKLUND TRANSFORMATIONS AND NONLINEAR EQUATIONS

and

$$\rho_{+x} - \rho_{-x} - \phi_{-x} - \phi_{+x} + 2i\lambda' - 2iv'\exp(\phi_+ - \phi_-) = 0. \quad (1.488)$$

Now, the latter relation shows that

$$\rho_{+xx} - \rho_{-xx} = \phi_{-xx} + \phi_{+xx} + 2iv'[\exp(\phi_+ - \phi_-)](\phi_{+x} - \phi_{-x}),$$

whence, on use of (1.486) and (1.487) to eliminate ρ_{+xx} and ρ_{-xx}, we obtain

$$i(\phi_+ - \phi_-)_t + i(\lambda - 2\lambda')(\phi_+ - \phi_-)_x + (\phi_+ + \phi_-)_{xx}$$
$$+ 2iv'(\phi_+ - \phi_-)_x \exp(\phi_+ - \phi_-) + (\phi_+ - \phi_-)_x(\phi_+ + \phi_-)_x = 0. \quad (1.489)$$

If U is now introduced according to

$$U := \phi_+ - \phi_- = \log(\bar{f}g/f\bar{g}), \quad (1.490)$$

then since, by virtue of (1.473) and (1.475),

$$iG\left[\frac{\partial}{\partial x}\log\left(\frac{\bar{f}g}{f\bar{g}}\right)\right] = iG\left[\frac{\partial}{\partial x}\log\left(\frac{\bar{f}}{f}\right) - \frac{\partial}{\partial x}\log\left(\frac{\bar{g}}{g}\right)\right]$$
$$= \lambda i \log\left(\frac{\bar{f}g}{f\bar{g}}\right) - \frac{\partial}{\partial x}\log\left(\frac{f\bar{f}}{g\bar{g}}\right),$$

it follows that

$$(\phi_+ + \phi_-)_x = -\lambda i U + iG[U_x], \quad (1.491)$$

$$(\phi_+ + \phi_-)_{xx} = -\lambda i U_x + iG[U_{xx}]. \quad (1.492)$$

Substitution of (1.490)–(1.492) into (1.489) now produces the "modified" Joseph equation

$$U_t - 2\lambda' U_x + 2v'e^U U_x + G[U_{xx}] + U_x G[U_x] - \lambda U U_x = 0. \quad (1.493)$$

Thus, if the product forms $f, \bar{f}, g, \bar{g}$ are subject to the conditions (1.477)–(1.479), then U as given by (1.490) is a solution of (1.493). N-soliton solutions of the modified Joseph equation have been generated via this procedure by Nakamura [286].

It is noted that Chen et al. [287] and Satsuma et al. [288] have also constructed Bäcklund transformations for Joseph's equation. In both of these papers, differential-difference operators were used to advantage in an interesting manner. It is to the important subject of Bäcklund transformations as applied to nonlinear differential-difference equations that we turn in the next two sections.

1.17 BÄCKLUND TRANSFORMATIONS OF NONLINEAR LATTICE EQUATIONS IN THEIR CONTINUUM APPROXIMATION. THE KONNO–SANUKI TRANSFORMATION

Toda [289], in an analysis of the longitudinal vibration of a chain of masses interconnected by nonlinear springs, introduced a nonlinear differential-difference equation which, remarkably, may be shown to possess stable N-soliton solutions (Hirota [290], Flashka [291]). Chen and Liu [292] and Wadati and Toda [293] later derived a Bäcklund transformation and an associated permutability theorem for the Toda lattice equation whereby the multi-soliton solution may be generated. Since that time, a number of papers have appeared in which Bäcklund transformations have been constructed for other nonlinear lattice equations. Thus, Wadati [294] early derived a Bäcklund transformation for a lattice which, in the continuum approximation, is modeled by a nonlinear evolution equation which is a combined form of the Korteweg–deVries and modified Korteweg–deVries equations. On the other hand, Konno and Sanuki [295] obtained, via the inverse formalism of Section 1.11, a Bäcklund transformation for a nonlinear lattice under a weak dislocation potential. Bäcklund transformations for discretised versions of the sine-Gordon and Korteweg–deVries equations have been obtained by Orfanidis [296, 297] and Hirota [298, 299].

Here, the Konno–Sanuki Bäcklund transformation for a *continuum* approximation to a nonlinear lattice equation is recorded. In the next section, the Bäcklund transformation for the nonlinear differential-difference equation associated with the *discrete* Toda lattice is described. Moreover, it is shown that the Hirota bilinear operator formulation of this Bäcklund transformation links the Toda lattice equation to other important nonlinear lattice equations.

The Konno–Sanuki Transformation

Konno *et al.* [300] introduced a nonlinear one-dimensional lattice in which the equation of motion of the nth atom is of the form

$$\frac{\partial^2 u_n}{\partial T^2} = u_{n+1} - 2u_n + u_{n-1}$$

$$+ \frac{1}{24}\{(u_{n+1} - u_n)^3 - (u_n - u_{n-1})^3\} - 2\alpha\varepsilon^4 \sin(u_n), \quad (1.494)$$

where u_n is a dimensionless displacement of the nth atom and ε is a measure of the size of the displacement.

In the case of disturbances of long wavelength compared to the spacing of the atoms in the lattice, a continuum approximation may be adopted. Thus, if a stretching

$$\xi_n = \varepsilon(nh - \bar{\lambda}T), \qquad \tau = \varepsilon^3 T \qquad (1.495)$$

is introduced, where h is a lattice spacing and $\bar{\lambda}$ is the group velocity, the displacement $u_{n\pm 1}$ ($\tau, \xi_{n\pm 1}$) of the $(n \pm 1)$th particle may be expanded as a Taylor series in the small parameter εh. It is shown in [300] that retention of terms to $O(\varepsilon^4)$ in this development leads to the nonlinear evolution equation

$$\bar{\lambda}\frac{\partial^2 u}{\partial \xi \, \partial \tau} + \frac{1}{16} h^4 \left[\frac{\partial u}{\partial \xi}\right]^2 \frac{\partial^2 u}{\partial \xi^2} + \frac{h^4}{24}\frac{\partial^4 u}{\partial \xi^4} - \alpha \sin u = 0, \qquad (1.496)$$

where, in the continuum approximation, ξ_n is replaced by $\xi = \varepsilon(X, \lambda T)$ with $X = nh$. Note that when $\alpha = 0$, the modified Korteweg–deVries equation in u_ξ is retrieved, while neglect of the terms in h^4 leads to the sine-Gordon equation. Moreover, introduction of the variables

$$x = (24)^{1/4}\xi/h, \qquad t = h\tau/[(24)^{1/4}\bar{\lambda}]$$

reduces (1.496) to the form

$$u_{xt} + \tfrac{3}{2}u_x^2 u_{xx} + u_{xxxx} - \alpha \sin u = 0, \qquad (1.497)$$

for which a Bäcklund transformation was later derived by Konno and Sanuki [295]. This result is now derived via the inverse scattering formalism outlined earlier in Section 1.11.

Thus, return to the AKNS system (1.271) with the specializations

$$\begin{aligned}
A &= -4\lambda^3 - \tfrac{1}{2}\lambda(u_x)^2 + (\alpha/4\lambda)\cos u, \\
B &= \tfrac{1}{2}u_{xxx} + \lambda u_{xx} + 2\lambda^2 u_x + \tfrac{1}{4}u_x^3 + (\alpha/4\lambda)\sin u, \\
C &= -\tfrac{1}{2}u_{xxx} + \lambda u_{xx} - 2\lambda^2 u_x - \tfrac{1}{4}u_x^3 + (\alpha/4\lambda)\sin u, \\
r &= -q = \tfrac{1}{2}u_x,
\end{aligned} \qquad (1.498)$$

leads to the nonlinear evolution equation (1.497). Now the Riccati equations (1.280) and (1.281), associated with the inverse scattering formulation of (1.497), are

$$\begin{aligned}
\partial \Gamma/\partial x &= 2\lambda\Gamma - \tfrac{1}{2}u_x - \tfrac{1}{2}u_x \Gamma^2, \\
\partial \Gamma/\partial t &= 2[-4\lambda^3 - \tfrac{1}{2}\lambda(u_x)^2 + (\alpha/4\lambda)\cos u]\Gamma \\
&\quad + \tfrac{1}{2}u_{xxx} + \lambda u_{xx} + 2\lambda^2 u_x + \tfrac{1}{4}u_x^3 + (\alpha/4\lambda)\sin u \\
&\quad - [-\tfrac{1}{2}u_{xxx} + \lambda u_{xx} - 2\lambda^2 u_x - \tfrac{1}{4}u_x^3 + (\alpha/4\lambda)\sin u]\Gamma^2,
\end{aligned} \qquad (1.499)$$

and, following the Konno–Wadati procedure as introduced in [299], it is seen that the pair of equations (1.499) is invariant under the transformations

$$\Gamma' = \Gamma^{-1}, \tag{1.500}$$

$$u' = u + 4\tan^{-1}\Gamma. \tag{1.501}$$

Elimination of Γ between (1.501) and (1.499) now produces an auto-Bäcklund transformation for the nonlinear equation (1.497), namely,

$$u_x + u'_x = -4\lambda \sin\tfrac{1}{2}(u-u'), \tag{1.502}$$

$$u_t - u'_t = 2[C-B] - 2[C+B]\cos\tfrac{1}{2}(u-u') + 4A\sin\tfrac{1}{2}(u-u'). \tag{1.503}$$

The permutability theorem based on the spatial part (1.502) of this Bäcklund transformation may now be generated in the manner earlier indicated for the sine-Gordon equation. Thus, if μ_1 and μ_2 are solutions of the nonlinear equation (1.497) generated via (1.502), (1.503) with the Bäcklund parameters λ_1 and λ_2, respectively, and starting solution u_0, then (1.497) admits the second-generation solution $\phi = \mathbb{B}_{\lambda_1}\mathbb{B}_{\lambda_2}u_0 = \mathbb{B}_{\lambda_2}\mathbb{B}_{\lambda_1}u_0$ given by

$$\phi = 4\tan^{-1}\left[\left(\frac{\lambda_1+\lambda_2}{\lambda_1-\lambda_2}\right)\tan\left\{\frac{u_1-u_2}{4}\right\}\right] + u_0. \tag{1.504}$$

Additional solutions of (1.497) may now be constructed iteratively in the usual manner by means of the Bianchi lattice.

If we set $u = u_0 = 0$ in the Bäcklund relations (1.502) and (1.503), integration produces the single-kink solutions

$$u_i = -4\tan^{-1}\left[\frac{c_i}{2\lambda_i}e^{s_i}\right], \tag{1.505}$$

where

$$s_i = 2A_i t + 2\lambda_i x, \tag{1.506}$$

$$A_i = -4\lambda_i^3 + (\alpha/4\lambda_i), \tag{1.507}$$

corresponding to the eigenvalues (Bäcklund parameters)λ_i, $i = 1, 2$. Associated with these single-kink solutions are the single solitons given by

$$u_{i,x} = -4\lambda_i \operatorname{sech}[s_i + \ln\{c_i/2\lambda_i\}]. \tag{1.508}$$

Interestingly, relation (1.507) indicates the existence of a critical eigenvalue at

$$\lambda_C = (\alpha/16)^{1/4}, \tag{1.509}$$

which is a result of the competing effects of the anharmonic and dislocation potentials. Thus, the propagation velocities

$$v_i = (16\lambda_i^4 - \alpha)/4\lambda_i^2 \tag{1.510}$$

are positive for $\lambda_i > \lambda_C$ and negative for $\lambda_i < \lambda_C$.

108   1. BÄCKLUND TRANSFORMATIONS AND NONLINEAR EQUATIONS

Substitution of single-kink solutions u_1 and u_2 into the permutability relation (1.504) generates a two-kink solution ϕ of the nonlinear evolution equation (1.497). The associated two-soliton solution is given by ϕ_x. These double-kink and double-soliton solutions are characterized by the parameter α together with λ_i and c_i, $i = 1, 2$. Three distinctive kinds of propagation emerge and are here illustrated in Figs. 1.25–1.30 below as reproduced from the paper by Konno et al. [300].

Case I is shown in Figs. 1.25 and 1.26. Thus, Fig. 1.25 depicts the two-step nature of the double-kink solution, while Fig. 1.26 shows the head-on collision of the two-component solitary waves in the spatial derivative of the two-kink solution.

Figures 1.27 and 1.28 illustrate the double-kink and associated double-soliton solutions in a typical Case II. It is noted that one of the solitary waves in the latter instance remains stationary, apart from the phase shift it undergoes due to the impingement of the second pulse.

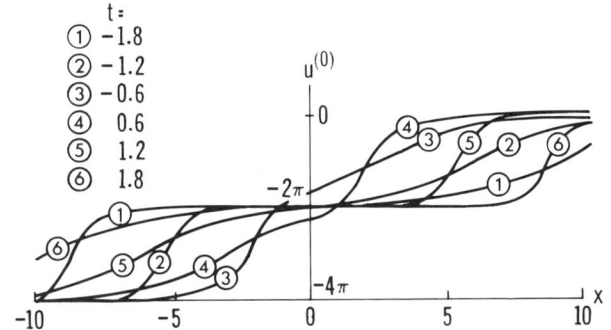

Fig. 1.25 Case I: The propagation of two kinks. (Konno et al. [300].)

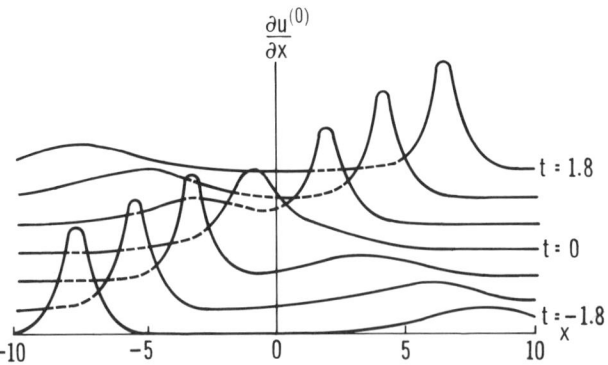

Fig. 1.26 Case I: The collision of two solitary waves with amplitudes (1.0, 4.0) and velocities (-3.75, 3.75). (Konno et al. [300].)

1.17 NONLINEAR LATTICE EQUATIONS

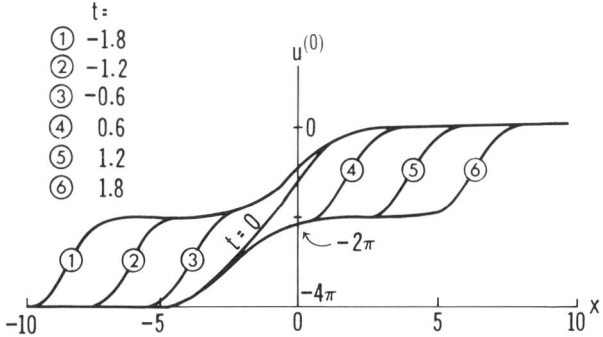

Fig. 1.27 Case II: The propagation of two kinks. (Konno et al. [300].)

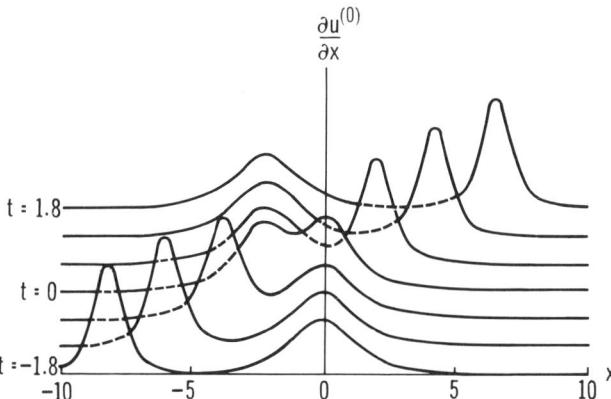

Fig. 1.28 Case II: The collision of two solitary waves: one with amplitude 4.0 and velocity 3.75, and the other with amplitude 2.0 and nonrunning. (Konno et al. [300].)

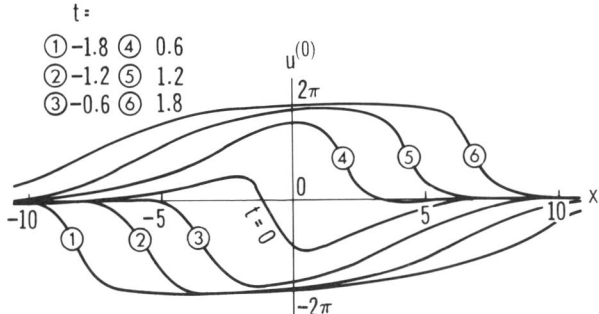

Fig. 1.29 Case III: The propagation of a kink and antikink. (Konno et al. [300].)

110   1. BÄCKLUND TRANSFORMATIONS AND NONLINEAR EQUATIONS

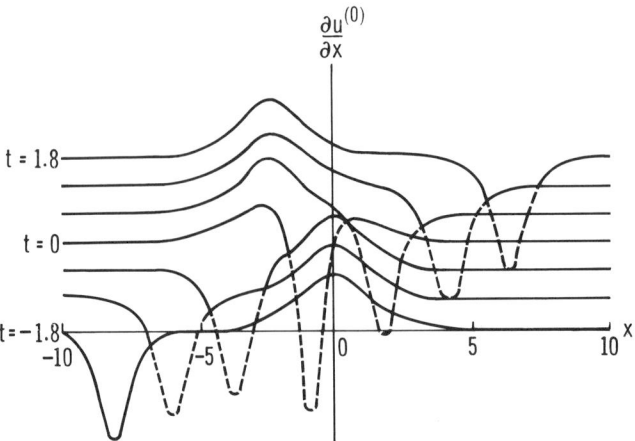

Fig. 1.30 Case III: The collision of two solitary waves with amplitudes (1.0, −4.0) and velocities (0.0, 3.75). (Konno *et al.* [300].)

Finally, typical Case III double-kink and double-soliton solutions are indicated in Figs. 1.29 and 1.30, respectively. Thus, Fig. 1.29 depicts the propagation of a kink and antikink,[†] while Fig. 1.30 shows the collision process of two solitons which have amplitudes of opposite sign at $t \to \pm \infty$.

This concludes our discussion of the Konno–Sanuki transformation for the continuum model of an anharmonic lattice under weak dislocation potential. In the next section, we turn to the derivation of Bäcklund transformations for the original nonlinear differential-difference equations descriptive of pulse propagation in certain lattices.

1.18 BÄCKLUND TRANSFORMATIONS OF NONLINEAR DIFFERENTIAL-DIFFERENCE EQUATIONS. THE TODA AND ASSOCIATED LATTICES

The nonlinear lattices discussed in this section consist of a one-dimensional chain of N particles, each of unit mass, interconnected by nonlinear springs. The equation of motion of the nth such particle is

$$\ddot{y}_n = -\phi'\{y_n - y_{n-1}\} + \phi'\{y_{n+1} - y_n\}, \qquad \ddot{y}_n := \partial^2 y_n/\partial t^2, \qquad (1.511)$$

where y_n is the displacement of the nth particle from its equilibrium position and $\phi(r)$ is the potential energy of the spring, where r denotes its elongation over its natural length. In the case of the Wadati lattice [294], for instance,

---
[†] A kink with negative amplitude is termed an *antikink*.

1.18 NONLINEAR DIFFERENTIAL-DIFFERENCE EQUATIONS

a potential was assumed in the form

$$\phi(r) = \alpha r^2 + \beta r^3 + \gamma r^4, \tag{1.512}$$

where $\alpha, \beta, \gamma \geq 0$ are constants. Accordingly, substitution in (1.511) yields

$$\begin{aligned}\ddot{y}_n =\; & 2\alpha\{(y_{n+1} - y_n) - (y_n - y_{n-1})\} \\ & + 3\beta\{(y_{n+1} - y_n)^2 - (y_n - y_{n-1})^2\} \\ & + 4\gamma\{(y_{n+1} - y_n)^3 - (y_n - y_{n-1})^3\},\end{aligned} \tag{1.513}$$

which, in the long wavelength continuum approximation adopted by Wadati, reduces to a mixed Korteweg–deVries–modified-Korteweg–deVries form

$$u_t + 6\lambda u u_x + 6\mu u^2 u_x + u_{xxx} = 0. \tag{1.514}$$

A Bäcklund transformation and associated permutability theorem were constructed by Wadati [294] for this nonlinear evolution equation via the inverse scattering formulation.[†] However, rather than pursue such continuum approximations further, our interest here lies in the construction of Bäcklund transformations for the original nonlinear differential-difference equations descriptive of *discrete* lattices. Specifically, our concern will be with the Toda and associated lattices.

In the case of the Toda lattice, the potential energy function is

$$\phi(r) = e^{-r} + r - 1, \tag{1.515}$$

so that the equation of motion (1.511) becomes

$$\ddot{y}_n = \exp\{-(y_n - y_{n-1})\} - \exp\{-(y_{n+1} - y_n)\}, \tag{1.516}$$

that is,

$$\ddot{r}_n = 2e^{-r_n} - e^{-r_{n+1}} - e^{-r_{n-1}} \tag{1.517}$$

in terms of the relative displacement of adjacent particles

$$r_n = y_n - y_{n-1}. \tag{1.518}$$

Wadati and Toda [293] introduced an auto-Bäcklund transformation for the *Toda lattice* equation (1.517) in the form

$$\dot{y}_n - \dot{y}'_{n-1} = \beta[\exp\{-(y'_n - y_n)\} - \exp\{-(y'_{n-1} - y_{n-1})\}], \tag{1.519}$$

$$\dot{y}'_n - \dot{y}_n = \beta^{-1}[\exp\{-(y_{n+1} - y'_n)\} - \exp\{-(y_n - y'_{n-1})\}], \tag{1.520}$$

[†] Nakamura and Hirota [415] have obtained a Bäcklund transformation for Wadati's equation by means of the Hirota bilinear operator formalism.

where β is a Bäcklund parameter. Thus, it is readily verified that if $r_n = y_n - y_{n-1}$ is a solution of (1.517), then so also is $r'_n = y'_n - y'_{n-1}$, and conversely. Alternatively, (1.519) and (1.520) imply that

$$\dot{y}_n - \beta \exp\{-(y'_n - y_n)\} - \beta^{-1} \exp\{-(y_n - y'_{n-1})\}$$
$$= \dot{y}_{n-1} - \beta \exp\{-(y'_{n-1} - y_{n-1})\} - \beta^{-1} \exp\{-(y_{n-1} - y'_{n-2})\};$$

whence by descent and imposition of the boundary conditions

$$y_n, y'_n \to \text{const} \quad \text{as} \quad |n| \to \infty, \tag{1.521}$$

it is evident that

$$\dot{y}_n = \beta[\exp\{-(y'_n - y_n)\} - c] + \beta^{-1}[\exp\{-(y_n - y'_{n-1})\} - c^{-1}], \tag{1.522}$$

where

$$c = \exp\{-(y'_{-\infty} - y_{-\infty})\}.$$

Similarly,

$$\dot{y}'_n = \beta[\exp\{-(y'_n - y_n)\} - c] + \beta^{-1}[\exp\{-(y_{n+1} - y'_n)\} - c^{-1}], \tag{1.523}$$

and the relations (1.522) and (1.523) represent a Bäcklund transformation for the Toda lattice equation (1.516), subject to the boundary conditions (1.521).

A permutability theorem is readily established for the Toda lattice. Thus in the notation indicated in the Bianchi diagram, Fig. 1.31, predicated on the assumption of a solution $y_n^{(12)} = \mathbb{B}_{\beta_1}\mathbb{B}_{\beta_2}y_n^{(0)} = \mathbb{B}_{\beta_2}\mathbb{B}_{\beta_1}y_n^{(0)}$, the Bäcklund relations (1.519) and (1.520) yield

$$\begin{aligned}
y_n^{(0)} - y_{n-1}^{(1)} &= \beta_1[\exp\{-(y_n^{(1)} - y_n^{(0)})\} - \exp\{-(y_{n-1}^{(1)} - y_{n-1}^{(0)})\}], \\
y_n^{(0)} - y_{n-1}^{(2)} &= \beta_2[\exp\{-(y_n^{(2)} - y_n^{(0)})\} - \exp\{-(y_{n-1}^{(2)} - y_{n-1}^{(0)})\}], \\
y_n^{(1)} - y_{n-1}^{(12)} &= \beta_2[\exp\{-(y_n^{(12)} - y_n^{(1)})\} - \exp\{-(y_{n-1}^{(12)} - y_{n-1}^{(1)})\}], \\
y_n^{(2)} - y_{n-1}^{(12)} &= \beta_1[\exp\{-(y_n^{(12)} - y_n^{(2)})\} - \exp\{-(y_{n-1}^{(12)} - y_{n-1}^{(2)})\}].
\end{aligned} \tag{1.524}$$

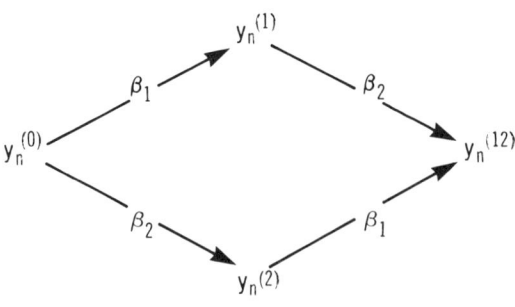

Fig. 1.31

1.18 NONLINEAR DIFFERENTIAL-DIFFERENCE EQUATIONS

On introduction of the boundary conditions

$$y^{(0)}_{-\infty} = \gamma^{(0)}, \qquad y^{(1)}_{-\infty} = \gamma^{(1)},$$
$$y^{(2)}_{-\infty} = \gamma^{(2)}, \qquad y^{(12)}_{-\infty} = \gamma^{(12)}, \qquad (1.525)$$
$$y^{(0)}_{-\infty} + y^{(12)}_{-\infty} = y^{(1)}_{-\infty} + y^{(2)}_{-\infty},$$

and elimination of the time derivatives in (1.524), a permutability theorem is obtained in the form (Wadati and Toda [293][†])

$$\exp(y_n^{(12)} - \gamma^{(12)} + y_{n+1}^{(0)} - \gamma^{(0)})$$
$$= \left\{ \frac{z_1 \exp(y_n^{(2)} - \gamma^{(2)}) - z_2 \exp(y_n^{(1)} - \gamma^{(1)})}{z_1 \exp(y_{n+1}^{(2)} - \gamma^{(2)}) - z_2 \exp(y_{n+1}^{(1)} - \gamma^{(1)})} \right\} \qquad (1.526)$$
$$\times \exp(y_{n+1}^{(1)} - \gamma^{(1)} + y_{n+1}^{(2)} - \gamma^{(2)}),$$

where

$$z_1 = \beta_1 \exp(\gamma^{(0)} - \gamma^{(1)}) = \beta_1 \exp(\gamma^{(2)} - \gamma^{(12)}),$$
$$z_2 = \beta_2 \exp(\gamma^{(0)} - \gamma^{(2)}) = \beta_2 \exp(\gamma^{(1)} - \gamma^{(12)}). \qquad (1.527)$$

It may readily be verified that $y_n^{(12)}$, as given by (1.526), is indeed a solution of the Toda lattice equation (1.516).

Interestingly, the Hirota bilinear operator formalism may be adduced to show that the Toda lattice and a bilinear representation of its Bäcklund transformation are associated with a number of other important lattices (Hirota and Satsuma [301, 302]). Thus, if

$$V_n = e^{-r_n} - 1 \qquad (1.528)$$

is introduced into the Toda lattice equation (1.517), it becomes

$$\ddot{r}_n = 2V_n - V_{n+1} - V_{n-1}; \qquad (1.529)$$

whence on the further substitution,

$$V_n = (\ddot{f}_n/f_n) - (\dot{f}_n^2/f_n^2), \qquad (1.530)$$

and subsequent integration, (1.529) yields

$$\ddot{f}_n f_n - \dot{f}_n^2 + f_n^2 - f_{n+1} f_{n-1} = 0. \qquad (1.531)$$

In terms of Hirota's bilinear operators, the nonlinear differential-difference equation (1.531) adopts the compact form

$$[D_t^2 - 4\sinh^2\{\tfrac{1}{2} D_n\}]f_n \circ f_n = 0, \qquad (1.532)$$

[†] Chen and Liu [292] independently derived a Bäcklund transformation and associated permutability theorem for the Toda lattice equation.

where

$$e^{(\varepsilon D_x + \delta D_t)} a(x,t) \circ b(x,t) := a(x+\varepsilon, t+\delta) b(x-\varepsilon, t-\delta). \quad (1.533)$$

Hirota and Satsuma [301, 302] constructed an auto-Bäcklund transformation for (1.532) given by the relations

$$D_t f'_n \circ f_n + 2\alpha \sinh\{\tfrac{1}{2} D_n\} g'_n \circ g_n = 0, \quad (1.534)$$

$$D_t g'_n \circ g_n + 2\alpha^{-1} \sinh\{\tfrac{1}{2} D_n\} f'_n \circ f_n = 0, \quad (1.535)$$

$$[\beta_1 \sinh\{\tfrac{1}{2} D_n\} + \cosh\{\tfrac{1}{2} D_n\}] g'_n \circ g_n = f'_n f_n, \quad (1.536)$$

$$[\beta_2 \sinh\{\tfrac{1}{2} D_n\} + \cosh\{\tfrac{1}{2} D_n\}] f'_n \circ f_n = g'_n g_n, \quad (1.537)$$

where α, β_1, β_2 are arbitrary parameters subject to the single constraint

$$\beta_1^2 - 1 = \alpha^2 (\beta_2^2 - 1). \quad (1.538)$$

Thus, it may be verified that if $f_n(g_n)$ is any solution of (1.533), then $f'_n(g'_n)$, defined through the relations (1.534)–(1.537), also satisfies the bilinear form (1.532) of the Toda lattice equation. Indeed, the Bäcklund transformation defined by (1.534)–(1.537) may be used to generate important nonlinear lattice equations closely related to the Toda equation. To this end, new variables

$$\hat{x}_n = \log\{f'_n f_n\}, \qquad \hat{y}_n = \log\{g'_n g_n\},$$
$$\hat{u}_n = \log\{f'_n / f_n\}, \qquad \hat{v}_n = \log\{g'_n / g_n\}, \quad (1.539)$$

are introduced into (1.534)–(1.537), and on elimination of $\hat{x}_n$ and $\hat{y}_n$ from the resulting equations, we obtain

$$\dot{u}_n = [\alpha^{-1}(\beta_1^2 - 1)\tfrac{1}{4} u_n^2 + \beta_1 u_n + \alpha][v_{n-1/2} - v_{n+1/2}],$$
$$\dot{v}_n = [\alpha(\beta_2^2 - 1)\tfrac{1}{4} v_n^2 + \beta_2 v_n + \alpha^{-1}][u_{n-1/2} - u_{n+1/2}], \quad (1.540)$$

where $u_n := \hat{u}_n$, $v_n := \hat{v}_n$. The coupled system (1.540) contains certain well-known nonlinear equations as special instances, namely,

(i) $\beta_1 = \beta_2 = 1$ This case corresponds to the nonlinear network equations descriptive of a Volterra system (Daikoku et al. [200], Hirota and Satsuma [303]).

(ii) $\beta_1 = \beta_2 = 0$, $\alpha = 1$ In this case, the self-dual nonlinear network equations are retrieved (Hirota [304]).

N-soliton solutions of the system (1.540), and hence of the above nonlinear lattice equations, are now readily generated via the Hirota technique. The results have been set out by Hirota and Satsuma in an extensive supplement devoted to the subject of nonlinear lattices [302].

1.19 BÄCKLUND TRANSFORMATIONS IN GENERAL RELATIVITY 115

Finally, note that Hirota has recently produced a further series of papers on nonlinear lattices and their Bäcklund transformations [298, 299, 305–307]. In particular, Bäcklund transformations were constructed via the bilinear operator formalism for a discrete-time Toda equation and a discretized Liouville equation. The reader is referred to the original publications for details.

1.19 BÄCKLUND TRANSFORMATIONS IN GENERAL RELATIVITY. THE ERNST EQUATIONS. NEUGEBAUER'S PERMUTABILITY THEOREM

In recent years, much research has been aimed at the construction of exact solutions of the stationary axisymmetric vacuum Einstein field equations which could represent the gravitational field of a spinning mass. A number of alternative approaches have been proposed for the generation of such solutions which preserve asymptotic flatness. Thus, Kinnersley and others [358–368] developed a formalism based on the infinite-dimensional Geroch group [369, 370] and discovered two independent infinite-dimensional subgroups which preserve asymptotic flatness, namely the Kinnersley–Chitre groups [361] and the Hoenselaers–Kinnersley–Xanthopoulos groups [363]. On the other hand, Cosgrove [371] investigated groups outside the Geroch group which preserve asymptotic flatness. Such a Lie group Q was subsequently used to construct the three-parameter Cosgrove–Tomimatsu–Sato solution [372–375].

Independently of the above investigations on continuous transformation groups, Bäcklund transformations had been discovered by Harrison [376], Belinskii and Zakharov [377], and Neugebauer [378], whereby solutions of the axially symmetric stationary Einstein field equations could be generated.

In [379] it was observed that Cosgrove's continuous groups are essentially the same as Neugebauer's Bäcklund transformations I_1 and I_2. Moreover, it was shown that Harrison's Bäcklund transformation may be decomposed into the products $I_1 I_2$ and $I_2 I_1$ and that Neugebauer's permutability theorem may be used to obtain a composition theorem for the Harrison transformations. Thus, the Neugebauer Bäcklund transformations provide an important framework for the unification of diverse solution-generating techniques of the axially symmetric stationary Einstein equations. The connection between the Neugebauer–Bäcklund transformations and those other methods has been set forth recently in an extensive review by Cosgrove [380]. Here, we first derive the Ernst equation and then proceed to the construction of Neugebauer's Bäcklund transformations and an associated permutability theorem. The latter is then used to derive the Kerr-NUT metric from Minkowski space–time.

1. BÄCKLUND TRANSFORMATIONS AND NONLINEAR EQUATIONS

A stationary, axially symmetric space–time is one which admits a two-dimensional isometry group with one timelike and one spacelike Killing vector, respectively. For such space–times, coordinates may be selected such that the metric tensor adopts the form

$$ds^2 = u[dt - w\,d\phi]^2 - u^{-1}[e^{2\delta}\,dx^1\,dx^2 + W^2\,d\phi^2], \quad (1.541)$$

where u, w, δ, W depend on x^1, x^2 only, and

$$W_{,12} = 0, \quad W_{,1} \neq 0, \quad W_{,2} \neq 0. \quad (1.542)$$

In terms of the cylindrical coordinates (ρ, z, ϕ, t), the metric is given by

$$ds^2 = u[dt - w\,d\phi]^2 - u^{-1}[e^{2\delta}(d\rho^2 + dz^2) + W^2\,d\phi^2], \quad (1.543)$$

where $x^1 = \rho + iz$, $x^2 = \rho - iz$.

In the system (1.541), the vacuum field equations

$$R_{ij} = 0$$

become

$$u[u_{,12} + \tfrac{1}{2}W^{-1}u_{,1}W_{,2} + \tfrac{1}{2}W^{-1}u_{,2}W_{,1}] - u_{,1}u_{,2} + W^{-2}u^4 w_{,1}w_{,2} = 0, \quad (1.544)$$

$$w_{,12} - \tfrac{1}{2}W^{-1}w_{,1}W_{,2} - \tfrac{1}{2}W^{-1}w_{,2}W_{,1} + u^{-1}[u_{,1}w_{,2} + u_{,2}w_{,1}] = 0, \quad (1.545)$$

$$\delta_{,1} = \tfrac{1}{2}W_{,1}^{-1}[W_{,11} + \tfrac{1}{2}Wu^{-2}u_{,1}^2 - \tfrac{1}{2}W^{-1}u^2 w_{,1}^2], \quad (1.546)$$

$$\delta_{,2} = \tfrac{1}{2}W_{,2}^{-1}[W_{,22} + \tfrac{1}{2}Wu^{-2}u_{,2}^2 - \tfrac{1}{2}W^{-1}u^2 w_{,2}^2], \quad (1.547)$$

where the integrability of (1.546) and (1.547) follows from equations (1.544) and (1.545).

In order to solve the system (1.544)–(1.547), we choose a real function $W(x^1, x^2)$ which satisfies (1.542) and then solve (1.544) and (1.545) for $u(x^1, x^2)$ and $w(x^1, x^2)$, respectively; $\delta(x^1, x^2)$ is then given by the pair of equations (1.546) and (1.547).

A more tractable system is obtained if it is noticed that the field equation (1.545) may be written as

$$\frac{\partial}{\partial x^1}\{W^{-1}u^2 w_{,2}\} + \frac{\partial}{\partial x^2}\{W^{-1}u^2 w_{,1}\} = 0;$$

whence a function $v(x^1, x^2)$ exists such that

$$v_{,1} = -iW^{-1}u^2 w_{,1}, \quad v_{,2} = iW^{-1}u^2 w_{,2}. \quad (1.548)$$

If the Ernst potential ε given by

$$\varepsilon = u + iv \quad (1.549)$$

1.19 BÄCKLUND TRANSFORMATIONS IN GENERAL RELATIVITY

is now introduced, then it is readily verified that the real and imaginary parts of the *Ernst equation*

$$(\text{Re}\,\varepsilon)W^{-1}[2W\varepsilon_{,12} + W_{,2}\varepsilon_{,1} + W_{,1}\varepsilon_{,2}] = 2\varepsilon_{,1}\varepsilon_{,2} \qquad (1.550)$$

constitute the Einstein field equations (1.544) and (1.545).

In terms of the Neugebauer field variables (M_i, N_i), $i = 1, 2, 3$, introduced according to

$$\begin{aligned} M_1 &= \varepsilon_{,1}/(\varepsilon + \bar{\varepsilon}), & M_2 &= \bar{\varepsilon}_{,1}/(\varepsilon + \bar{\varepsilon}), & M_3 &= W_{,1}/W, \\ N_1 &= \bar{\varepsilon}_{,2}/(\varepsilon + \bar{\varepsilon}), & N_2 &= \varepsilon_{,2}/(\varepsilon + \bar{\varepsilon}), & N_3 &= W_{,2}/W, \end{aligned} \qquad (1.551)$$

the Ernst equation (1.550) together with potential equation (1.542) are equivalent to the system

$$M_{i,2} = C_i^{kl} M_k N_l, \qquad N_{i,1} = C_i^{kl} N_k M_l, \qquad (1.552)$$

where the nonvanishing C_i^{kl} are given by

$$\begin{aligned} C_1^{11} &= C_2^{22} = C_3^{33} = -C_1^{12} = -C_2^{21} = -1, \\ C_1^{32} &= C_1^{13} = C_2^{31} = C_2^{23} = -\tfrac{1}{2}. \end{aligned} \qquad (1.553)$$

The Neugebauer Bäcklund transformations I_1 and I_2 may now be introduced in the context of the system (1.552). They map a known solution $(\mathring{M}_i, \mathring{N}_i)$ of the system to a new solution (M_i, N_i). A new associated solution set $\{u, v, w, \delta, W\}$ of the Einstein field equations (1.544)–(1.548) is then readily generated. Thus, if the solution set $\{\mathring{u}, \mathring{v}, \mathring{w}, \mathring{\delta}, \mathring{W}\}$ is known, the solution pairs $(\mathring{M}_i, \mathring{N}_i)$, $i = 1, 2, 3$, may be constructed via the relations (1.551). Substitution of the new solution pairs (M_i, N_i) into, in turn, equations (1.551), (1.548), and (1.546)–(1.547) generates, on integration, the new solution set $\{u, v, w, \delta, W\}$.

(i) **The Neugebauer–Bäcklund Transformation I_1** The Neugebauer–Bäcklund transformation $I_1(\alpha, \gamma)$ is given by

$$\begin{aligned} \frac{\partial \alpha}{\partial x^1} &= \alpha(\alpha - 1)\mathring{M}_1 + (\alpha - \gamma)\mathring{M}_2 + \frac{1}{2}\alpha(\gamma - 1)\mathring{M}_3, \\ \frac{\partial \alpha}{\partial x^2} &= (\alpha - 1)\mathring{N}_1 + \frac{\alpha}{\gamma}(\alpha - \gamma)\mathring{N}_2 + \frac{\alpha}{2\gamma}(\gamma - 1)\mathring{N}_3, \end{aligned} \qquad (1.554)$$

and

$$\frac{\partial \gamma}{\partial x^1} = \gamma(\gamma - 1)\mathring{M}_3, \qquad \frac{\partial \gamma}{\partial x^2} = (\gamma - 1)\mathring{N}_3, \qquad (1.555)$$

where
$$M_i = \alpha_i^k \overset{\circ}{M}_k, \qquad N_i = \alpha_i^{-1k} \overset{\circ}{N}_k,$$

$$(\alpha_i^k) = \begin{bmatrix} \alpha & 0 & 0 \\ 0 & \beta & 0 \\ 0 & 0 & \gamma \end{bmatrix}, \qquad \alpha\beta = \gamma, \tag{1.556}$$

and α_i^{-1k} is the inverse of α_i^k.

The integrability conditions for (1.554) and (1.555) follow from the fact that the pairs $(\overset{\circ}{M}_i, \overset{\circ}{N}_i)$, $i = 1, 2, 3$, satisfy the system (1.552). It is noted that the solution (α, γ) of (1.554) and (1.555) contains two complex constants of integration.

(ii) **The Neugebauer–Bäcklund Transformation I_2** The Neugebauer–Bäcklund transformation $I_2(\alpha, \gamma)$ is defined by

$$I_2(\alpha, \gamma) = S \circ I_1(\alpha, \gamma) \circ S, \tag{1.557}$$

where S is the *symmetry operator* which maps the set $(\overset{\circ}{M}_i, \overset{\circ}{N}_i)$ to the set (M_i, N_i) in accordance with the relations

$$M_i = \mu_i^k \overset{\circ}{M}_k, \qquad N_i = \nu_i^k \overset{\circ}{N}_k,$$

$$(\mu_i^k) = \begin{bmatrix} 0 & -1 & \tfrac{1}{2} \\ -1 & 0 & \tfrac{1}{2} \\ 0 & 0 & 1 \end{bmatrix}, \qquad (\nu_i^k) = \begin{bmatrix} -1 & 0 & \tfrac{1}{2} \\ 0 & -1 & \tfrac{1}{2} \\ 0 & 0 & 1 \end{bmatrix}. \tag{1.558}$$

This transformation, which satisfies $S^{-1} = S$ will, in general, send a real solution of the Einstein field equations to a complex one, the original solution $(\overset{\circ}{u}, \overset{\circ}{v}, \overset{\circ}{w}, \overset{\circ}{W})$ being mapped to the solution $(u, v, w, W) = (\overset{\circ}{u}^{-1}\overset{\circ}{W}, i\overset{\circ}{w}, -i\overset{\circ}{v}, \overset{\circ}{W})$. If there is no rotation, so that $\overset{\circ}{w} = \overset{\circ}{\psi} = 0$, the operator S generates a real solution. For example, Minkowski space–time $ds^2 = dt^2 - d\rho^2 - dz^2 - \rho^2 d\phi^2$ is mapped to the nonflat metric $ds^2 = \rho\, dt^2 - \rho^{-1}(\rho^{1/2} d\rho^2 + \rho^{1/2} dz^2 + \rho^2 d\phi^2)$ by S.

We record the following important properties of $I_1(\alpha, \gamma)$ and $I_2(\alpha, \gamma)$:

$$I_a(\alpha_2, \gamma_2) \circ I_a(\alpha_1, \gamma_1) = I_a(\alpha_1\alpha_2, \gamma_1\gamma_2), \tag{1.559}$$

$$I_a^{-1}(\alpha, \gamma) = I_a(\alpha^{-1}, \gamma^{-1}), \qquad a = 1, 2. \tag{1.560}$$

The Bäcklund transformations may now be used to construct an infinite sequence of axially symmetric vacuum solutions of Einstein's equations via purely algebraic procedures by appeal to the following permutability theorem:

THEOREM 1.13 If $(\overset{\circ}{M}_i, \overset{\circ}{N}_i)$, $i = 1, 2, 3$, is a solution of the system (1.552) and $I_1(\alpha, \gamma)$ is an arbitrary Neugebauer–Bäcklund transformation, then there

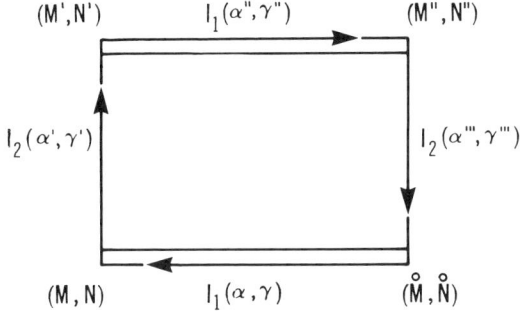

Fig. 1.32 The Bianchi-type diagram for the Neugebauer permutability theorem.

exist three Bäcklund transformations $I_2(\alpha', \gamma')$, $I_1(\alpha'', \gamma'')$, and $I_2(\alpha''', \gamma''')$ such that

$$I_2(\alpha''', \gamma''') \circ I_1(\alpha'', \gamma'') \circ I_2(\alpha', \gamma') \circ I_1(\alpha, \gamma)(\mathring{M}_i, \mathring{N}_i) = (\mathring{M}_i, \mathring{N}_i), \quad (1.561)$$

where

$$\begin{aligned}
\alpha' &= (\alpha - \gamma)/\gamma(\alpha - 1), & \gamma' &= \gamma^{-1}, \\
\alpha'' &= \gamma/\alpha, & \gamma'' &= \gamma, \\
\alpha''' &= (\alpha - 1)/(\alpha - \gamma), & \gamma''' &= \gamma^{-1}.
\end{aligned} \quad (1.562)$$

∎

The above result is readily demonstrated on appeal to the definitions of I_1 and I_2. It may be conveniently represented by a Bianchi-type diagram wherein a double line represents a transformation of type I_1 and a single line a transformation of type I_2 (Fig. 1.32).

Let us now suppose that a solution $(\mathring{M}, \mathring{N})$ of the system of field equations (1.552) is known and that (1.554) and (1.555) have been solved for (α_A, γ_A), $A = 1, 2$, where (α_1, γ_1), (α_2, γ_2) denote solutions with different integration constants. Reference to Fig. 1.33 shows that the permutability theorem provides for the construction of a second-generation solution

$$I_2(\alpha^{(2)}, \gamma^{(2)}) \circ I_1(\alpha_1, \gamma_1)(\mathring{M}_i, \mathring{N}_i), \quad (1.563)$$

where

$$\alpha^{(2)} = (\gamma_1 \alpha_2 - \alpha_1 \gamma_2)/\gamma_1(\alpha_2 - \alpha_1), \quad \gamma^{(2)} = \gamma_2/\gamma_1. \quad (1.564)$$

The Neugebauer procedure is illustrated here by using a second generation solution to construct the Kerr-NUT metric from Minkowski space–time (Neugebauer and Kramer [381]). Thus, if we start with the solution (1.544)–(1.547)

$$\mathring{u} = 1, \quad \mathring{w} = \mathring{\delta} = 0, \quad \mathring{W} = \tfrac{1}{2}(x^1 + x^2),$$

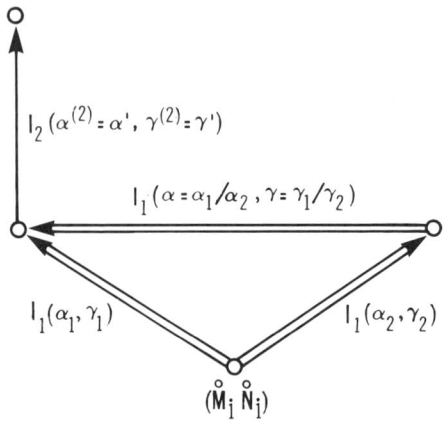

Fig. 1.33 Construction of second-generation solutions $I_2(\alpha^{(2)}, \gamma^{(2)}) \circ I_1(\alpha_1, \gamma_1)(\overset{\circ}{M}_i, \overset{\circ}{N}_i)$ via the Neugebauer permutability theorem.

corresponding to Minkowski space–time, then (1.551) shows that

$$\overset{\circ}{\mathbf{M}} = \overset{\circ}{\mathbf{N}} = \begin{bmatrix} 0 \\ 0 \\ (x^1 + x^2)^{-1} \end{bmatrix},$$

where

$$\overset{\circ}{\mathbf{M}} = \begin{bmatrix} \overset{\circ}{M}_1 \\ \overset{\circ}{M}_2 \\ \overset{\circ}{M}_3 \end{bmatrix}, \quad \overset{\circ}{\mathbf{N}} = \begin{bmatrix} \overset{\circ}{N}_1 \\ \overset{\circ}{N}_2 \\ \overset{\circ}{N}_3 \end{bmatrix}.$$

Application of the symmetry operator S to the set $(\overset{\circ}{M}_i, \overset{\circ}{N}_i)$ and use of the resulting set (M_i, N_i) to solve (1.554) and (1.555) yield

$$\alpha_A = \gamma_A^{1/2} \left[\frac{b_A + \gamma_A^{1/2}}{1 + b_A \gamma_A^{1/2}} \right], \quad A = 1, 2, \tag{1.565}$$

$$\gamma_A = -\left[\frac{x^2 + ik_A}{x^1 - ik_A} \right], \quad A = 1, 2, \tag{1.566}$$

where b_A and k_A, $A = 1, 2$, are complex constants of integration.

A straightforward calculation now shows that the image of Minkowski space–time under the Bäcklund transformation

$$I_1(\alpha^{(2)}, \gamma^{(2)}) I_2(\alpha, \gamma) = I_1(\alpha^{(2)}, \gamma^{(2)}) S I_1(\alpha, \gamma) S \tag{1.567}$$

1.19 BÄCKLUND TRANSFORMATIONS IN GENERAL RELATIVITY

is given by

$$\mathbf{M} = (x^1 + x^2)^{-1} \begin{bmatrix} \dfrac{A(\gamma_1 - 1)\gamma_2^{1/2}}{2B(b_1 + \gamma_1^{1/2})} \\ \dfrac{B(\gamma_1 - 1)\gamma_2^{1/2}}{2A(1 + b_1\gamma_1^{1/2})} \\ \gamma_2 \end{bmatrix}, \quad (1.568)$$

$$\mathbf{N} = (x^1 + x^2)^{-1} \begin{bmatrix} \dfrac{B(1 - \gamma_1)b_1}{2A(\gamma_1\gamma_2)^{1/2}(b_1 + \gamma_1^{1/2})} \\ \dfrac{A(1 - \gamma_1)}{2B(\gamma_1\gamma_2)^{1/2}(1 + b_1\gamma_1^{1/2})} \\ \gamma_2^{-1} \end{bmatrix}, \quad (1.569)$$

where

$$A = b_1\gamma_2^{1/2}(\gamma_1 - 1) + b_1 b_2(\gamma_1 - \gamma_2) - b_2\gamma_1^{1/2}(\gamma_2 - 1) \quad (1.570)$$

and

$$B = -b_2\gamma_2^{1/2}(\gamma_1 - 1) + (\gamma_2 - \gamma_1) + b_1\gamma_1^{1/2}(\gamma_2 - 1). \quad (1.571)$$

The metric obtained by solution of (1.551) for W and ε using the set (M_i, N_i) given by (1.568) and (1.569) will be real if $M_i = \bar{N}_i$, and this condition holds if $k_A = \bar{k}_A$ and $\bar{b}_A = b_A^{-1}$. It follows that

$$W(x^1, x^2) = (x^1 + x^2)/[(x^1 - ik_2)(x^2 + ik_2)],$$

while, if the coordinate change

$$w^1 = \rho + iz = 2/(x^1 - ik_2), \quad (1.573)$$
$$w^2 = \rho - iz = 2/(x^2 + ik_2) \quad (1.574)$$

is introduced, the Ernst potential is obtained in the form

$$\varepsilon = \frac{e^{-i\phi}r_1 + e^{i\phi}r_2 - 2(m + il)\cos\phi}{e^{-i\phi}r_1 + e^{i\phi}r_2 + 2(m + il)\cos\phi}, \quad (1.575)$$

where

$$r_A^2 = \rho^2 + (z - K_A)^2, \quad K_1 = \frac{k_1 k_2}{k_1 - k_2}, \quad K_2 = k_2, \quad (1.576)$$

and the original constants of integration b_A, k_A are given by

$$b_1 = e^{2i\phi}, \quad b_2 = -\sqrt{\frac{m-il}{m+il}}\, e^{i\phi},$$
$$K_1 = -\sqrt{m^2+l^2}\cos\phi + z_0, \quad K_2 = \sqrt{m^2+l^2}\cos\phi + z_0. \quad (1.577)$$

The Ernst potential (1.575) corresponds to the Kerr-NUT solution in Weyl coordinates where m and l are the mass and NUT parameters, respectively. The Kerr parameter is $m\sin\phi$, while z_0 denotes the position of the rotating body on the z axis.

The above procedure may be iterated to generate a lattice whereby an infinite sequence of solutions of the system (1.552) may be constructed without further integration. Thus, iteration can be used to construct exact solutions of the Einstein field equations corresponding not only to a Tomimatsu–Sato solution [382] but also to a superposition of Schwarzschild solutions [383, 384]. Furthermore, particular Neugebauer–Bäcklund transformations have recently been employed to generate new solutions of the Einstein equations both from the Van Stockhum metric and from Weyl's class of static vacuum solutions (Herlt [385, 386]). In particular, Darmois and Reissner–Nordström solutions were thereby constructed.

CHAPTER 2

A Local Jet-Bundle Formulation of Bäcklund Transformations

2.1 PRELIMINARIES

In this chapter, a geometric formulation of an important class of Bäcklund transformations is presented. We consider the case in which the defining equations of the Bäcklund transformations may be solved to express the first derivatives of the new dependent variables as functions of the old independent and dependent variables, their derivatives of some finite order, and the new dependent variables themselves. The formalism is based on the idea of a Bäcklund map [48, 407]. It has the advantage that it allows consideration of Bäcklund transformations of equations without restriction as to order or to the number of dependent or independent variables.

We shall demonstrate how a one-parameter symmetry group of a differential equation allows the introduction of a parameter into its Bäcklund transformation. Further, we shall note how a certain factorization of Bäcklund maps permits one to identify the presence of a Lie algebra (such as $\mathfrak{SL}$ (2, $\mathbb{R}$)) which, in turn, may be used to construct an associated linear scattering problem. Finally, the Wahlquist–Estabrook procedure for the construction

124   2. A JET-BUNDLE FORMULATION OF BÄCKLUND TRANSFORMATIONS

of Bäcklund maps will be presented. This method, based on the notion of pseudopotentials, is more economical than the traditional Clairin approach. Moreover, it clarifies the roles played by the various Lie groups associated with differential equations which possess Bäcklund transformations.

Here, a jet-bundle formulation of Bäcklund transformations is set forth. In general terms, the utility of jet bundles in this connection is due to the fact that they have local coordinates which adopt the roles of the field variables and their derivatives; in this context, these derivatives may be legitimately regarded as independent quantities.[†]

We begin the discussion with some definitions. Let M and N denote spaces of independent variables, $M = \mathbb{R}^m$ and $N = \mathbb{R}^n$ say,[‡] with coordinates x^a, $a = 1, \ldots, m$, and z^A, $A = 1, \ldots, n$, respectively. Let $C^\infty(M, N)$ denote the smooth maps from M to N. Then, if $f \in C^\infty(M, N)$ is defined at $\mathbf{x} \in M$, it follows that f is determined by n coordinate functions $z^A := f^A(\mathbf{x})$. If $g \in C^\infty(M, N)$ is another map defined at $\mathbf{x}$, then f and g are said to be k-*equivalent* at $\mathbf{x}$ iff f and g have the same Taylor expansion to order k at $\mathbf{x}$. Thus, f and g are k-equivalent at $\mathbf{x}$ iff

$$f^A(\mathbf{x}) = g^A(\mathbf{x}), \qquad A = 1, \ldots, n,$$
$$\partial_{a_1 \cdots a_j} f^A(\mathbf{x}) = \partial_{a_1 \cdots a_j} g^A(\mathbf{x}), \qquad A = 1, \ldots, n, \quad j = 1, \ldots, k, \qquad (2.1)$$

where ∂_a denotes the derivative $\partial/\partial x^a$, while $\partial_{a_1 \cdots a_j}$ denotes repeated differentiation:

$$\partial_{a_1 \cdots a_j} := \partial_{a_1} \circ \partial_{a_2} \circ \cdots \circ \partial_{a_j}.$$

Here and subsequently, the lower case indices $a_1, a_2, \ldots$ range over $1, \ldots, m$. The k-equivalence class of f at $\mathbf{x}$ is called the k-*jet of* f *at* $\mathbf{x}$ and is denoted by $j_\mathbf{x}^k f$. The collection of all k jets $j_\mathbf{x}^k f$ as $\mathbf{x}$ ranges over M and f ranges over $C^\infty(M, N)$ is called the k-*jet bundle of maps from* M *to* N and is denoted by $J^k(M, N)$. Thus

$$J^k(M, N) := \bigcup_{\mathbf{x} \in M, \, f \in C^\infty(M,N)} j_\mathbf{x}^k f. \qquad (2.2)$$

It follows from the definitions that if $\mathbf{p}$ is a "point" in $J^k(M, N)$, then $\mathbf{p} = j_\mathbf{x}^k f$ for some $\mathbf{x} \in M$ and $f \in C^\infty(M, N)$ and hence that $\mathbf{p}$ is determined uniquely by the numbers

$$x^a, z^A, z^A_{a_1 \cdots a_j}, \qquad a = 1, \ldots, m, \quad A = 1, \ldots, N, \quad j = 1, \ldots, k, \qquad (2.3)$$

[†] Other geometric approaches to the theory of Bäcklund transformations have been given by Gardner [442] and Payne [357].

[‡] The subsequent treatment can be formulated equally well if M and N are smooth manifolds of dimension m and n, respectively (Pirani *et al.*, [48]).

2.1 PRELIMINARIES

where x^a, z^A are the coordinates of $\mathbf{x}$ and $f(\mathbf{x})$, respectively, and

$$z^A_{a_1 \cdots a_j} = \partial_{a_1} \cdots \partial_{a_j} f^A(\mathbf{x}).$$

The latter quantities are, by definition, independent of the choice of f in the equivalence class $j^k_\mathbf{x} f$. Conversely, any collection of numbers x^a, z^A, $z^A_{a_1 \cdots a_j}$, $j = 1, \ldots, k$, with the $z^A_{a_1 \cdots a_j}$ symmetric in the indices $a_1, \ldots, a_j$ determines a point of $J^k(M, N)$.†

Example 2.1 Let $M = \mathbb{R}^2$, $N = \mathbb{R}$, and choose coordinates x^1, x^2 on M and z^1 on N. Then, if f and $g \in C^\infty(M, N)$ are given by

$$f(x^1, x^2) = (x^1)^2 + (x^2)^2, \qquad g(x^1, x^2) = 0,$$

f and g belong to the same 1-jet at $(0, 0)$ but not to the same 2-jet.

The 2-jet bundle $J^2(M, N)$ may be identified with $\mathbb{R}^8$ with the coordinates $x^1, x^2, z^1, z^1_1, z^1_2, z^1_{11}, z^1_{12}, z^1_{22}$. ∎

Note that two maps which are k-equivalent at $x \in M$ are also j-equivalent for all $j \leq k$. This fact allows us to define the *canonical projection maps* π^k_{k-r} from $J^k(M, N)$ to $J^{k-r}(M, N)$, $r = 0, \ldots, k-1$, by

$$\pi^k_{k-r} : j^k_\mathbf{x} f \to j^{k-r}_\mathbf{x} f. \tag{2.4}$$

It is sometimes convenient to identify $J^0(M, N)$ with $M \times N$ and to define π^k_0 by

$$\pi^k_0 : j^k_\mathbf{x} f \to (\mathbf{x}, f(\mathbf{x})). \tag{2.5}$$

There are two further canonical projection maps which will prove useful in what follows, namely, the *source and target maps* $\alpha : J^k(M, N) \to M$ and $\beta : J^k(M, N) \to N$ defined by

$$\alpha : j^k_\mathbf{x} f \to \mathbf{x}, \qquad \beta : j^k_\mathbf{x} f \to f(\mathbf{x}), \tag{2.6}$$

respectively.

A map $h : M \to J^k(M, N)$ which satisfies $\alpha \circ h = \mathrm{id}_M$, where id_M is the identity map on M, is called a *cross section of the source map* α. An important example of such a cross section is the *k-jet extension* of a map $f \in C^\infty(M, N)$ denoted by $j^k f$ and defined by

$$j^k f : \mathbf{x} \to j^k_\mathbf{x} f.$$

In the case $k = 0$, $j^k f$ is just the graph of f.

Example 2.2 If $M = \mathbb{R}^2$ and $N = \mathbb{R}$ with the same coordinates as in Example 2.1, then the 1-jet extension $j^1 f$ of $f \in C^\infty(\mathbb{R}^2, \mathbb{R})$ is defined by

$$j^1 f : (x^1, x^2) \to (x^1, x^2, f(x^1, x^2), \partial_1 f(x^1, x^2), \partial_2 f(x^1, x^2)). \quad \blacksquare$$

---
† The k-jet bundle $J^k(M, N)$ with $M = \mathbb{R}^m$, $N = \mathbb{R}^n$ may, accordingly, be identified with $\mathbb{R}^{\dim J^k}$ with the coordinates (2.3).

2.2 CONTACT STRUCTURES

Those cross sections of α that are k-jet extensions of maps from M to N may be conveniently characterized in terms of differential forms on $J^k(M, N)$.[†]

We first consider the case $k = 1$ and define differential 1-forms θ^A, $A = 1, \ldots, n$, on $J^1(M, N)$ by

$$\theta^A := dz^A - z_a^A \, dx^a. \tag{2.7}$$

Since z^A and z_a^A are coordinates on $J^1(M, N)$, the 1-forms θ^A are not identically zero. There are, however, certain privileged submanifolds on $J^1(M, N)$ on which the θ^A vanish. The ones of interest for the present investigation arise as images of cross sections of α. Thus if $g: M \to J^1(M, N)$ is a cross section of α and is given in coordinates by

$$g: \mathbf{x} \to (x^b, g^A(x^b), g_a^A(x^b)),$$

then the image of g is the m-dimensional submanifold $\text{im}(g) \subset J^1(M, N)$ determined by the constraint equations

$$z^A = g^A(x^b), \qquad z_a^A = g_a^A(x^b), \tag{2.8}$$

(see Fig. 2.1).

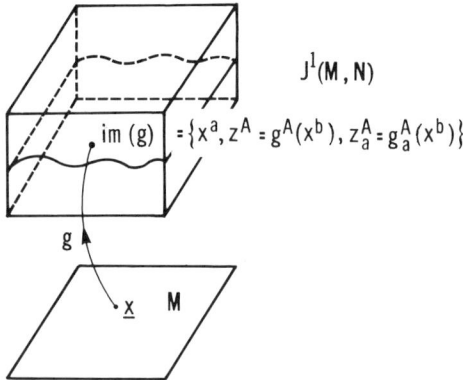

Fig. 2.1 A schematic representation of $\text{im}(g)$, $g: M \to J^1(M, N)$.

It follows from (2.7) and (2.8) that on the submanifold $\text{im}(g)$, θ^A is given by

$$\theta^A \Big|_{\text{im}(g)} = dg^A - g_a^A \, dx^a = \left(\frac{\partial g^A}{\partial x^a} - g_a^A \right) dx^a,$$

[†] The reader unfamiliar with the exterior differential calculus is referred to Appendixes II–III or, for a more complete account, to Flanders [122].

2.2 CONTACT STRUCTURES

and this 1-form vanishes iff

$$g_a^A = \partial g^A/\partial x^a. \tag{2.9}$$

But (2.9) holds iff g is the 1-jet extension of the map $\beta \circ g: M \to N$ since $\beta \circ g$ is given in coordinates by

$$\beta \circ g: \mathbf{x} \to g^A(x^a).$$

The same result may be rephrased in terms of the so-called pull-back maps (see Appendix II). Thus, if we consider the 1-forms $g^*\theta^A$ on M, we have

$$g^*\theta^A = dg^A - g_a^A dx^a = \left(\frac{\partial g^A}{\partial x^a} - g_a^A\right) dx^a,$$

so that $g^*\theta^A = 0$ iff $g_a^A = \partial g^A/\partial x^a$ and this, in turn, holds iff

$$g = j^1(\beta \circ g). \tag{2.10}$$

To summarize then, if g is a cross section of α, then g is the 1-jet extension of $\beta \circ g: M \to N$ iff $g^*\theta^A = 0$. The 1-forms θ^A are called *contact forms* and the set of all finite linear combinations of the 1-forms θ^A over the ring of C^∞ functions on $J^1(M, N)$ is called the *first-order contact* module and is denoted by $\Omega^1(M, N)$.[†] Thus

$$\Omega^1(M, N) := \{\sum f_A \theta^A \mid f_A \in C^\infty(J^1(M, N), \mathbb{R})\}. \tag{2.11}$$

The set of 1-forms $\{\theta^A\}$ is called the *standard basis* for $\Omega^1(M, N)$.

It is a simple matter to generalize the above discussion to the case $k > 1$ by introducing the 1-forms θ^A, $\theta^A_{a_1 \cdots a_l}$, $l = 1, \ldots, k-1$, according to

$$\begin{aligned}
\theta^A &:= dz^A - z_a^A dx^b, \\
\theta^A_{a_1} &:= dz^A_{a_1} - z^A_{a_1 b} dx^b, \\
&\vdots \\
\theta^A_{a_1 \cdots a_{k-1}} &= dz^A_{a_1 \cdots a_{k-1}} - z^A_{a_1 \cdots a_{k-1} b} dx^b.
\end{aligned} \tag{2.12}$$

It may be readily verified that if $g: M \to J^k(M, N)$ is a cross section of the source map α, then

$$g^*\theta^A_{a_1 \cdots a_l} = 0, \quad l = 1, \ldots, k-1 \quad \text{iff} \quad g = j^k(\beta \circ g).$$

Thus, if g is given in coordinates by

$$g: \mathbf{x} \to (x^a, g^A(x^b), g^A_{a_1 \cdots a_j}(x^b)), \quad j = 1, \ldots, k,$$

it follows that

$$g^*\theta^A = 0 \quad \text{iff} \quad g_a^A = \partial_a g^A,$$

[†] See Adamson [123] for an introduction to module theory. The notion of contact module used here is based on that of Johnson [124].

and then iteratively that

$$g^*\theta^A_{a_1\cdots a_l} = 0 \quad \text{iff} \quad g^A_{a_1\cdots a_{l+1}} = \partial_{a_1\cdots a_{l+1}} g^A, \quad (2.13)$$
$$l = 1, \ldots, k-1,$$

The collection $\Omega^k(M, N) := \{\sum f_A \theta^A + \sum f^{a_1\cdots a_j}_A \theta^A_{a_1\cdots a_j}\}$ is called the kth-order contact module.

One may equally well characterize k-jet extensions in terms of differential m-forms on $J^k(M, N)$ or, indeed, in terms of differential p-forms with $1 \leq p \leq m$. This characterization is presented here for the cases $p = m$ and $p = m - 1$ since these arise in the later discussion of the Wahlquist–Estabrook procedure.

Let w be the volume m-form on M given by

$$w := dx^1 \wedge dx^2 \wedge \cdots \wedge dx^m,$$

and define $(m-1)$-forms w_a, $a = 1, \ldots, m$, and $(m-2)$-forms w_{ab}, $a < b$, $1, \ldots, m$, by

$$w_a := \partial_a \,\lrcorner\, w \qquad (2.14)$$

and

$$w_{ab} := \partial_a \,\lrcorner\, w_b, \qquad (2.15)$$

respectively. Here $\lrcorner$ denotes the interior product of differential forms and vector fields (see Appendix II).

The modules of *contact m-forms* $\Omega^k_{(m)}(M, N)$ and *contact $(m-1)$-forms* $\Omega^k_{(m-1)}(M, N)$ are defined by

$$\Omega^k_{(m)}(M, N) := \{\theta \wedge w_a \mid \theta \in \Omega^k(M, N)\} \qquad (2.16)$$

and

$$\Omega^k_{(m-1)}(M, N) := \{\theta \wedge w_{ab} \mid \theta \in \Omega^k(M, N)\}, \qquad (2.17)$$

respectively.

Suppose now that g is a cross section of α. It follows that

$$g^*(\theta \wedge w_a) = 0 \quad \text{iff} \quad g^*\theta = 0,$$

whence

$$g^*\Omega^k_{(m)}(M, N) = 0 \quad \text{iff} \quad g^*\Omega^k(M, N) = 0.$$

Accordingly, if g is a cross section of α, then

$$g^*\Omega^k_{(m)}(M, N) = 0 \quad \text{iff} \quad g = j^k(\beta \circ g), \qquad (2.18)$$

and a similar calculation shows that

$$g^*\Omega^k_{(m-1)}(M, N) = 0 \quad \text{iff} \quad g = j^k(\beta \circ g). \qquad (2.19)$$

2.3 PARTIAL DIFFERENTIAL EQUATIONS ON JET BUNDLES

A general system of kth order partial differential equations in n unknowns u^A may be written in the form

$$F_\lambda(x^a, u^A, \ldots, \partial_{a_1 \cdots a_k} u^A) = 0, \qquad \lambda = 1, \ldots, r, \tag{2.20}$$

and such a system determines a submanifold of the k-jet bundle $J^k(M, N)$, namely, the submanifold R^k given by the constraint equations

$$F_\lambda(x^a, z^A, z^A_a, \ldots, z^A_{a_1 \cdots a_k}) = 0, \qquad \lambda = 1, \ldots, r. \tag{2.21}$$

This submanifold, being a geometric object, provides a very convenient means whereby transformation properties of the system (2.20) may be analyzed.

In what follows, we shall refer to R^k as a kth-*order differential equation*. A *solution of* R^k is defined to be a map $f \in C^\infty(M, N)$ such that $\text{im}(j^k f) \subset R^k$ (see Fig. 2.2). Thus, if f is a solution of R^k and if f is given in coordinates by $z^A = f^A(x^b)$, then it follows that

$$F_\lambda(x^a, f^A, \partial_a f^A, \ldots, \partial_{a_1 \cdots a_k} f^A) = 0, \qquad \lambda = 1, \ldots, r,$$

and f is a solution of the system (2.20).

Example 2.3 Let $M = \mathbb{R}^2$, $N = \mathbb{R}$, and choose coordinates x^a, z, z_a, z_{ab} on $J^2(M, N)$.

The sine-Gordon equation

$$\frac{\partial^2 u}{\partial x^1 \, \partial x^2} = \sin u,$$

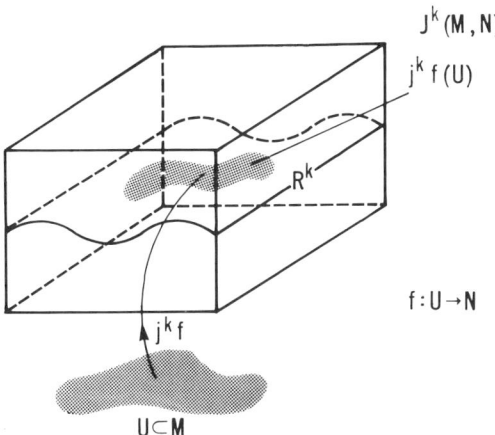

Fig. 2.2 A schematic representation of solutions f of R^k.

determines the submanifold R^2 of $J^2(M, N)$ given by
$$z_{12} - \sin z = 0,$$
and a solution of R^2 is a map $f \in C^\infty(M, N)$ such that if $\mathbf{x} \in M$, then $j^2 f(\mathbf{x}) \in R^2$, that is,
$$\partial_{12} f(x^1, x^2) - \sin\{f(x^1, x^2)\} = 0. \quad\blacksquare$$

To summarize then, a system of partial differential equations may be replaced by a submanifold of an appropriate jet bundle. This point of view is the basis of the formulation of Bäcklund maps and Bäcklund transformations that follows. Both linear and nonlinear systems of the type (2.20) are treated. In the simplest case, when the system is linear with constant coefficients, the associated submanifold R^k is a subspace of $J^k(M, N)$, where it is recalled that here M and N are taken to be Euclidean m and n spaces, respectively, so that $J^k(M, N)$ may be identified with $\mathbb{R}^{\dim J^k}$. Thus, the functions F_λ which serve to define R^k are linear in each of the coordinates so that if $\mathbf{p}$ and $\mathbf{p}'$ are points in R^k, so also is $a\mathbf{p} + b\mathbf{p}'$ where $a, b \in \mathbb{R}$. In the case of nonlinear systems, the associated submanifold of $J^k(M, N)$ is *not* a subspace. These points are illustrated by the following:

Example 2.4 Let $M = \mathbb{R}^2$, $N = \mathbb{R}$, and consider, in turn, the Klein–Gordon equation
$$\frac{\partial^2 u}{\partial x^1 \partial x^2} = \alpha u \tag{2.22}$$
and the sine-Gordon equation
$$\frac{\partial^2 u}{\partial x^1 \partial x^2} = \sin u. \tag{2.23}$$
These correspond to the submanifolds R and $R' \subset \mathbb{R}^8$ given by
$$z_{12} = \alpha z \tag{2.24}$$
and
$$z_{12} = \sin z, \tag{2.25}$$
respectively.

If $\mathbf{p}$ and $\mathbf{p}'$ are points in $\mathbb{R}^8$ with coordinates $(x^a, z, z_a, z_{11}, z_{12} = \alpha z, z_{22})$ and $(x'^a, z', z'_a, z'_{11}, z'_{12} = \alpha z', z'_{22})$, respectively, then since
$$a z_{12} + b z'_{12} = a\alpha z + b\alpha z' = \alpha(az + bz')_{12},$$
it is readily seen that $a\mathbf{p} + b\mathbf{p}'$ belongs to the submanifold (2.24) associated with the Klein–Gordon equation. Accordingly, (2.24) is a subspace of $\mathbb{R}^8$.

2.3 PARTIAL DIFFERENTIAL EQUATIONS ON JET BUNDLES

On the other hand, if **p** and **p**' are points in R' with coordinates $(x^a, z, z_a, z_{11}, z_{12} = \sin z, z_{22})$ and $(x'^a, z', z'_a, z'_{11}, z'_{12} = \sin z', z'_{22})$, then since

$$az_{12} + bz'_{12} = a \sin z + b \sin z' \neq \sin(az + bz'), \quad a, b \in \mathbb{R},$$

examination of the z_{12} coordinates of $a\mathbf{p} + b\mathbf{p}'$ shows that, in general, $a\mathbf{p} + b\mathbf{p}'$ does not lie in R', so that R' is not a subspace of $\mathbb{R}^8$. ∎

At this point, it is convenient to introduce the important notion of prolongation of systems of differential equations. Thus it is often useful to consider a system together with its derivatives and to characterize the new system in terms of jets. In order to describe the jet bundle submanifolds corresponding to the system, we introduce the *total derivative operators* $D_a^{(k)}$ defined on $J^k(M, N)$ for $k \geq 1$ by

$$D_a^{(k)} := \frac{\partial}{\partial x^a} + z_a^A \frac{\partial}{\partial z^A} + \cdots + z_{aa_1 \cdots a_{k-1}}^A \frac{\partial}{\partial z_{a_1 \cdots a_{k-1}}^A}. \tag{2.26}$$

For notational convenience, repeated total derivatives $D_{a_1}^{(k)} \circ D_{a_2}^{(k)} \circ \cdots \circ D_{a_p}^{(k)}$ will be designated by $D_{a_1 \cdots a_p}^{(k)}$. The *l*th *prolongation of* R^k is the submanifold $R^{k+l} \subset J^{k+l}(M, N)$ defined by the constraint equations

$$\begin{aligned} F_\lambda &= 0, \\ D_a^{(k+l)} F_\lambda &= 0, \\ &\vdots \\ D_{a_1 \cdots a_l}^{(k+l)} F_\lambda &= 0. \end{aligned} \quad \lambda = 1, \ldots, r. \tag{2.27}$$

It follows from (2.27) that if f is a solution of R^{k+l}, then

$$\begin{aligned} F_\lambda(x^a, f^A, \ldots, \partial_{a_1 \cdots a_k} f^A) &= 0, \\ \partial_a F_\lambda(x^a, f^A, \ldots, \partial_{a_1 \cdots a_k} f^A) &= 0, \\ &\vdots \\ \partial_{a_1 \cdots a_l} F_\lambda(x^a, f^A, \ldots, \partial_{a_1 \cdots a_k} f^A) &= 0. \end{aligned} \tag{2.28}$$

Example 2.5 Consider the sine-Gordon equation as in Example 2.3. The total derivatives on $J^3(M, N)$ are given by

$$D_a^{(3)} = \partial_a + z_a \frac{\partial}{\partial z} + z_{ab} \frac{\partial}{\partial z_b} + z_{abc} \frac{\partial}{\partial z_{bc}},$$

and consequently, the first prolongation of the sine-Gordon equation is given by

$$z_{12} - \sin z = 0, \quad D_a^{(3)}(z_{12} - \sin z) = 0,$$

that is,

$$z_{12} - \sin z = 0, \quad z_{112} - z_1 \cos z = 0, \quad z_{122} - z_2 \cos z = 0. \quad \blacksquare$$

2. A JET-BUNDLE FORMULATION OF BÄCKLUND TRANSFORMATIONS

The following properties of the total derivative as introduced in (2.26) will prove important in the context of prolongations of R^k:

(a) If $\theta \in \Omega^k(M, N)$, then $D_a^{(k)} \lrcorner\, \theta = 0$.
(b) If $\phi: J^{k-1}(M, N) \to \mathbb{R}$, then

$$d(\pi_{k-1}^{k*}\phi) \equiv D_a^{(k)}(\pi_{k-1}^{k*}\phi)\, dx^a \quad \mathrm{mod}\, \Omega^k(M, N),$$

where congruence "$\mathrm{mod}\, \Omega^k(M, N)$" means that 1-forms lying in $\Omega^k(M, N)$ are discarded.

Example 2.6 Consider the function $\phi: J^1(\mathbb{R}^2, \mathbb{R}) \to \mathbb{R}$ defined by

$$\phi:(x^a, z, z_a) \to zz_1.$$

Then $\pi_1^{2*}\phi: J^2(\mathbb{R}^2, \mathbb{R}) \to \mathbb{R}$ is given by

$$\pi_1^{2*}\phi:(x^a, z, z_a, z_{ab}) \to zz_1,$$

so that

$$d(\pi_1^{2*}\phi) = z_1\, dz + z\, dz_1 \quad \text{and} \quad D_a^{(2)}(\pi_1^{2*}\phi) = z_1 z_a + zz_{1a}.$$

Hence if $\theta, \theta_1 \in \Omega^2(M, N)$ are introduced according to

$$\theta := dz - z_a\, dx^a, \qquad \theta_1 := dz_1 - z_{1a}\, dx^a,$$

then it follows that

$$d(\pi_1^{2*}\phi) = (z_1 z_a + zz_{1a})\, dx^a + z_1 \theta + z\theta_1$$
$$= D_a^{(2)}(\pi_1^{2*}\phi)\, dx^a + z_1\theta + z\theta_1. \quad\blacksquare$$

2.4 FIBERED PRODUCTS OF JET BUNDLES

If $M = \mathbb{R}^m$, $N = \mathbb{R}^n$, $N' = \mathbb{R}^{n'}$, then the subset of $J^k(M, N) \times J^l(M, N')$ which consists of all pairs of points $(\mathbf{p}, \mathbf{p}')$, $\mathbf{p} \in J^k(M, N)$ and $\mathbf{p}' \in J^l(M, N')$, such that $\alpha(\mathbf{p}) = \alpha(\mathbf{p}')$ is called the *fibered product* of $J^k(M, N)$ and $J^l(M, N')$ and is denoted by $J^k(M, N) \underset{M}{\times} J^l(M, N')$. Thus

$$J^k(M, N) \underset{M}{\times} J^l(M, N') := \{(\mathbf{p}, \mathbf{p}') \in J^k(M, N) \times J^l(M, N')\,|\, \alpha(\mathbf{p}) = \alpha(\mathbf{p}')\},$$

(2.29)

and points in the fibered product have the form $(j_x^k f, j_x^l g)$ for some $f \in C^\infty(M, N)$ and $g \in C^\infty(M, N')$.

If local coordinates $x^a, z^A, \ldots, z^A_{a_1 \cdots a_k}$ and $x^a, y^\mu, \ldots, y^\mu_{a_1 \cdots a_l}$ are chosen on $J^k(M, N)$ and $J^l(M, N')$, respectively, then it follows that there is a one-to-one correspondence between $J^k(M, N) \underset{M}{\times} J^l(M, N')$ and the set of points $(x^a, z^A, \ldots, z^A_{a_1 \cdots a_k}, y^\mu, \ldots, y^\mu_{a_1 \cdots a_l})$.

2.4 FIBERED PRODUCTS OF JET BUNDLES

The important concepts of projection maps $\pi^{k,l}: J^k(M,N) \times_M J^l(M,N') \to J^{k-1}(M,N) \times_M J^{l-1}(M,N')$ and modules $\Omega^{k,l}(M,N,N')$ on $J^k(M,N) \times_M J^l(M,N')$ are defined, in turn, by

$$\pi^{k,l}: (j_x^k f, j_x^l g) \to (j_x^{k-1} f, j_x^{l-1} g), \tag{2.30}$$

and

$$\Omega^{k,l}(M,N,N') := \text{pr}_1^* \Omega^k(M,N) + \text{pr}_2^* \Omega^l(M,N') \tag{2.31}$$

where pr_i represents projection on the ith factor, while total derivative operators $D_a^{k,l}$ may defined on $J^k(M,N) \times_M J^l(M,N')$ by

$$D_a^{k,l} := \frac{\partial}{\partial x^a} + z_a^A \frac{\partial}{\partial z^A} + \cdots + z_{aa_1 \cdots a_{k-1}}^A \frac{\partial}{\partial z_{a_1 \cdots a_{k-1}}^A}$$
$$+ y_a^\mu \frac{\partial}{\partial y^\mu} + \cdots + y_{aa_1 \cdots a_{l-1}}^\mu \frac{\partial}{\partial y_{a_1 \cdots a_{l-1}}^\mu}. \tag{2.32}$$

It is readily verified that

(a) $D_a^{k,l} \lrcorner \theta = 0$, $\forall \theta \in \Omega^{k,l}(M,N,N')$,
(b) If $\phi: J^{k-1}(M,N) \times_M J^{l-1}(M,N') \to \mathbb{R}$, then

$$d(\pi^{k,l*}\phi) \equiv D_a^{k,l}(\pi^{k,l*}\phi) \, dx^a \quad \text{mod } \Omega^{k,l}(M,N,N').$$

Example 2.7 Let $M = \mathbb{R}^2$, $N = N' = \mathbb{R}$, and take coordinates x^a, z, y on M, N, and N', respectively. Then, on $J^2(M,N) \times_M J^1(M,N')$, $\Omega^{2,1}$ is generated by the 1-forms

$$\theta = dz - z_b \, dx^b, \quad \theta_a = dz_a - z_{ab} \, dx^b, \quad \text{and} \quad \bar{\theta} = dy - y_b \, dx^b,$$

while the total derivative operators $D_a^{2,1}$ are given by

$$D_a^{2,1} = \frac{\partial}{\partial x^a} + z_a \frac{\partial}{\partial z} + z_{ab} \frac{\partial}{\partial z_b} + y_a \frac{\partial}{\partial y}.$$

If ϕ is a function from $J^1(M,N) \times_M J^0(M,N')$ to $\mathbb{R}$, then since $J^1(M,N) \times_M J^0(M,N')$ has coordinates (x^a, z, z_a, y), it follows that

$$d(\pi^{2,1*}\phi) = \left(dx^a \frac{\partial}{\partial x^a} + dz \frac{\partial}{\partial z} + dz_a \frac{\partial}{\partial z_a} + dy \frac{\partial}{\partial y} \right)(\pi^{2,1*}\phi)$$
$$= D_a^{2,1}(\pi^{2,1*}\phi) \, dx^a + \left(\theta \frac{\partial}{\partial z} + \theta_a \frac{\partial}{\partial z_a} + \bar{\theta} \frac{\partial}{\partial y} \right)(\pi^{2,1*}\phi)$$
$$\equiv D_a^{2,1}(\pi^{2,1*}\phi) \, dx^a \quad \text{mod } \Omega^{2,1}. \blacksquare$$

134   2. A JET-BUNDLE FORMULATION OF BÄCKLUND TRANSFORMATIONS

Finally, note that if f and g are maps $f \in C^\infty(M, N)$ and $g \in C^\infty(M, N')$, then there is a natural map from M to $J^k(M, N) \underset{M}{\times} J(M, N')$, namely,

$$j^k f \underset{M}{\times} j^l g := (j^k f \times j^l g) \circ \Delta,$$

where Δ is the diagonal map

$$\Delta : x \to (x, x).$$

2.5 BÄCKLUND MAPS

In this section, we introduce jet-bundle transformations known as Bäcklund maps, which reproduce features common to many of the Bäcklund transformations set forth in Chapter 1. Indeed, the concept of Bäcklund map as formulated by Pirani and Robinson [47] is sufficiently general to provide an important unification of the inverse scattering, Wahlquist–Estabrook, and classical approaches to Bäcklund transformations. Bäcklund maps have the feature of being maps of finite-dimensional spaces that admit a simple geometric characterization. As such they may be used to give a geometric formulation of Bäcklund transformations as described in the sequel.

Our aim is to develop a natural jet-bundle framework for Bäcklund transformations of the type

$$y_a^\mu = \psi_\mu^a(x^a, z^A, \ldots, z^A_{a_1 \cdots a_k}; y^\nu),$$
$$a = 1, \ldots, m, \quad A = 1, \ldots, n, \quad \nu = 1, \ldots, n', \qquad (2.33)$$
$$x'^a = x^a,$$

without restriction as to dimension. In accordance with previous notation, let M and N be the spaces of the original independent and dependent variables x^a, z^A, respectively, and let $N' = \mathbb{R}^{n'}$ be taken as the space of the n' new dependent variables y^ν. Let ψ be a map from $J^k(M, N) \times N' = J^k(M, N) \underset{M}{\times} J^0(M, N') \to J^1(M, N)$ which is such that the diagrams of Figs. 2.3 and 2.4 commute, where, as before, pr_i denotes projection on the ith factor and α, β

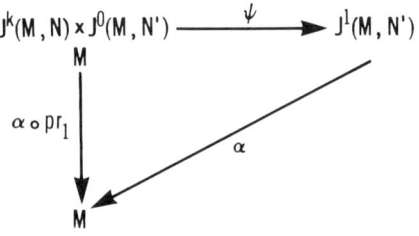

Fig. 2.3

2.5 BÄCKLUND MAPS

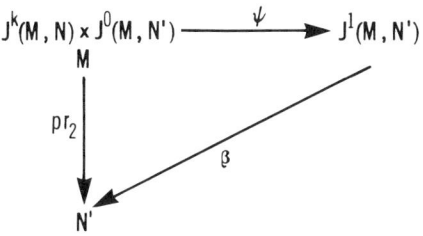

Fig. 2.4

are the source and target projection maps. Thus we require that

$$\alpha \circ \psi = \alpha \circ \mathrm{pr}_1 \tag{2.34}$$

and

$$\beta \circ \psi = \mathrm{pr}_2. \tag{2.35}$$

If $\mathbf{p} \in J^k(M,N) \underset{M}{\times} J^0(M,N')$ and $\mathbf{p}' = \psi(\mathbf{p})$, then $\mathbf{p} = (j^k_{\mathbf{x}} f, \mathbf{y})$ and $\mathbf{p}' = j^1_{\mathbf{x}'} g$, where $j^k_{\mathbf{x}} f \in J^k(M,N)$, $\mathbf{y} \in N'$, and $j^1_{\mathbf{x}'} g \in J^1(M,N')$. Hence, from (2.34),

$$(\alpha \circ \psi)(\mathbf{p}) = (\alpha \circ \mathrm{pr}_1)(\mathbf{p}),$$

so that

$$\alpha(j^1_{\mathbf{x}'} g) = \alpha(j^k_{\mathbf{x}} f),$$

which implies that

$$\mathbf{x}' = \mathbf{x}. \tag{2.36}$$

Further, from (2.35),

$$(\beta \circ \psi)(\mathbf{p}) = \mathrm{pr}_2(\mathbf{p}),$$

which implies that

$$\beta(j^1_{\mathbf{x}'} g) = \mathbf{y},$$

whence

$$g(\mathbf{x}') = \mathbf{y}. \tag{2.37}$$

Thus, in terms of coordinates x^a, z^A, ..., $z^A_{a_1 \cdots a_k}$, y^ν and x'^a, y'^μ, y'^μ_a on the domain and codomain of ψ, respectively, it follows from (2.36) and (2.37) that

$$x'^a = x^a, \tag{2.38}$$

and

$$y'^\mu = y^\mu. \tag{2.39}$$

In view of (2.38) and (2.39), it remains only to specify the coordinates y'^μ_a as functions of x^a, z^A, ..., $z^A_{a_1 \cdots a_k}$, y^ν in order to determine completely the map ψ. In other words, ψ is prescribed by mn' functions $\psi^\mu_a(x^b, z^A, \ldots, z^A_{a_1 \cdots a_k}, y^\nu)$ with

$$y'^\mu_a = \psi^\mu_a. \tag{2.40}$$

2. A JET-BUNDLE FORMULATION OF BÄCKLUND TRANSFORMATIONS

If the map ψ is given and f, g are maps $f \in C^\infty(M, N)$, $g \in C^\infty(M, N')$, then we shall be concerned with two ways of constructing out of f, g, and ψ a map from M to $J^1(M, N')$, namely, $j^1 g$ and $\psi \circ (j^k f \underset{M}{\times} j^0 g)$, respectively. In local coordinates, these maps are given by

$$x'^a = x^a, \quad y'^\mu = g^\mu(x^b), \quad y'^\mu_a = \partial_a g^\mu(x^b)$$

and

$$x'^a = x^a, \quad y'^\mu = g^\mu(x^b),$$

$$y'^\mu_a = \psi^\mu_a(x^b, f^A(x^b), \ldots, \partial_{a_1 \cdots a_k} f^A(x^b), g^\nu(x^b)),$$

respectively. These representations will not generally agree, in that it will *not* usually be the case that

$$\psi^\mu_a(x^b, f^A, \ldots, \partial_{a_1 \cdots a_k} f^A, g^\nu) = \partial_a g^\mu. \tag{2.41}$$

In fact, (2.41) holds only if the *integrability conditions*

$$\partial_b \psi^\mu_a(x^b, f^A, \ldots, \partial_{a_1 \cdots a_k} f^A, g^\nu) = \partial_a \psi^\mu_b(x^b, f^A, \ldots, \partial_{a_1 \cdots a_k} f^A, g^\nu) \tag{2.42}$$

are satisfied.

We now present a geometrical formulation of the integrability conditions for ψ which arises naturally out of consideration of the pull-back under ψ of the contact module $\Omega^1(M, N')$. Since (2.42) involves the $(k+1)$th derivatives of the "old" dependent variables f^A and the first derivatives of the "new" dependent variables g^ν, the formulation is in terms of objects on a space that contains both $J^{k+1}(M, N)$ and $J^1(M, N')$, namely, $J^{k+1}(M, N) \underset{M}{\times} J^1(M, N')$. In fact, the integrability conditions for ψ will be seen to determine a submanifold $B^{k+1,1} \subset J^{k+1}(M, N) \underset{M}{\times} J^1(M, N')$.

Let $\{\theta'^\mu\}$ denote the standard basis for $\Omega^1(M, N')$ and let Σ_ψ be the exterior differential system on $J^{k+1}(M, N) \underset{M}{\times} J^1(M, N')$ generated by $\{\psi^* \theta'^\mu, \psi^* d\theta'^\mu\}$. Thus Σ_ψ is the *differential ideal* generated by the pull-back of $\Omega^1(M, N')$ under ψ. It is demonstrated in Appendix IV that there is a submanifold of $J^{k+1}(M, N) \underset{M}{\times} J^1(M, N')$, denoted here by $B^{k+1,1}(\psi)$, naturally associated with Σ_ψ. This is such that if $f \in C^\infty(M, N)$, $g \in C^\infty(M, N')$, then

$$(j^k f \underset{M}{\times} j^0 g)^* \Sigma_\psi = 0, \tag{2.43}$$

iff

$$\mathrm{im}(j^{k+1} f \underset{M}{\times} j^1 g) \subset B^{k+1,1}. \tag{2.44}$$

The constraint equations defining $B^{k+1,1}(\psi)$ are given by

$$y^\mu_a = \psi^\mu_a, \tag{2.45}$$

$$D^{k+1,1}_a \psi^\mu_b = D^{k+1,1}_b \psi^\mu_a, \tag{2.46}$$

2.5 BÄCKLUND MAPS

and it follows that a pair of maps $f \in C^\infty(M, N)$, $g \in C^\infty(M, N')$ satisfies (2.44) iff

$$\partial_a g^\mu(x^b) = \psi_a^\mu(x^b, f^A, \ldots, \partial_{a_1 \cdots a_k} f^A, g^v) \tag{2.47}$$

and

$$\partial_a \psi_b^\mu(x^c, f^A, \ldots, g^v) = \partial_b \psi_a^\mu(x^c, f^A, \ldots, g^v). \tag{2.48}$$

A pair (f, g) of such maps is called a *solution* of $B^{k+1, 1}$ in what follows. Note that the relations (2.45) are identical to the relations (2.40) which define the image of ψ on $J^1(M, N')$, while (2.46) may be regarded as a system of equations on $J^{k+1}(M, N)$ parametrized by the y^μ. If this system, which we may identify as $B^{k+1, 1} \cap J^{k+1}(M, N)$, is nonempty, then ψ will be called an *ordinary Bäcklund map* [47, 48]. The Bäcklund maps of present interest arise when $B^{k+1} \cap J^{k+1}(M, N)$ contains a system of differential equations R^{k+1}. Then the above procedure provides a representation of the constraint equations for R^{k+1} as integrability conditions for the map ψ, and in this case ψ is called an *ordinary Bäcklund map for* R^{k+1}.

Example 2.8 Let $M = \mathbb{R}^2$, $N = N' = \mathbb{R}$, and let ψ be the map

$$\psi : J^1(M, N) \times N' \to J^1(M, N'), \tag{2.49}$$

given by

$$\begin{aligned}
x'^a &= x^a, \\
y' &= y, \\
y'_1 &= y_1 = \psi_1 := z_1 + 2e^t \sin \tfrac{1}{2}(y + z), \\
y'_2 &= y_2 = \psi_2 := -z_2 + 2e^{-t} \sin \tfrac{1}{2}(y - z),
\end{aligned} \tag{2.50}$$

where t is a real parameter. Here the coordinates are x^a on M, z on N, and y' on N'.

In this case, $B^{2, 1}$ is the submanifold of $J^2(M, N) \underset{M}{\times} J^1(M, N')$ given by

$$y_a = \psi_a$$

together with

$$D^{2,1}_{[a} \psi_{b]} = 0,^\dagger$$

that is,†

$$\left\{\frac{\partial}{\partial x^1} + z_1 \frac{\partial}{\partial z} + z_{11} \frac{\partial}{\partial z_1} + z_{12} \frac{\partial}{\partial z_2} + y_1 \frac{\partial}{\partial y}\right\} \psi_2$$
$$- \left\{\frac{\partial}{\partial x^2} + z_2 \frac{\partial}{\partial z} + z_{21} \frac{\partial}{\partial z_1} + z_{22} \frac{\partial}{\partial z_2} + y_2 \frac{\partial}{\partial y}\right\} \psi_1 = 0. \tag{2.51}$$

---

† Here and in what follows, square brackets denote antisymmetrization of the enclosed indices.

Insertion into (2.51) of ψ_1 and ψ_2 as given by (2.50) now yields
$$-z_{12} + e^{-t}(y_1 - z_1)\cos\tfrac{1}{2}(y-z) - [z_{21} + e^t(y_2+z_2)\cos\tfrac{1}{2}(y+z)] = 0,$$
that is,
$$-2z_{12} + 2\sin\tfrac{1}{2}(y+z)\cos\tfrac{1}{2}(y-z) - 2\sin\tfrac{1}{2}(y-z)\cos\tfrac{1}{2}(y+z) = 0,$$
whence
$$z_{12} - \sin z = 0. \tag{2.52}$$

Thus ψ is an ordinary Bäcklund map for R^2, where R^2 is the system consisting simply of the sine-Gordon equation (2.52). ∎

Example 2.9 (Pirani *et al.* [48]) Let ψ be a map of the form
$$\psi: J^1(M, N) \times N' \to J^1(M, N')$$
given, in the notation of [48], by
$$\begin{aligned} x'^a &= x^a, \quad y'^A = y^A, \\ y'^A_b &= \psi^A_b(x^a, z^\mu, z^\mu_a, y^B) = \bar{A}^{Ac}_{\rho b} z^\rho_c + \bar{B}^A_{\rho b} z^\rho + \bar{C}^A_{Bb} y^B + \bar{D}^A_b, \end{aligned} \tag{2.53}$$
where the matrices $\bar{A}$, $\bar{B}$, $\bar{C}$, and $\bar{D}$ are dependent on the x^a. Here, the coordinates are x^a on $M = \mathbb{R}^n$, z^μ on $N = \mathbb{R}^m$, and y'^A on $N = N'$.

It is readily shown that subject to the conditions
$$\bar{A}^{Ac}_{\rho b} = \bar{A}^A_\rho \delta^c_b, \; ^\dagger \tag{2.54}$$
$$\partial_d \bar{C}^A_{Bb} - \partial_b \bar{C}^A_{Bd} + \bar{C}^A_{Db} \bar{C}^D_{Bd} - \bar{C}^A_{Dd} \bar{C}^D_{Bb} = 0, \tag{2.55}$$
the integrability conditions (2.46) yield
$$\alpha^B_{\rho[b} z^\rho_{a]} + \beta^B_{\rho[ab]} z^\rho + \zeta^B_{[ab]} = 0, \tag{2.56}$$
with
$$\alpha^B_{\rho b} := \partial_b(\bar{G}^B_A \bar{A}^A_\rho) - \bar{G}^B_A \bar{B}^A_{\rho b}, \tag{2.57}$$
$$\beta^B_{\rho ab} := \partial_b(\bar{G}^B_A \bar{B}^A_{\rho a}), \tag{2.58}$$
$$\zeta^B_{ab} := \partial_b(\bar{G}^B_A \bar{D}^A_a), \tag{2.59}$$
where it has been noted that the condition (2.55) implies that the matrix $\bar{C}$ is such that
$$\bar{C}^A_{Bb} = \bar{G}^C_B \partial_b g^A_C. \tag{2.60}$$

Here $[g^A_C]$ is an arbitrary nonsingular matrix with entries dependent on x^a and $[\bar{G}^C_B]$ is its inverse.

---
† δ^c_b is the usual Kronecker symbol.

2.6 BÄCKLUND TRANSFORMATIONS DETERMINED BY BÄCKLUND MAPS 139

If the matrix $[\bar{A}_\rho^A]$ is nonsingular and $[a_A^\rho]$ denotes its inverse, then subject to the additional conditions

$$\beta_{\sigma[ab]}^B = \alpha_{\sigma[b}^B a_{|A|}^\rho \bar{B}_{\rho|a]}^A, \tag{2.61}$$

the integrability conditions (2.56) imply the system R^2 given by

$$\alpha_{\rho[b}^B a_{|A|}^\rho (y_{a]}^A - \bar{C}_{|B|a]}^A y^B - \bar{D}_{a]}^A) + \zeta_{[ab]}^B. \tag{2.62}$$

Accordingly, subject to the conditions (2.54) and (2.57)–(2.61), the map ψ given by (2.53) is an ordinary Bäcklund map for the system (2.62).

The specialization of the above result to $M = N = \mathbb{R}^2$ leads to the linear Bäcklund transformations introduced by Loewner [64]. In subsequent chapters these are used extensively both in gasdynamics and elasticity. ∎

2.6 BÄCKLUND TRANSFORMATIONS DETERMINED BY BÄCKLUND MAPS

In this section, we describe the way in which Bäcklund maps may be used to determine Bäcklund transformations. The idea is to consider submanifolds $B^{k+r,r} \subset J^{k+r}(M,N) \underset{M}{\times} J^r(M,N')$, $r = 1, \ldots$, constructed by analogy with the prolongations of a differential equation, and to use ψ to define maps ψ^r from these submanifolds to $J^{r+1}(M,N')$. One then obtains a Bäcklund transformation if the image of some ψ^r is a differential equation.

Thus, let $B^{k+r,r}$ be the submanifold of $J^{k+r}(M,N) \underset{M}{\times} J^r(M,N')$ defined by the constraint equations

$$\begin{aligned} y_a^\mu - \psi_a^\mu &= 0, & D_{[a}^{k+r,r}\psi_{b]}^\mu &= 0, \\ D_{a_1 \cdots a_j}^{k+r,r}(y_a^\mu - \psi_a^\mu) &= 0, & & \\ D_{a_1 \cdots a_j}^{k+r,r}(D_{[a}^{k+r,r}\psi_{b]}^\mu) &= 0, & j &= 1, \ldots, r. \end{aligned} \tag{2.63}$$

We may now introduce maps

$$\psi^r : B^{k+r,r} \to J^{r+1}(M,N')$$

given in coordinates by

$$\begin{aligned} x'^a &= x^a, & y'^\mu &= y^\mu, & y_a'^\mu &= \psi_a^\mu, \\ y_{aa_1 \cdots a_j}'^\mu &= D_{a_1 \cdots a_j}^{k+r,r}\psi_a^\mu, & j &= 1, \ldots, r. \end{aligned} \tag{2.64}$$

Note that since the domain of ψ^r is restricted to $B^{k+r,r}$, the conditions (2.63) are satisfied, and thus $y_{aa_1 \cdots a_j}'^\mu$ is symmetric in the indices $aa_1 \cdots a_j$ and ψ^r is well defined. If for some r, the image of the map ψ^r is contained in a system of differential equations $R' \subset J^{r+1}(M,N')$, then the correspondence between R^{k+1} and R' will be called the *Bäcklund transformation determined by the Bäcklund map* ψ. Suppose now that this is the case and that $f \in C^\infty(M,N)$, $g \in C^\infty(M,N')$ are such that the pair (f,g) is a solution of $B^{k+1,1}$, that is,

140   2. A JET-BUNDLE FORMULATION OF BÄCKLUND TRANSFORMATIONS

$\operatorname{im}(j^{k+1}f \underset{M}{\times} j^1 g) \subset B^{k+1,1}$. A routine calculation in coordinates shows that

$$(\psi^r \circ (j^{k+r}f \underset{M}{\times} j^r g))^* \Omega^{r+1}(M, N') = 0,$$

and, consequently, that the map

$$\psi^r \circ (j^{k+r}f \underset{M}{\times} j^r g) : M \to J^{r+1}(M, N')$$

is the $(r+1)$-jet of $\beta \circ \psi^r \circ (j^{k+r}f \underset{M}{\times} j^r g)$. But

$$\beta \circ \psi^r \circ (j^{k+r}f \underset{M}{\times} j^r g) = g,$$

so that

$$\psi^r \circ (j^{k+r}f \underset{M}{\times} j^r g) = j^{r+1} g. \tag{2.65}$$

Accordingly, since $\operatorname{im} \psi^r \subset R'$, it follows that whenever (f, g) is a solution of $B^{k+1,1}$, g is a solution of R'. In other words, if ψ determines a Bäcklund transformation between R^{k+1} and R' and if f is a solution of R^{k+1}, then a solution g of R' may be obtained by solving the first-order system

$$\frac{\partial g^\mu}{\partial x^a} = \psi_a^\mu(x^b, f^A, \ldots, \partial_{a_1 \cdots a_j} f^A, g^\nu). \tag{2.66}$$

Example 2.10 *The Korteweg–deVries Potential Equation*   Let $M = \mathbb{R}^2$, $N = \mathbb{R}$, and $N' = \mathbb{R}$. The Korteweg–deVries potential equation

$$u_2 + u_{111} - 6(u_1)^2 = 0$$

determines the submanifold of $J^3(M, N)$ given by

$$z_2 + z_{111} - 6(z_1)^2 = 0. \tag{2.67}$$

Wahlquist and Estabrook [36] constructed a Bäcklund map for (2.67), namely, the map

$$\psi : J^2(M, N) \underset{M}{\times} J^0(M, N') \to J^1(M, N')$$

given by

$$\begin{aligned} y'_1 &= -z_1 - 2t + (y-z)^2, \\ y'_2 &= -z_2 + 4\{4t^2 + 2tz_1 - 2t(y-z)^2 \\ &\quad + z_1^2 + z_1(y-z)^2 + z_{11}(y-z)\}, \end{aligned} \tag{2.68}$$

where t is a Bäcklund parameter.

The integrability condition, obtained by substitution into (2.46) yields

$$(y - z)(z_{111} + z_2 - 6z_1^2) = 0,$$

and this equation defines $B^{3,1}(\psi)$. Since the Korteweg–deVries potential equation is thus contained in $B^{3,1}(\psi)$, it follows that ψ is a Bäcklund map for (2.67).

The map ψ^1 is given by (2.68) together with

$$y'_{11} = -z_{11} + 2(y-z)\{-2z_1 - 2t + (y-z)^2\},$$
$$y'_{12} = -z_{12} + 2(y-z)\{16t^2 + 8tz_1 - 2z_1^2 + z_{111} - z_2\}$$
$$\quad + 8(y-z)^2 z_{11} + 8(y-z)^3(z_1 - 2t),$$
$$y'_{22} = -z_{22} + 8tz_{12} + 8z_1 z_{12} + 4z_{112}(y-z) + 4z_{12}(y-z)^2$$
$$\quad + 4\{z_{11} + (2z_1 - 4t)(y-z)\}$$
$$\quad \times \{-2z_1 + 4[4t^2 + 2tz_1 - 2t(y-z)^2 + z_1^2$$
$$\quad + z_1(y-z)^2 + z_{11}(y-z)]\}.$$

A further prolongation shows that ψ^2 has as its image the submanifold of $J^3(M, N')$ given by

$$y'_2 + y'_{111} - 6(y'_1)^2 = 0.$$

Thus, the map ψ determines an auto-Bäcklund transformation for the Korteweg–deVries potential equation. ∎

Example 2.11 *The AKNS System for the Sine-Gordon Equation* The AKNS linear scattering equations for the sine-Gordon equation may be generated from a Bäcklund map in the following manner:

Let $M = \mathbb{R}^2$, $N = \mathbb{R}$, $N' = \mathbb{R}^2$, and let ψ be the map $\psi: J^1(M, N) \times_M J^0(M, N') \to J^1(M, N')$ given by

$$y'^1_1 = \psi^1_1 := \tfrac{1}{2}\lambda(y^1 \cos z + y^2 \sin z),$$
$$y'^2_1 = \psi^2_1 := \tfrac{1}{2}\lambda(y^1 \sin z - y^2 \cos z),$$
$$y'^1_2 = \psi^1_2 := \tfrac{1}{2}(\lambda^{-1} y^1 - y^2 z_2), \qquad (2.69)$$
$$y'^2_2 = \psi^2_2 := \tfrac{1}{2}(y^1 z_2 - \lambda^{-1} y^2).$$

It is readily verified that ψ is a Bäcklund map for the sine-Gordon equation and the corresponding system (2.66) may be written as

$$\partial_1 \begin{bmatrix} g^1 \\ g^2 \end{bmatrix} = \frac{\lambda}{2} \begin{bmatrix} \cos f & \sin f \\ \sin f & -\cos f \end{bmatrix} \begin{bmatrix} g^1 \\ g^2 \end{bmatrix},$$

$$\partial_2 \begin{bmatrix} g^1 \\ g^2 \end{bmatrix} = \frac{1}{2} \begin{bmatrix} \lambda^{-1} & -f_2 \\ f_2 & -\lambda^{-1} \end{bmatrix} \begin{bmatrix} g^1 \\ g^2 \end{bmatrix}$$

where f is any solution of the sine-Gordon equation. This is precisely the AKNS system for the sine-Gordon equation as set out in Section 1.11. ∎

Observe that for both of the above Bäcklund maps, the functions ψ^μ_a of equation (2.40) factorize according to

$$\psi^\mu_a = \Psi^\lambda_a(x^b, z^A, \ldots, z^A_{a_1 \cdots a_k}) X^\mu_\lambda(y^v).$$

142 2. A JET-BUNDLE FORMULATION OF BÄCKLUND TRANSFORMATIONS

The functions $X_\lambda^\mu(y^v)$ may be used to define vector fields on N', namely, the vector fields X_λ given by

$$X_\lambda := X_\lambda^\mu(y^v)\partial/\partial y^\mu.$$

If these X_λ constitute part of a basis for a finite-dimensional Lie algebra **G**, then there is a Lie group naturally associated with the Bäcklund map ψ, namely, the Lie group G with Lie algebra **G**. Moreover, in this case, R^{k+1} may be expressed as the Maurer–Cartan structure equations for the Lie group G (Crampin [408]).

Example 2.12 In the case of the Bäcklund map (2.69), the vector fields X_λ are given by

$$X_1 = y^1\,\partial_1 - y^2\,\partial_2, \quad X_2 = y^2\,\partial_1, \quad X_3 = y^1\,\partial_2,$$

and these form a representation for the three-dimensional Lie algebra $\mathfrak{Sl}(2,\mathbb{R})$. Accordingly, the Lie group $\mathrm{Sl}(2,\mathbb{R})$ may be associated with the Bäcklund map ψ. ∎

To conclude this section, we describe how the notion of Bäcklund map may be extended to incorporate constraint equations. Thus we consider the case in which p equations of the type (2.40), namely,

$$y_a^\mu = \psi_a^\mu(x^b, z^A, \ldots, z^A_{a_1 \cdots a_k}, y^v) \quad \mu = 1, \ldots, p, \tag{2.70}$$

are augmented by the n'-p constraint equations

$$\Phi^\mu(x^b, z^A, \ldots, z^A_{a_1 \cdots a_k}, y^v) = 0 \quad \mu = p+1, \ldots, n',$$

here assumed to be solvable in the form

$$y^\mu = \phi^\mu(x^b, z^A, \ldots, z^A_{a_1 \cdots a_k}) \quad \mu = p+1, \ldots, n'. \tag{2.71}$$

The approach now is to adjoin to (2.70) the system obtained by operating on (2.71) with D_a^k, namely,

$$y_a^\mu = D_a^k \phi^\mu \quad \mu = p+1, \ldots, n', \tag{2.72}$$

and to use (2.70) together with (2.72) to define a map $\psi: J^k(M,N) \times_M J^0(M,N') \to J^0(M,N')$. Thus ψ is the map given in coordinates by

$$\begin{aligned} x'^a &= x^a, \quad y'^\mu = y^\mu \\ y_a'^\mu &= \psi_a^\mu, \quad \mu = 1, \ldots, p \\ y_a'^\mu &= D_a^k \phi^\mu, \quad \mu = p+1, \ldots, n'. \end{aligned} \tag{2.73}$$

Let C denote the submanifold of $J^k(M,N) \times_M J^0(M,N')$ given by (2.71), and let C' denote the submanifold of $J^{k+1}(M,N) \times_M J^1(M,N')$ given by (2.71) and (2.72). In view of (2.73), we call the map $\psi|_C$ a Bäcklund map for a system of equations R^{k+1} if

$$B^{k+1,1}(\psi) \supset R^{k+1} \cap C'. \tag{2.74}$$

Example 2.13 *Pohlmeyer's Bäcklund Transformation for the Yang Equations* As noted in Section 1.14, Pohlmeyer [401] has recently constructed a Bäcklund transformation from Yang's equations that was derived from the Bäcklund transformation for the nonlinear σ-models [255]. That this transformation in fact determines a Bäcklund map may be seen as follows: Let $M = \mathbb{R}^4$ and $N = \mathbb{R}^4 = N'$, and let M, N, N' have, in turn, the coordinates $x^1, \ldots, x^4, z^1, \ldots, z^4$, and $y^1, \ldots, y^4$. Let $\mathbf{z}$ and $\mathbf{y}$ denote the vectors with components z^A and y^A, respectively. Then Yang's equations may be written in vector form on $J^2(M, N)$ as the system R^2 given by

$$\mathbf{z}_{12} + \mathbf{z}_{34} + (\mathbf{z}_1 \cdot \mathbf{z}_2 + \mathbf{z}_3 \cdot \mathbf{z}_4)\mathbf{z}$$
$$+ i[\mathbf{z}; \mathbf{z}_2; \mathbf{z}_1] + i[\mathbf{z}; \mathbf{z}_4; \mathbf{z}_3] = 0 \quad \mathbf{z} \cdot \mathbf{z} = 1, \tag{2.75}$$

where $\cdot$ denotes the Minkowski scalar product $\mathbf{z} \cdot \mathbf{z} = z_A z^A$ with $z_A := g_{AB} z^B$ and

$$[g_{AB}] = \mathrm{diag}[1, -1, -1, -1],$$

while $[\mathbf{u}; \mathbf{v}; \mathbf{w}]$ is the triple vector product of $\mathbf{u}$, $\mathbf{v}$, and $\mathbf{w}$, namely, the vector with components given by

$$[\mathbf{u}; \mathbf{v}; \mathbf{w}]^A = \varepsilon^{ABCD} u_B v_C w_D.$$

Pohlmeyer's Bäcklund transformation may be written as a Bäcklund map $\psi|_C$ for R^2, where $\psi : J^1(M, N) \times_M J^0(M, N') \to J^1(M, N')$ is given in vector notation by

$$\begin{aligned}
\mathbf{y}'_1 &= -\mathbf{z}_4 - (\mathbf{z}_4 \cdot \mathbf{y})\mathbf{y}, & \mathbf{y}'_2 &= -\mathbf{z}_3 - (\mathbf{z}_3 \cdot \mathbf{y})\mathbf{y}, \\
\mathbf{y}'_3 &= \mathbf{z}_1 + (\mathbf{z}_1 \cdot \mathbf{y})\mathbf{y}, & \mathbf{y}'_4 &= \mathbf{z}_2 + (\mathbf{z}_2 \cdot \mathbf{y})\mathbf{y},
\end{aligned} \tag{2.76}$$

and the constraint submanifold $C \subset J^1(M, N) \times_M J^0(M, N')$ is given by[†]

$$\mathbf{z} \cdot \mathbf{z} = 1, \quad \mathbf{z} \cdot \mathbf{y} = 0, \quad \mathbf{y} \cdot \mathbf{y} = -1. \tag{2.77}$$

2.7 SYMMETRIES OF DIFFERENTIAL EQUATIONS AND ONE-PARAMETER FAMILIES OF BÄCKLUND MAPS

In this section we indicate how symmetries of a system of differential equations R^{k+1} may be combined with a Bäcklund map for R^{k+1} to produce another Bäcklund map for R^{k+1} [48]. In the case when R^{k+1} admits

---

[†] The correspondence between the present notation and that of Pohlmeyer is

$$(x^1, x^2, x^3, x^4, z^1, z^2, z^3, z^4) \leftrightarrow (y, \bar{y}, z, \bar{z}, q^1, q^2, q^3, q^4).$$

a one-parameter group of symmetries, this procedure leads to a one-parameter family of Bäcklund maps. This result generalizes the so-called "Theorem of Lie" for the sine-Gordon equation whereby the parameter is inserted in the Bäcklund relations by conjugation of the parameter-free Bianchi transformation with the Lie transformation $(x^1, x^2) \to (ax^1, x^2/a)$ (see Eisenhart [168]). The general construction is an important one, since the parameter so introduced plays the role of the eigenvalue in the associated inverse scattering problems, and moreover, is intrinsic to both the permutability theorems and the method of generation of conservation laws (Wadati, *et al.* [310], Shadwick [407]). In what follows, we use the notation of Appendixes V and VI to which the reader is referred for the relevant definitions.

Suppose that (Φ, ϕ) is a symmetry of R^{k+1}, ψ is a Bäcklund map for R^{k+1}, and (Φ', ϕ) is any diffeomorphism of $J^0(M, N')$. A diffeomorphism $p^{k,0}(\Phi \times \Phi')$ of $J^k(M, N) \underset{M}{\times} J^0(M, N')$ may be constructed from the prolongations of Φ and Φ' according to

$$p^{k,0}(\Phi \times \Phi') := p^k\Phi \times p^0\Phi'.$$

If we now introduce a map $\tilde{\psi}: J^k(M, N) \underset{M}{\times} J^0(M, N') \to J^1(M, N')$ by

$$\tilde{\psi} := p^1\Phi'^{-1} \circ \psi \circ p^{k,0}(\Phi \times \Phi'), \tag{2.78}$$

it is shown in Appendix VI that the integrability conditions for $\tilde{\psi}$ are satisfied on $R^{k+1} \times \operatorname{im} \tilde{\psi}$, that is,

$$R^{k+1} \times \operatorname{im} \tilde{\psi} \subset B^{k+1,1}(\tilde{\psi}),$$

so that $\tilde{\psi}$ is also a Bäcklund map for R^{k+1}.

Given a one-parameter group of symmetries (Φ_t, ϕ_t) of R^{k+1} and any other one-parameter group of diffeomorphisms (Φ'_t, ϕ_t) of $J^0(M, N')$, the same procedure may be adopted to obtain the one-parameter family of Bäcklund maps $\tilde{\psi}_t$ given by

$$\tilde{\psi}_t := p^1\Phi'^{-1}_t \circ \psi \circ p^{k,0}(\Phi_t \times \Phi'_t). \tag{2.79}$$

Example 2.14 *The Theorem of Lie for the* Sine-*Gordon Equation* Let $M = \mathbb{R}^2$, $N = \mathbb{R} = N'$, and let ψ be the Bäcklund map for the sine-Gordon equation given by

$$y'_1 = z_1 + 2\sin\left\{\frac{y+z}{2}\right\}, \qquad y'_2 = -z_2 + 2\sin\left\{\frac{y-z}{2}\right\}.$$

This is the Bäcklund map corresponding to the Bianchi transformation for the sine-Gordon equation (Eisenhart [168]).

Let (Φ_t, ϕ_t) be the one-parameter group of Lie symmetries of the sine-Gordon equation given by

$$(x^1, x^2, z) \to (e^t x^1, e^{-t} x^2, z).$$

2.8 THE WAHLQUIST–ESTABROOK PROCEDURE

Let (Φ'_t, ϕ_t) be the one-parameter group of diffeomorphisms of $J^0(M, N')$ given by

$$(x'^1, x'^2, y') \to (e^t x'^1, e^{-t} x'^2, y').$$

Then, the maps $p^{1,0}(\Phi_t \times \Phi'_t)$ and $p^1 \Phi'^{-1}_t$ are given by

$$(x^1, x^2, z, z_1, z_2, y) \to (e^t x^1, e^{-t} x^2, z, e^{-t} z_1, e^t z_2, y)$$

and

$$(x'^1, x'^2, y', y'_1, y'_2) \to (e^{-t} x'^1, e^t x'^2, y', e^t y'_1, e^{-t} y'_2),$$

respectively, so that the maps $\psi_t : J^1(M, N) \underset{M}{\times} J^0(M, N') \to J^1(M, N')$ with $\psi_t = p^1 \Phi'_{-t} \circ \psi \circ p^{1,0}(\Phi_t \times \Phi'_t)$ have the coordinate representation

$$x''^a = x^a, \qquad y'' = y,$$

$$y''_1 = e^t y'_1 = e^t [e^{-t} z_1 + 2 \sin\{\frac{y+z}{2}\}],$$

$$y''_2 = e^{-t} y'_2 = e^{-t} [-e^t z_2 + 2 \sin\{\frac{y-z}{2}\}].$$

This is the Bäcklund map with an incorporated "Bäcklund parameter" e^t as set out previously in Example 2.8. ∎

Example 2.15 *A One-parameter Family of Bäcklund Maps for the Yang Equations* It is readily verified that the Yang equations (2.75) admit the one-parameter group of symmetries (Φ_t, ϕ_t) given by

$$(x^1, x^2, x^3, x^4, \mathbf{z}) \to (e^t x^1, e^{-t} x^2, x^3, x^4, \mathbf{z}).$$

Thus, if we let (Φ'_t, ϕ_t) be given by

$$(x'^1, x'^2, x'^3, x'^4, \mathbf{y}') \to (e^t x^1, e^{-t} x^2, x^3, x^4, \mathbf{y}'),$$

then the Bäcklund map $\psi|_C$ of Example 2.13 gives rise to the one-parameter family of Bäcklund maps $\psi_t|_C$ given by

$$\mathbf{y}'_1 = e^t[-\mathbf{z}_4 - (\mathbf{z}_4 \cdot \mathbf{y})\mathbf{y}], \qquad \mathbf{y}'_2 = e^{-t}[-\mathbf{z}_3 - (\mathbf{z}_3 \cdot \mathbf{y})\mathbf{y}],$$
$$\mathbf{y}'_3 = e^{-t}[\mathbf{z}_1 + (\mathbf{z}_1 \cdot \mathbf{y})\mathbf{y}], \qquad \mathbf{y}'_4 = e^t[\mathbf{z}_2 + (\mathbf{z}_2 \cdot \mathbf{y})\mathbf{y}]. \quad ∎ \qquad (2.80)$$

2.8 THE WAHLQUIST–ESTABROOK PROCEDURE

We now turn to the problem of construction of Bäcklund maps. The method of Clairin as outlined by Lamb [35] is readily translated into the language of jet bundles. Wahlquist and Estabrook [409, 410] devised a more economical method than that of Clairin based on the notion of pseudopotentials. This has subsequently been applied in the construction of Bäcklund transformations and Lax formulations (Dodd and Gibbon [244, 411]). The Wahlquist–Estabrook procedure has the advantage that it highlights the roles of the various Lie groups associated with differential equations that admit Bäcklund transformations. Moreover, the method has a natural formulation within a jet-bundle context. Here, we present a generalization of the Wahlquist–Estabrook procedure to the case of m independent variables (Pirani et al. [48], Shadwick [407]).[†]

The technique adopted takes as its starting point an exterior differential system Σ^k of m-forms on $J^k(M, N)$ associated with a quasilinear system $R^{k+1} \subset J^{k+1}(M, N)$.[‡] Let $N' = \mathbb{R}^n$ and choose coordinates x'^a, y'^μ, and y'^μ_a on $J^1(M, N')$. Then a basis for the contact $(m-1)$-forms on $J^1(M, N')$ is given by $\theta'^\mu \wedge w'_{ab}$, where

$$\theta'^\mu := dy'^\mu - y'^\mu_a dx'^a \qquad (2.81)$$

and for $a < b = 1, 2, \ldots, m$, w'_{ab} is the $(m-2)$-form defined by

$$w'_{ab} = \frac{\partial}{\partial x'^b} \lrcorner \frac{\partial}{\partial x'^a} \lrcorner (dx'^1 \wedge \cdots \wedge dx'^m). \qquad (2.82)$$

A short computation shows that

$$\theta'^\mu \wedge w'_{ab} = dy'^\mu \wedge w'_{ab} + y'^\mu_a w'_b - y'^\mu_b w'_a. \qquad (2.83)$$

Suppose now that the map $\psi: J^k(M, N) \underset{M}{\times} J^0(M, N') \to J^1(M, N')$ is given in coordinates by

$$\begin{aligned} x'^a &= x^a, \qquad y'^\mu = y^\mu, \\ y'^\mu_a &= \psi^\mu_a(x^b, z^A, \ldots, z^A_{a_1 \cdots a_k}, y^\nu). \end{aligned} \qquad (2.84)$$

Let Ω' denote the module of contact $(m-1)$-forms on $J^1(M, N')$. It is readily shown that the requirement

$$\psi^* d\Omega' \subset \mathscr{I}(\Sigma^k, \psi^*\Omega') \qquad (2.85)$$

implies that ψ is a Bäcklund map for R^{k+1}. Now, the condition (2.85) holds only if

$$\pi^{k+1,1*}\psi^* d\Omega' \subset \pi^{k+1,1*}\mathscr{I}(\Sigma^k, \psi^*\Omega'), \qquad (2.86)$$

where $\pi^{k+1,1}$ is the canonical projection

$$\pi^{k+1,1}: J^{k+1}(M, N) \underset{M}{\times} J^1(M, N') \to J^k(M, N) \underset{M}{\times} J^0(M, N'). \qquad (2.87)$$

But since $\pi^{k+1,1*}\mathscr{I}(\Sigma^k, \psi^*\Omega')$ is generated by $\Omega^{k+1,1}$ and Fw, the condition

---

[†] The method has its roots in Cartan's theory of exterior differential systems [49].
[‡] We present the technique in the case for which R^{k+1} is given by a single equation $F = 0$. extension to systems is obvious.

2.8 THE WAHLQUIST–ESTABROOK PROCEDURE

(2.85) obtains only if

$$\pi^{k+1,1}*\psi^* \, d\Omega' \subset \mathscr{I}(\Omega^{k+1,1}, Fw). \tag{2.88}$$

Thus, since Ω' is generated by $\{\theta'^\mu \wedge w'_{ab}\}$, a necessary condition for (2.85) to be satisfied is that

$$\pi^{k+1,1}*\psi^*(d\theta'^\mu \wedge w'_{ab}) \in \mathscr{I}(\Omega^{k+1,1}, Fw). \tag{2.89}$$

But on application of (2.83), a routine calculation shows that

$$\pi^{k+1,1}*\psi^*(d\theta'^\mu \wedge w'_{ab}) \equiv D^{k+1,1}_{[a}\psi^\mu_{b]}w \quad \bmod \mathscr{I}(\Omega^{k+1,1});$$

whence, from (2.88),

$$D^{k+1,1}_{[a}\psi^\mu_{b]}w \in \mathscr{I}(\Omega^{k+1,1}, Fw). \tag{2.90}$$

The only m-forms in $\mathscr{I}(\Omega^{k+1,1}, Fw)$ that are linearly dependent on w are multiples of Fw, so we conclude that (2.85) holds only if $D^k_{[a}\psi^\mu_{b]}$ is some multiple of F. Thus, the integrability conditions for ψ are satisfied on R^{k+1} and ψ is a Bäcklund map for R^{k+1}.

The Wahlquist–Estabrook procedure requires, essentially, that the map ψ be determined by solving for the functions ψ^μ_a as given by the conditions

$$d\psi^*\theta'^\mu \wedge w_{ab} \in \mathscr{I}(\Sigma^k, \{\psi^*\theta'^\nu \wedge w_{cd}\}). \tag{2.91}$$

In practice, it is usually convenient to seek a differential subideal $\Sigma \subset \Sigma^k$ that has R^{k+1} as its associated equation. Such a subideal is said to be *effective for* R^{k+1}. Thus if Σ is effective for R^{k+1} and one finds a map $\psi: J^k(M, N) \underset{M}{\times} N' \to J^1(M, N')$ such that

$$\psi^*(\Omega', d\Omega') \subset \mathscr{I}(\Sigma, \psi^*\Omega'), \tag{2.92}$$

then since $\mathscr{I}(\Sigma, \psi^*\Omega') \subset \mathscr{I}(\Sigma^k, \psi^*\Omega')$, it follows that ψ is a Bäcklund map for R^{k+1}.

The advantage of working with the subideal Σ is that a reduction in the number of variables in the equations for ψ may be achieved thereby. This reduction can be described in terms of characteristic vector fields for Σ (see Appendix III). Specifically, if it is assumed that every characteristic vector field for Σ is vertical over M, that is,

$$X \lrcorner \, dx^a = 0, \quad a = 1, \ldots, m, \quad \forall x \in \mathrm{Char}(\Sigma),$$

then it may be verified that the condition

$$\psi^* d\Omega' \subset \mathscr{I}(\Sigma, \psi^*\Omega'),$$

requires that

$$X(\psi^\mu_a) = 0.^\dagger \tag{2.93}$$

† An alternative construction is given in [48].

It is the additional equations (2.93) for ψ_a^μ that allow a reduction in the number of variables to be considered. The procedure is illustrated by the following example:

Example 2.16 *The Wahlquist–Estabrook Procedure for the Sine-Gordon Equation* Let $M = \mathbb{R}^2$, $N = \mathbb{R}$, and $N' = \mathbb{R}$. The exterior system Σ^1 on $J^1(M, N)$ generated by the 2-forms

$$\begin{aligned}\eta_1 &:= dz \wedge dx^1 + z_2\, dx^1 \wedge dx^2, \\ \eta_2 &:= dz \wedge dx^2 - z_1\, dx^1 \wedge dx^2, \\ \eta_3 &:= dz_1 \wedge dx^1 + \sin z\, dx^1 \wedge dx^2, \\ \eta_4 &:= dz_2 \wedge dx^2 - \sin z\, dx^1 \wedge dx^2,\end{aligned} \qquad (2.94)$$

has as its associated equation

$$z_{12} = \sin z.$$

There are three effective subideals of Σ^1, namely Σ_1 generated by $\{\eta_1, \eta_4\}$, Σ_2 generated by $\{\eta_2, \eta_3\}$, and Σ_3 generated by $\{\eta_1, \eta_2, \eta_3 + \eta_4\}$. It is readily verified that if (Φ, ϕ) is the diffeomorphism of $J^0(M, N)$ given by

$$(x^1, x^2, z) \to (x^2, x^1, z),$$

then $p^1\Phi$ is given by

$$(x^1, x^2, z, z_1, z_2) \to (x^2, x^1, z, z_2, z_1),$$

and further, that $p^1\Phi^*\Sigma_2 = \Sigma_1$. Thus, there are essentially only two effective subideals of Σ^1 and we may, without loss of generality, restrict attention to Σ_1 and Σ_3.

A short calculation shows that $\text{Char}(\Sigma_1) = \{\partial/\partial z_1\}$ and $\text{Char}(\Sigma_3) = \{0\}$. Thus, we may expect a reduction in the number of variables to be considered if we apply the Wahlquist–Estabrook procedure to Σ_1.[†] To do this, we must find functions ψ_1 and ψ_2 on $J^1(M, N) \underset{M}{\times} J^0(M, N')$ which satisfy (2.91), that is

$$d(dy - \psi_1\, dx^1 - \psi_2\, dx^2) \in \mathscr{I}(\Sigma_1, dy - \psi_a\, dx^a),$$

together with the additional requirement

$$\frac{\partial}{\partial z_1}\psi_a = 0.$$

Thus, functions f, g, and a 1-form ξ must exist such that

$$d\psi_1 \wedge dx^1 + d\psi_2 \wedge dx^2 = f\eta_1 + g\eta_4 + \xi \wedge \zeta,$$

[†] For an analysis of the Wahlquist–Estabrook procedure applied to the effective ideal Σ_3, the reader is referred to Shadwick [171].

2.8 THE WAHLQUIST–ESTABROOK PROCEDURE

where $\zeta := dy - \psi_a dx^a$ and the ψ_a are independent of z_1. Consequently, if $\partial \psi_a / \partial x^b = 0$, then

$$\frac{\partial \psi_1}{\partial z} dz \wedge dx^1 + \frac{\partial \psi_1}{\partial z_2} dz_2 \wedge dx^1 + \frac{\partial \psi_2}{\partial z} dz \wedge dx^2 + \frac{\partial \psi_2}{\partial z_2} dz_2 \wedge dx^2$$

$$+ \left(\psi_1 \frac{\partial \psi_2}{\partial y} - \psi_2 \frac{\partial \psi_1}{\partial y} \right) dx^1 \wedge dx^2$$

$$= f\eta_1 + g\eta_4 + \xi \wedge \zeta;$$

whence, on use of (2.94) and comparison of the terms in $dz \wedge dx^1$, $dz \wedge dx^2$, $dz_2 \wedge dx^1$, and $dz_2 \wedge dx^2$, the additional equations

$$\frac{\partial \psi_1}{\partial z_2} = 0, \quad \frac{\partial \psi_2}{\partial z} = 0, \quad \frac{\partial \psi_1}{\partial z} = f, \quad \frac{\partial \psi_2}{\partial z_2} = g, \quad (2.95)$$

for the ψ_a are obtained, so that $\psi_1 = \psi_1(z, y)$ and $\psi_2 = \psi_2(z_2, y)$. Finally, a comparison of the terms in $dx^1 \wedge dx^2$ and use of (2.95) shows that

$$\psi_1 \frac{\partial \psi_2}{\partial y} - \psi_2 \frac{\partial \psi_1}{\partial y} = z_2 \frac{\partial \psi_1}{\partial z} - \sin z \frac{\partial \psi_2}{\partial z_2}. \quad (2.96)$$

Solutions of (2.96) are sought for which ψ_2 is linear in z_2, so that

$$\psi_2(z_2, y) = X_1(y) + X_2(y) z_2 \quad (2.97)$$

for functions $X_i(y)$, $i = 1, 2$, to be determined. Substitution of (2.97) into (2.96) yields

$$\psi_1 \left(\frac{\partial X_1}{\partial y} + \frac{\partial X_2}{\partial y} z_2 \right) - (X_1 + X_2 z_2) \frac{\partial \psi_1}{\partial y} = z_2 \frac{\partial \psi_1}{\partial z} - \sin z \, X_2,$$

whence

$$\psi_1 \frac{\partial X_2}{\partial y} - X_2 \frac{\partial \psi_1}{\partial y} = \frac{\partial \psi_1}{\partial z} \quad \text{and} \quad \psi_1 \frac{\partial X_1}{\partial y} - X_1 \frac{\partial \psi_1}{\partial y} = -\sin z \, X_2. \quad (2.98)$$

A solution of (2.98) is given by

$$\psi_1 = a \sin z \, \bar{X}_3(y) + a \cos z \, \bar{X}_1(y),$$
$$\psi_2 = a^{-1} \bar{X}_1(y) + \bar{X}_2(y) z_2, \quad (2.99)$$

provided that $\bar{X}_1, \bar{X}_2, \bar{X}_3$ are such that

$$\bar{X}_1 = [\bar{X}_2, \bar{X}_3], \quad \bar{X}_2 = [\bar{X}_1, \bar{X}_3], \quad \bar{X}_3 = [\bar{X}_1, \bar{X}_2],$$

where $[\bar{X}_i, \bar{X}_j] := \bar{X}_i(\partial \bar{X}_j / \partial y) - \bar{X}_j(\partial \bar{X}_i / \partial y)$. These equations are the commutation relations for $\mathfrak{SL}(2, \mathbb{R})$, so that a solution ψ_1, ψ_2 is obtained by substitution into (2.99) of any one-dimensional representation of $\mathfrak{SL}(2, \mathbb{R})$

150   2. A JET-BUNDLE FORMULATION OF BÄCKLUND TRANSFORMATIONS

in this basis. Thus, for example, if we take

$$\bar{X}_1 = y, \quad \bar{X}_2 = -\tfrac{1}{2}(1 + y^2), \quad \bar{X}_3 = \tfrac{1}{2}(1 - y^2),$$

the Bäcklund map

$$\begin{aligned} y'_1 &= \psi_1 = \tfrac{1}{2}(1 - y^2)\sin z + ay\cos z, \\ y'_2 &= \psi_2 = a^{-1}y - \tfrac{1}{2}(1 + y^2)z_2 \end{aligned} \quad (2.100)$$

is obtained. The system (2.100) is equivalent to the Riccati form of the AKNS scattering equations. ∎

In conclusion, we note that the choice of the effective ideal can limit the transformations retrievable from the Wahlquist–Estabrook procedure. Thus, for instance, the Bäcklund map which defines the classical Bäcklund transformation for the sine-Gordon equation, cannot be obtained by applying the procedure to the Σ_1 of the previous example. It is, on the other hand, the only nontrivial map, in the case $N' = \mathbb{R}$ and $\partial \psi_a / \partial x^b = 0$, if the starting point is the effective ideal Σ_3. For further discussion and applications of the Wahlquist–Estabrook technique, the reader is referred to the work of Pirani et al. [48], Harrison [376], Morris [412], and Shadwick [407].

CHAPTER 3

Bäcklund Transformations in Gasdynamics, Nonlinear Heat Conduction, and Magnetogasdynamics

3.1 THE RECIPROCAL RELATIONS AND THE HAAR TRANSFORMATION

In 1928, Haar [51], in a paper devoted to adjoint variational problems, established a remarkable invariance property of the gasdynamic equations. Subsequently, Bateman [53] introduced an associated but less restricted class of invariant transformations known as the reciprocal relations. The application of the latter in the approximation of subsonic gas flows was later noted by Tsien [54].

That the adjoint and reciprocal transformations are both of the Bäcklund-type may be readily seen in terms of Martin's formulation of the gasdynamic equations [138, 172]. However, it is also interesting to note that both types

3. GASDYNAMICS, HEAT CONDUCTION, AND MAGNETOGASDYNAMICS

of transformation may be shown to lie within a more general class of Bäcklund transformations subsequently introduced by Loewner [64, 65]. The appropriate specializations which lead to the adjoint and reciprocal relations are indicated in the discussion of Loewner's work later in this chapter. Here the reciprocal transformations which follow are generated as a Bäcklund transformation of Martin's system of equations.

Martin's Equations

The steady two-dimensional flow of an inviscid thermally nonconducting gas, subject to no external force, is governed by the following system of equations (Yih [405])

$$u^j \frac{\partial \rho}{\partial x^j} + \rho \frac{\partial u^j}{\partial x^j} = 0, \qquad j = 1, 2 \tag{3.1}$$

$$\rho u^j \frac{\partial u^i}{\partial x^j} + \frac{\partial p}{\partial x^i} = 0, \qquad i, j = 1, 2 \tag{3.2}$$

$$u^j \frac{\partial s}{\partial x^j} = 0,$$

to which must be adjoined an equation of state

$$\rho = \rho(p, s), \qquad (\partial p / \partial \rho)_s > 0. \tag{3.4}$$

Here, p, ρ, s, and u^i, $i = 1, 2$, denote, in turn, the gas pressure, gas density, specific entropy, and components of the velocity field **u**. In irrotational flow, the additional condition

$$\frac{\partial u^2}{\partial x^1} - \frac{\partial u^1}{\partial x^2} = 0 \tag{3.5}$$

holds.

Multiplication of (3.1) by u^i and addition to (3.2) allows the replacement of the latter by

$$\frac{\partial}{\partial x^j}(\rho u^i u^j) + \frac{\partial p}{\partial x^i} = 0, \qquad i, j = 1, 2, \tag{3.6}$$

which, together with the continuity equation (3.1), permits the introduction of $\bar{\xi}(x^a)$, $\bar{\eta}(x^a)$, $\psi(x^a)$, $a = 1, 2$, such that

3.1 RECIPROCAL RELATIONS AND THE HAAR TRANSFORMATION

$$d\bar{\xi} = \rho u^1 u^2 \, dx^1 - (p + \rho(u^1)^2) \, dx^2,$$
$$d\bar{\eta} = (p + \rho(u^2)^2) \, dx^1 - \rho u^1 u^2 \, dx^2, \qquad (3.7)$$
$$d\psi = \rho u^2 \, dx^1 - \rho u^1 \, dx^2.^\dagger$$

If we now set

$$\xi = \bar{\xi} + px^2, \qquad \eta = \bar{\eta} - px^1, \qquad (3.8)$$

then the Pfaffian system (3.7) becomes

$$d\xi = u^1 \, d\psi + x^2 \, dp, \qquad d\eta = u^2 \, d\psi - x^1 \, dp, \qquad d\psi = \rho u^2 \, dx^1 - \rho u^1 \, dx^2, \quad (3.9)$$

and the introduction of p and ψ as independent variables instead of the x^a is suggested.[‡] In terms of p and ψ, $(3.9)_{1,2}$ show that

$$x^1 = -\eta_p, \qquad x^2 = \xi_p, \qquad u^1 = \xi_\psi, \qquad u^2 = \eta_\psi, \qquad (3.10)$$

while, $(3.9)_3$ then requires that $\xi(p,\psi)$ and $\eta(p,\psi)$ satisfy Martin's equations

$$\xi_\psi \xi_{p\psi} + \eta_\psi \eta_{p\psi} + \rho^{-1} = 0,$$
$$\xi_\psi \xi_{pp} + \eta_\psi \eta_{pp} = 0. \qquad (3.11)$$

Note that the isentropic condition (3.3) shows that $s = s(\psi)$, whence, the constitutive law (3.4) implies that $\rho = \rho(p,\psi)$. Specification of the latter characterizes the system (3.11).

A Monge–Ampère equation associated with the system (3.11) is readily generated. Thus, $(3.10)_2$ and $(3.11)_2$ show that if q is the gas speed given by $q = (\mathbf{u} \cdot \mathbf{u})^{1/2}$, then

$$\eta_\psi = (q^2 - \xi_\psi^2)^{1/2}, \qquad (3.12)$$
$$\eta_{pp} = -\xi_\psi \xi_{pp}/(q^2 - \xi_\psi^2)^{1/2}, \qquad (3.13)$$

and elimination of η produces a Monge–Ampère equation for $\xi(p,\psi)$, namely,

$$2qq_p \xi_\psi \xi_{\psi p} - qq_\psi \xi_\psi \xi_{pp} + q^3 q_{pp}$$
$$- (qq_{pp} + q_p^2)\xi_\psi^2 + q^2(\xi_{\psi\psi}\xi_{pp} - \xi_{\psi p}^2) = 0. \qquad (3.14)$$

Bäcklund transformations of this equation may now be generated by a direct approach to the Martin system (3.11) as indicated below for the reciprocal relations.

[†] $\bar{\xi}$ and $\bar{\eta}$ are termed the lift and drag functions, respectively [53].
[‡] The class of gasdynamic flows for which the streamlines are isobars must thereby be excluded. The geometry of this restricted set of flows has been delineated by Martin [172].

The Reciprocal Relations

It is readily verified that the system (3.11) is invariant under the multi-parameter class of Bäcklund transformations

$$\xi'_{p'} = \beta_1^{-1}[\xi - (p + \beta_2)\xi_p], \qquad \beta_1 \neq 0,$$

$$\xi'_{\psi'} = \frac{-\beta_1 \xi_\psi}{[p + \beta_2]}, \qquad p + \beta_2 \neq 0,$$

$$\eta'_{p'} = \beta_1^{-1}[\eta - (p + \beta_2)\eta_p], \qquad (3.15)$$

$$\eta'_{\psi'} = \frac{-\beta_1 \eta_\psi}{[p + \beta_2]},$$

$$p' = \beta_4 - \frac{\beta_1^2 \beta_2}{[p + \beta_2]}, \qquad \psi' = \beta_3 \psi, \quad \beta_3 \neq 0.$$

where β_i, $i = 1, \ldots, 4$, are constants and the new density ρ' is given by

$$\rho' = \beta_3 \rho (p + \beta_2)/[p + \beta_2 + \rho q^2]. \qquad (3.16)$$

Use of relations (3.10) reveals that (3.15) and (3.16) correspond to the reciprocal transformations introduced originally by Bateman [53] and subsequently extended to rotational flow by Power and Smith [60], who obtained their result in the following form:

THEOREM 3.1 The system of gasdynamic equations (3.1)–(3.3) is invariant under the multiparameter class of transformations

$$\{u^j, p, \rho, s, x^i\} \to \{u'^j, p', \rho', s', x'^i\}$$

given by

$$u'^j = \frac{-\beta_1 u^j}{[p + \beta_2]}, \qquad j = 1, 2, \quad \beta_1 \neq 0,$$

$$p' = \beta_4 - \frac{\beta_1^2 \beta_3}{[p + \beta_2]}, \qquad \beta_3 \neq 0,$$

$$\rho' = \frac{\beta_3 \rho (p + \beta_2)}{[p + \beta_2 + \rho q^2]}, \qquad s' = s, \qquad (3.17)$$

$$dx'^1 = \beta_1^{-1}[-(p + \beta_2 + \rho(u^2)^2)\,dx^1 + \rho u^1 u^2 \,dx^2],$$
$$dx'^2 = \beta_1^{-1}[\rho u^1 u^2 \,dx^1 - (p + \beta_2 + \rho(u^1)^2)\,dx^2],$$

subject to the requirement $0 < |J(x'^1, x'^2; x^1, x^2)| < \infty$, so that

$$0 < |(p + \beta_2)(p + \beta_2 + \rho q^2)| < \infty. \qquad (3.18)$$

The Haar Relations

The Haar class of invariant transformations is of a type similar to the reciprocal relations but is restricted to irrotational flow. Haar's result is as follows:

THEOREM 3.2 The system of homentropic gasdynamic equations (3.1)–(3.2) augmented by the irrotational condition (3.5) is invariant under the multiparameter class of transformations $\{u^j, p, \rho, x^i\} \to \{\bar{u}^j, \bar{p}, \bar{\rho}, \bar{x}^i\}$ given by

$$\bar{u}^j = \frac{\beta_1 \rho u^j}{[p + \beta_2 + \rho q^2]}, \qquad j = 1, 2, \quad \beta_1 \neq 0,$$

$$\bar{p} = \beta_4 + \frac{\beta_1^2 \beta_3}{[p + \beta_2 + \rho q^2]}, \qquad \beta_3 \neq 0,$$

$$\bar{\rho} = \frac{\beta_3(p + \beta_2 + \rho q^2)}{\rho[p + \beta_2]},$$

$$d\bar{x}^1 = \beta_1^{-1}[\rho u^1 u^2 \, dx^1 - (p + \beta_2 + \rho(u^1)^2) \, dx^2],$$
$$d\bar{x}^2 = \beta_1^{-1}[(p + \beta_2 + \rho(u^2)^2) \, dx^1 - \rho u^1 u^2 \, dx^2],$$

(3.19)

subject again to the constraint (3.18) imposed by the condition

$$0 < |J(\bar{x}^1, \bar{x}^2; x^1, x^2)| < \infty.$$

The above Haar relations are readily established via the Loewner formalism as introduced later in this chapter.

3.2 PROPERTIES OF THE RECIPROCAL RELATIONS. INVARIANCE OF THE EQUATION OF STATE

We now proceed to establish certain invariance properties of the reciprocal relations. The first such result is incorporated in the following:

THEOREM 3.3 (Power and Smith [60]) Under the reciprocal transformations (3.17)

(i) $\left.\begin{matrix}\text{irrotational}\\\text{rotational}\end{matrix}\right\}$ flows are mapped to $\left.\begin{matrix}\text{irrotational}\\\text{rotational}\end{matrix}\right\}$ flows,

(ii) $\left.\begin{matrix}\text{subsonic}\\\text{supersonic}\end{matrix}\right\}$ regions are mapped to $\left.\begin{matrix}\text{subsonic}\\\text{supersonic}\end{matrix}\right\}$ regions,

subject to the assumption of uniform stagnation pressure and of a Prim-type equation of state

$$\rho = P(p)S(s). \qquad (3.20)$$

Proof (i) The reciprocal relations (3.17)$_{1,5,6}$ may be combined to produce the relationship

$$\frac{\partial u^2}{\partial x^1} - \frac{\partial u^1}{\partial x^2} = \beta_1^{-1}(p+\beta_2)(p+\beta_2+\rho q^2)\left\{\frac{\partial u'^2}{\partial x'^1} - \frac{\partial u'^1}{\partial x'^2}\right\}, \qquad (3.21)$$

between the vorticities in the original and reciprocal flows. In view of condition (3.18), the result follows.

(ii) The equations of motion (3.2) and of energy (3.3) together produce the Bernoulli integral

$$q^2 + 2h(p,s) = 2h(p_0,s), \qquad (3.22)$$

where h is the specific enthalpy, given by

$$h(p,s) = \int_0^p \frac{d\sigma}{\rho(\sigma,s)}, \qquad (3.23)$$

and p_0 is the stagnation pressure. The latter is constant along any individual streamline, that is,

$$u^j \, \partial p_0/\partial x^j = 0. \qquad (3.24)$$

The reciprocal enthalpy may be introduced similarly according to

$$h'(p',s') = \int_0^{p'} \frac{d\sigma'}{\rho'(\sigma',s')}, \qquad (3.25)$$

while the reciprocal equations of motion and energy produce the reciprocal Bernoulli integral

$$q'^2 + 2h'(p',s') = 2h'(p'_0,s'), \qquad (3.26)$$

where p'_0 is the reciprocal stagnation pressure.

The equation of state (3.4) is henceforth specialized to be of the Prim product type (3.20). In particular, this includes the constitutive law for a perfect gas, namely,

$$\rho = A p^{1/\gamma} e^{-s/c_p}, \qquad (3.27)$$

where A is a constant, $\gamma = c_p/c_v$ and c_p, c_v are the specific heats at constant pressure and volume, respectively.

With the Prim law (3.20), the specific enthalpy (3.23) adopts the form

$$h(p,s) = \Pi(p)/S(s), \qquad (3.28)$$

where

$$\Pi(p) = \int_0^p \frac{d\sigma}{P(\sigma)}, \qquad (3.29)$$

3.2 RECIPROCAL RELATIONS; INVARIANCE OF THE EQUATION OF STATE

and (3.22) becomes
$$q^2 S(s) + 2\Pi(p) = 2\Pi(p_0). \tag{3.30}$$
Use of the reciprocal relation (3.16), together with (3.20) and (3.30), yields
$$\rho' = \left[\frac{\beta_3 P(p)(p + \beta_2)}{[p + \beta_2 + 2\{\Pi(p_0) - \Pi(p)\}P(p)]}\right] S(s), \tag{3.31}$$
which together with the reciprocal relations for p' and s', as given in (3.17), determine the reciprocal equation of state. Thus, in the case of uniform stagnation pressure p_0, the reciprocal equation of state adopts the form
$$\rho' = P'(p')S'(s') \tag{3.32}$$
so that the Prim form is preserved under the reciprocal transformations. We proceed on this assumption.

The Mach numbers M and M' in the original and reciprocal flows are given by
$$M^2 = q^2/c^2 = q^2[(\partial p/\partial \rho)_s]^{-1}, \tag{3.33}$$
and
$$M'^2 = q'^2/c'^2 = q'^2[(\partial p'/\partial \rho')_{s'}]^{-1}, \tag{3.34}$$
respectively, where c, c' are the corresponding local speeds of sound. On use of the reciprocal relations $(3.17)_{1,2}$, together with the original and reciprocal Prim-type equations of state (3.20) and (3.32), it is readily shown that
$$1 - M'^2 = (1 - M^2)(p + \beta_2)^2/(p + \beta_2 + \rho q^2)^2. \tag{3.35}$$
Thus, $1 - M'^2$ and $1 - M^2$ have the same sign and the result follows.[†] ∎

Further properties of reciprocal relations were detailed by Power and Smith in [60]. Thus, it was shown, under an additional assumption, that shock lines are mapped to shock lines under the transformations. Moreover, in [60] and a subsequent paper by Rogers [61] the conditions under which the equation of state remains invariant were investigated. In this connection, the following result was established in [61]:

THEOREM 3.4 Prim-type equations of state
$$\rho = \beta_1[\Phi(\ln|\beta_1^{-1}(p + \beta_2)|) + \Phi'(\ln|\beta_1^{-1}(p + \beta_2)|)]^{-1} S(s), \tag{3.36}$$
where Φ is an even differentiable function, are invariant under reciprocal relations (3.17) with the specializations $\beta_2 = -\beta_4$, $\beta_3 = +1$.

[†] Regions in which $M < 1$ are called subsonic, while regions with $M > 1$ are termed supersonic.

3. GASDYNAMICS, HEAT CONDUCTION, AND MAGNETOGASDYNAMICS

Proof The equation of state is invariant provided that

$$P(p') = \mathbb{P}(p), \tag{3.37}$$

where

$$\mathbb{P}(p) = \beta_3 P(p)(p + \beta_2)/[p + \beta_2 + 2\{\Pi(p_0) - \Pi(p)\}P(p)]. \tag{3.38}$$

In view of relation (3.29), this condition yields

$$\frac{d\Pi(p)}{dp}(p + \beta_2) + 2[\Pi(p_0) - \Pi(p)] = \beta_3 \frac{d\Pi(p')}{dp'}(p + \beta_2),$$

whence, on integration,

$$\Pi(p) - \Pi(p_0) = \beta_1^{-2}(p + \beta_2)^2[\Pi(p') + \varepsilon], \tag{3.39}$$

where ε is a constant of integration. The choice $\varepsilon = -\Pi(p'_0)$ reduces the problem to the solution of the functional equation embodied in

$$\Pi(p) - \Pi(p_0) = \beta_1^{-2}(p + \beta_2)^2[\Pi(p') - \Pi(p'_0)],$$

$$\Pi(p) = \int_0^p \frac{d\sigma}{P(\sigma)}, \qquad \Pi(p') = \int_0^{p'} \frac{d\sigma}{P(\sigma)}, \tag{3.40}$$

$$p' = \beta_4 - \beta_1^2 \beta_3 / \{p + \beta_2\}.$$

If we now set $\beta_2 = -\beta_4$, $\beta_3 = -1$, together with

$$\Pi^*(p) := \Pi(p) - \Pi(p_0), \qquad p^* := p + \beta_2, \tag{3.41}$$

then $(3.40)_1$ reduces to

$$\Pi^*(p^* - \beta_2) = \beta_1^{-2} p^{*2} \Pi^*\left[\frac{\beta_1^2}{p^*} - \beta_2\right], \tag{3.42}$$

whence, we obtain the canonical form

$$\tilde{\Pi}(\tilde{p}) = \tilde{\Pi}(\tilde{p}^{-1}), \tag{3.43}$$

where

$$\tilde{\Pi}(p^*) := p^{*-1} \Pi^*(\beta_1 p^* - \beta_2), \qquad \tilde{p} := \beta_1^{-1} p^*. \tag{3.44}$$

The substitutions $\tilde{p} = e^u$, $\tilde{p} > 0$, $\tilde{p} = -e^u$, $\tilde{p} < 0$ (it has been stipulated that $\tilde{p} \neq 0$) show that a sufficient condition for (3.43) to be satisfied is that

$$\tilde{\Pi}(\tilde{p}) = \Phi(\ln|\tilde{p}|), \tag{3.45}$$

where Φ is an even function (required by earlier considerations to be differentiable).

3.2 RECIPROCAL RELATIONS; INVARIANCE OF THE EQUATION OF STATE

To show that (3.45) is the necessary form, we set $|\tilde{p}| \equiv e^{\ln|\tilde{p}|}$ and decompose $\tilde{\Pi}(\tilde{p})$ into even and odd parts. Then by virtue of (3.43),

$$\tilde{\Pi}(\tilde{p}) \equiv g(\ln|\tilde{p}|) = \tfrac{1}{2}[g(\ln|\tilde{p}|) + g(-\ln|\tilde{p}|)] + \tfrac{1}{2}[g(\ln|\tilde{p}|) - g(-\ln|\tilde{p}|)]$$
$$= \tfrac{1}{2}[g(-\ln|\tilde{p}|) + g(\ln|\tilde{p}|)] + \tfrac{1}{2}[g(-\ln|\tilde{p}|) - g(\ln|\tilde{p}|)]. \quad (3.46)$$

Thus,

$$g(\ln|\tilde{p}|) - g(-\ln|\tilde{p}|) = 0,$$

so that $\tilde{\Pi}(\tilde{p})$ is necessarily an even function of $\ln|\tilde{p}|$.

Hence, it has been shown that the most general solution of the system (3.40) with the specializations $\beta_2 = -\beta_4$, $\beta_3 = -1$ is

$$\Pi(p) - \Pi(p_0) = \beta_1^{-1}(p + \beta_2)\Phi(\ln|\beta_1^{-1}(p + \beta_2)|). \quad (3.47)$$

The required result now follows, namely, that Prim-type gas laws with

$$P(p) = \beta_1[\Phi(\ln|\beta_1^{-1}(p + \beta_2)|) + \Phi'(\ln|\beta_1^{-1}(p + \beta_2)|)]^{-1}, \quad (3.48)$$

are preserved by the present reciprocal relations. ∎

As an illustration, if we take

$$\Phi(\ln|\beta_1^{-1}(p + \beta_2)|) = \sum_{r=0}^{k} 2A_r \cosh\{r \ln|\beta_1^{-1}(p + \beta_2)|\}, \quad (3.49)$$

then

$$\Pi(p) = \Pi(p_0) + \beta_1^{-1}(p + \beta_2) \sum_{r=0}^{k} A_r(|\beta_1^{-r}(p + \beta_2)^r| + |\beta_1^r(p + \beta_2)^{-r}|), \quad (3.50)$$

whence, if $\beta_1^{-1}(p + \beta_2) > 0$, it follows that

$$P(p) = \left[\sum_{r=0}^{k} A_r\{\beta_1^{-r-1}(1+r)(p + \beta_2)^r + \beta_1^{r-1}(1-r)(p + \beta_2)^{-r}\}\right]^{-1}. \quad (3.51)$$

It is of interest to note that if we set

$$A_0 = [\beta_1/(2B)](A + \beta_2), \qquad A_1 = [-1/(2B)]\beta_1^2, \qquad A_r = 0, \qquad r \geq 2,$$

the invariance of the Kármán-Tsien type gas law

$$\rho = [B/(A - p)]S(s) \quad (3.52)$$

is established for the present transformations. The form (3.52) has been widely used to approximate the real gas law in subsonic potential flow since, with such an equation of state, the hodograph equations can be reduced to a

3.3 RECIPROCAL RELATIONS IN SUBSONIC GASDYNAMICS

Cauchy–Riemann system. The application of reciprocal relations in such a context is described in the next section.

In general, the equation of state is not invariant under the reciprocal relations. Since the classical theory of plane, incompressible, potential flow is well established, it is of interest to investigate the properties of the reciprocal flow when the flow in the original x^a plane is incompressible. The subsequent discussion of this aspect of the reciprocal relations proceeds under the assumption of homentropic, irrotational conditions.

If $\rho = \rho_0$ in the original flow, the Bernoulli integral (3.22) then yields

$$\rho_0 q^2 + 2p = 2p_0, \qquad (3.53)$$

so that the reciprocal density ρ' is given by

$$\rho' = \beta_3 \rho_0 (p + \beta_2)/[p + \beta_2 + 2\{p_0 - p\}]. \qquad (3.54)$$

Elimination of p between (3.54) and the reciprocal relation (3.17)$_2$ for p' now shows that the reciprocal equation of state is of the *Kármán–Tsien type*

$$p' = A - B/\rho' \qquad (3.55)$$

where

$$A = \beta_4 - \beta_1^2 \beta_3 / 2(p_0 + \beta_2), \qquad B = \beta_1^2 \beta_3^2 \rho_0 / 2(p_0 + \beta_2).$$

The importance of the Kármán–Tsien relation (3.55) in the approximation of subsonic flows of an adiabatic gas

$$p = C\rho^\gamma \qquad (3.56)$$

is well known (Tsien [54], von Mises [55], Shapiro [356], Power and Smith [125]). On the other hand, properties of supersonic flow of gases with an equation of state approximated by a Kármán–Tsien relation have been investigated by Coburn [68].

It is convenient at this point to proceed in terms of a hodograph system. Thus, by virtue of the continuity equation (3.1) and the irrotationality condition (3.5), stream and potential functions $\psi(x^a)$, $\phi(x^a)$ may be introduced in turn according to

$$\rho_0 (\rho q)^{-1} d\psi = -\sin\theta \, dx^1 + \cos\theta \, dx^2, \qquad (3.57)$$

$$q^{-1} d\phi = \cos\theta \, dx^1 + \sin\theta \, dx^2, \qquad (3.58)$$

where $qe^{i\theta} = u^1 + iu^2$. Hence, if $z = x^1 + ix^2$,

$$dz = q^{-1} e^{i\theta} [d\phi + i\rho_0 \rho^{-1} \, d\psi], \qquad (3.59)$$

3.3 RECIPROCAL RELATIONS IN SUBSONIC GASDYNAMICS

so that, on introduction of q, θ as independent variables, under the assumption

$$0 < |J(x^1, x^2; q, \theta)| < \infty, \tag{3.60}$$

it is seen that

$$\frac{\partial z}{\partial q} = q^{-1} e^{i\theta} \left[\frac{\partial \phi}{\partial q} + i \rho_0 \rho^{-1} \frac{\partial \psi}{\partial q} \right],$$

$$\frac{\partial z}{\partial \theta} = q^{-1} e^{i\theta} \left[\frac{\partial \phi}{\partial \theta} + i \rho_0 \rho^{-1} \frac{\partial \psi}{\partial \theta} \right]. \tag{3.61}$$

The integrability condition for the latter pair of equations now produces the well-known hodograph system (Shapiro [356])

$$\begin{bmatrix} \phi \\ \psi \end{bmatrix}_q = \begin{bmatrix} 0 & -\rho_0(1-M^2)/\rho q \\ \rho/\rho_0 q & 0 \end{bmatrix} \begin{bmatrix} \phi \\ \psi \end{bmatrix}_\theta, \tag{3.62}$$

where the $M(q)$, $\rho(q)$ relationships are provided by the prevailing equation of state together with the Bernoulli integral.

In subsonic flow, $M < 1$ and introduction of the variable s according to

$$s = \int_{q_0}^{q} [1 - M(\sigma)^2]^{1/2} \sigma^{-1} d\sigma, \tag{3.63}$$

reduces (3.62) to the elliptic canonical form

$$\begin{bmatrix} \phi \\ \psi \end{bmatrix}_s = \begin{bmatrix} 0 & -K^{1/2}(s) \\ K^{-1/2}(s) & 0 \end{bmatrix} \begin{bmatrix} \phi \\ \psi \end{bmatrix}_\theta, \tag{3.64}$$

where

$$K = \rho_0^2 (1 - M^2)/\rho^2, \tag{3.65}$$

and q_0 is a reference gas speed.

In particular, for the Kármán–Tsien gas law (3.55), use of the Bernoulli integral (3.22) in the reciprocal flow shows that

$$K' = 1, \tag{3.66}$$

whence the hodograph system (3.64) reduces to the Cauchy–Riemann equations

$$\phi'_s = -\psi'_\theta, \quad \psi'_s = \phi'_\theta, \tag{3.67}$$

and

$$\phi' + i\psi' = g(s - i\theta), \tag{3.68}$$

where g is an analytic function of its argument. In this manner, if the adiabatic gas law (3.56) is approximated by a Kármán–Tsien law of the form (3.55) by appropriate choice of the parameters A and B, then established complex-variable techniques may be adduced to approximate plane subsonic gas flows associated with classical hydrodynamic solutions.†

The Kármán–Tsien method may be regarded as an improvement over the Glauert–Prandtl theory for the analysis of small disturbances due to the introduction of solid bodies of small thickness ratio in subsonic uniform flow ([126], [54]). Thus, the Kármán–Tsien theory is not restricted to subsonic flow past thin bodies and accordingly has proved to be of considerable importance in aerodynamics, there being an extensive literature on the subject [416–422].

In Section 3.7, the reduction to canonical form of the hodograph system (3.62) in subsonic, transsonic, and supersonic flow is investigated in the broader context of matrix Bäcklund transformations. In particular, it will be seen that the Kármán–Tsien approximation is but one of a class of (p, ρ) relations for which (3.62) may be reduced to the Cauchy–Riemann equations in subsonic flow.

3.4 RECIPROCAL-TYPE TRANSFORMATIONS IN STEADY MAGNETOGASDYNAMICS

The governing equations of the flow of a thermally nonconducting gas of infinite electrical conductivity in the presence of a magnetic field are, in rationalized MKS units [74]

$$\frac{\partial \rho}{\partial t} + u^j \frac{\partial \rho}{\partial x^j} + \rho \frac{\partial u^j}{\partial x^j} = 0, \tag{3.69}$$

$$\rho \left[\frac{\partial u^i}{\partial t} + u^j \frac{\partial u^i}{\partial x^j} \right] + \frac{\partial \Pi_m}{\partial x^i} - \mu H^j \frac{\partial H^i}{\partial x^j} = 0, \quad i = 1, 2, 3, \tag{3.70}$$

$$\frac{\partial H^j}{\partial x^j} = 0, \tag{3.71}$$

$$\frac{\partial H^i}{\partial t} - \frac{\partial u^i}{\partial x^j} H^j + \frac{\partial H^i}{\partial x^j} u^j + H^i \frac{\partial u^j}{\partial x^j} = 0, \quad i = 1, 2, 3, \tag{3.72}$$

where $\Pi_m = p + \tfrac{1}{2}\mu H^i H^i$ is the total magnetic pressure, μ is the magnetic permeability (assumed constant), while u^i, H^i are the rectangular Cartesian components of the velocity and magnetic fields, respectively; p, ρ denote, in turn, the magnetogas pressure and density.

† In Chapter 5 a detailed application of the method is given for analogous antiplane deformations of nonlinear elastic materials. In that context, certain deformations of nonlinear elastic materials are linked to associated deformations of classical Hookean materials.

3.4 TRANSFORMATIONS IN STEADY MAGNETOGASDYNAMICS

In two-dimensional steady magnetogasdynamics, the system (3.69)–(3.72) can be written in the form (Rogers *et al.* [62])

$$\frac{\partial A_2^i}{\partial x^1} - \frac{\partial A_1^i}{\partial x^2} = 0, \qquad i = 1, \ldots, 4, \tag{3.73}$$

together with the relation

$$\mu(u^1 H^2 - u^2 H^1) = \alpha, \tag{3.74}$$

where α is a constant and

$$[A_j^i] = \begin{bmatrix} \rho u^1 u^2 - \mu^2 H^1 H^2 & -\{\Pi_m + \rho(u^1)^2 - \mu^2(H^1)^2\} \\ -\{\Pi_m + \rho(u^2)^2 - \mu^2(H^2)^2\} & \rho u^1 u^2 - \mu^2 H^1 H^2 \\ \rho u^2 & -\rho u^1 \\ \mu H^2 & -\mu H^1 \end{bmatrix}. \tag{3.75}$$

Equations such as (3.73) in the form of conservation laws in two independent variables may be conveniently expressed in terms of differential 1-forms. Thus, if we let M be $\mathbb{R}^2$ with coordinates x^1, x^2 and let N be $\mathbb{R}^{10}$ with coordinates u^1, u^2, Π_m, ρ, H^1, H^2, and w^i, $i = 1, \ldots, 4$, then the exterior differential system $\sum$ on $J^0(M, N)$ generated by the 0-form

$$h := \mu(u^1 H^2 - u^2 H^1) - \alpha, \tag{3.76}$$

and the 1-forms

$$\gamma^i := dw^i - A_a^i dx^a, \tag{3.77}$$

is equivalent to the system of equations (3.73)–(3.74).

A *solution* of $\sum$ is a map $f : M \to N$ such that

$$j^0 f^* \sum = 0, \tag{3.78}$$

while a generalized symmetry of $\sum$ is a diffeomorphism ϕ of $J^0(M, N)$ which satisfies

$$\phi^* \sum \subset \sum \tag{3.79}$$

(see Appendix V). If ϕ is such a diffeomorphism and f is a solution of $\sum$, then the map $\beta \circ \phi \circ j^0 f : M \to N$ will be another solution of $\sum$, provided $\phi \circ j^0 f = j^0 g$ for some map $g : M \to N$. Here, it is demonstrated that reciprocal-type generalized symmetries of $\sum$ exist, and that, except in a degenerate case, they map solutions to solutions.

Consider a diffeomorphism ϕ of $J^0(M, N)$ which is linear in the coordinates x^a, w^i, namely,

$$x'^a := \phi^* x^a = B_j^a w^j + C_b^a x^b, \tag{3.80}$$

$$w'^i := \phi^* w^i = b_j^i w^j + c_a^i x^a, \qquad a, b = 1, 2, \quad i, j = 1, \ldots, 4, \tag{3.81}$$

164   3. GASDYNAMICS, HEAT CONDUCTION, AND MAGNETOGASDYNAMICS

where B_j^a, C_b^a, b_j^i, and c_a^i are constants. The condition (3.79) may now be used to determine the remaining quantities

$$u'^i := \phi^* u^i, \qquad \Pi'_m := \phi^* \Pi_m,$$
$$\rho' := \phi^* \rho, \qquad H'^i := \phi^* H^i,$$

as follows:
From (3.80) and (3.81)

$$\phi^* \gamma^i = \phi^* dw^i - \phi^* A_a^i \phi^* dx^a$$
$$= b_j^i dw^j + c_a^i dx^a - \phi^* A_k^i \{B_j^k dw^j + C_a^k dx^a\},$$

and from (3.76), (3.77),

$$dw^j \equiv A_a^j dx^a \qquad \mod \sum,$$

so that

$$\phi^* \gamma^i \equiv \{b_j^i A_a^j + c_a^i - A'^i_k (B_j^k A_a^j + C_a^k)\} dx^a \qquad \mod \sum,$$

where

$$A'^i_k := \phi^* A_k^i = A_k^i \circ \phi.$$

Thus $\phi^* \gamma^i \in \sum$ iff

$$b_j^i A_a^j + c_a^i - A'^i_k \{B_j^k A_a^j + C_a^k\} = 0,$$

which condition provides eight equations for the six unknowns in terms of u^i, Π_m, ρ, and H^i together with the constants b_j^i, c_a^i, B_j^k, and C_a^k. The requirement

$$\phi^* h = \lambda h, \qquad (3.82)$$

where λ is a constant, provides a ninth equation. In general, it is to be expected that it will be necessary to impose three additional constraints in order to obtain symmetries of the present type. It is possible however, to reduce the number of constraints to two as is evidenced in the following illustration of the method.

The matrices $[b_j^i]$, $[c_j^i]$, $[B_j^i]$, and $[C_j^i]$ are taken to be of the forms

$$[b_j^i] = \begin{bmatrix} 0_2^2 & 0_2^2 \\ 0_2^2 & \Lambda(a_1, a_2) \end{bmatrix}, \qquad [c_j^i] = \begin{bmatrix} & c & & c \\ -c & & c \\ & & 0_2^2 \end{bmatrix},$$

$$[B_j^i] = \begin{bmatrix} b & -b & \\ & & 0_2^2 \\ b & b & \end{bmatrix}, \qquad [C_j^i] = [0_2^2], \qquad (3.83)$$

3.4 TRANSFORMATIONS IN STEADY MAGNETOGASDYNAMICS

where 0_2^2 is the 2×2 null matrix, and

$$\Lambda(a_1, a_2) = \begin{bmatrix} a_1 & 0 \\ 0 & a_2 \end{bmatrix},$$

while $[A_j^i]$ is written as

$$[A_j^i] = \begin{bmatrix} S & -U \\ -V & S \\ \rho u^2 & -\rho u^1 \\ \mu H^2 & -\mu H^1 \end{bmatrix}, \tag{3.84}$$

where

$$S := \rho u^1 u^2 - \mu^2 H^1 H^2, \quad U := \Pi_m + \rho(u^1)^2 - \mu^2(H^1)^2,$$
$$V := \Pi_m + \rho(u^2)^2 - \mu^2(H^2)^2,$$

and, in the above, a_i; $i = 1, 2$, b, and c are real constants. It follows that,

$$[B_k^i A_j^k + C_j^i] = b \begin{bmatrix} S+V & -(S+U) \\ S-V & S-U \end{bmatrix}, \tag{3.85}$$

whence,

$$\det\{B_k^i A_j^k + C_j^i\} = 2b^2\{S^2 - UV\} = -2b^2 J, \tag{3.86}$$

where

$$J = \Pi_m^2 + \Pi_m\{\rho q^2 - \mu^2 H^i H^i\} - \rho \alpha^2. \tag{3.87}$$

Thus,

$$[B_k^i A_j^k + C_j^i]^{-1} = -\frac{1}{2bJ} \begin{bmatrix} S-U & S+U \\ -(S-V) & S+V \end{bmatrix}, \tag{3.88}$$

where it is required that

$$0 < |J| < \infty, \tag{3.89}$$

while

$$[b_k^i A_j^k + c_j^i] = \begin{bmatrix} c & c \\ -c & c \\ a_1 \rho u^2 & -a_1 \rho u^1 \\ a_2 \mu H^2 & -a_2 \mu H^1 \end{bmatrix}, \tag{3.90}$$

so that

$$[A_j^{\prime i}] = -\frac{1}{2bJ}\begin{bmatrix} c(V-U) & c\{2S+U+V\} \\ -c\{2S-U-V\} & c\{V-U\} \\ -a_1\rho\{\Pi_m(u^1+u^2)+\mu\alpha(H^1-H^2)\} & -a_1\rho\{\Pi_m(u^1-u^2)-\mu\alpha(H^1+H^2)\} \\ -a_2\{\mu\Pi_m(H^1+H^2)+\rho\alpha(u^1-u^2)\} & -a_2\{\mu\Pi_m(H^1-H^2)-\rho\alpha(u^1+u^2)\} \end{bmatrix}.$$

(3.91)

In this case, the form of the matrices B_j^i and c_j^i has guaranteed that $A_1'^1 = A_2'^2$ so that the number of equations for u'^1, u'^2, Π_m', ρ', H'^1, H'^2 has been reduced to the seven relations obtained by comparison of (3.91) and the primed counterpart of (3.75), augmented by the constraint (3.82). Thus, in general, u'^1, u'^2, Π_m', ρ', H'^1, H'^2 may be expressed in terms of u^1, u^2, Π_m, ρ, H^1, and H^2 subject to two constraints on the latter six quantities.

Now, comparison of the (4,1) and (4,2) entries in (3.91) and $[A_j^{\prime i}]$ given by

$$[A_j^{\prime i}] = \begin{bmatrix} \rho'u'^1u'^2 - \mu^2 H'^1 H'^2 & -\{\Pi_m' + \rho'(u'^1)^2 - \mu^2(H'^1)^2\} \\ -\{\Pi_m' + \rho'(u'^2)^2 - \mu^2(H'^2)^2\} & \rho'u'^1u'^2 - \mu^2 H'^1 H'^2 \\ \rho'u'^2 & -\rho'u'^1 \\ \mu H'^2 & -\mu H'^1 \end{bmatrix}, \quad (3.92)$$

produces the required expressions for the new components H'^i of the magnetic field in terms of the original magnetogasdynamic variables, namely,

$$\mu H'^1 = -a_2[\Pi_m\mu(H^1 - H^2) - \rho\alpha(u^1 + u^2)]/2Jb,$$
$$\mu H'^2 = a_2[\Pi_m\mu(H^1 + H^2) + \rho\alpha(u^1 - u^2)]/2Jb,$$
(3.93)

or

$$\mu H' = -a_2[\Pi_m\mu(1-i)\bar{H} - \rho\alpha(1+i)\bar{q}]/2Jb, \quad (3.94)$$

where $H := H^1 + iH^2$, $q := u^1 + iu^2$ and similarly for the primed quantities.

On the other hand, comparison of the (3,1) and (3,2) entries of (3.91) and (3.92) yields

$$\rho'u'^1 = -a_1\rho[\Pi_m(u^1 - u^2) - \mu\alpha(H^1 + H^2)]/2Jb,$$
$$\rho'u'^2 = a_1\rho[\Pi_m(u^1 + u^2) + \mu\alpha(H^1 - H^2)]/2Jb,$$
(3.95)

or

$$\rho'q' = -a_1\rho[\Pi_m(1-i)\bar{q} - \mu\alpha(1+i)\bar{H}]/2Jb. \quad (3.96)$$

Now, the restriction (3.82) implies that

$$2i\rho'\alpha' = -\mu\{\rho'q'\bar{H}' - \rho'\bar{q}'H'\}, \quad (3.97)$$

where $\alpha' = \lambda\alpha$, so that, from (3.94) and (3.96) we obtain the expressions for

3.4 TRANSFORMATIONS IN STEADY MAGNETOGASDYNAMICS

the new density and new velocity components in the forms

$$\rho' = -a_1 a_2 \alpha \rho / 2b^2 \alpha' J, \tag{3.98}$$

and

$$q' = b\alpha'[\Pi_m(1-i)\bar{q} - \mu\alpha(1+i)\bar{H}]/a_2\alpha, \tag{3.99}$$

respectively.

Further, comparison of the (1,2) or (2,1) entries of (3.91) and (3.92) gives the new total magnetic pressure, namely,

$$\Pi'_m = a_1 \alpha' \rho [\Pi_m(u^1 + u^2) + \mu\alpha(H^1 - H^2)]^2 / 2a_2 \alpha J \\ + c[2\Pi_m + \rho(u^1 - u^2)^2 - \mu^2(H^1 - H^2)^2]/2bJ \\ + a_2^2[\Pi_m \mu(H^1 + H^2) + \rho\alpha(u^1 - u^2)]^2 / 4J^2 b^2. \tag{3.100}$$

Finally, the (1,1) entries of (3.91) and (3.92) yield

$$\rho' u'^1 u'^2 - \mu^2 H'^1 H'^2 = c\{\rho[(u^1)^2 - (u^2)^2] - \mu^2[(H^1)^2 - (H^2)^2]\}/2bJ, \tag{3.101}$$

while the (1,2) and (2,1) terms give, for Π'_m to be defined consistently,

$$\rho'[(u'^1)^2 - (u'^2)^2] - \mu^2[(H'^1)^2 - (H'^2)^2] = 2c[\rho u^1 u^2 - \mu^2 H^1 H^2]/bJ. \tag{3.102}$$

Conditions (3.101) and (3.102) may conveniently be combined to give the single complex constraint

$$\rho' q'^2 - \mu^2 H'^2 = ic[\rho \bar{q}^2 - \mu^2 \bar{H}^2]/bJ. \tag{3.103}$$

That solutions of Σ are mapped to solutions of Σ by ϕ is readily shown. Thus, it follows from (3.80) and (3.83) that

$$\phi^* dx^1 = b(dw^1 - dw^2), \tag{3.104}$$

$$\phi^* dx^2 = b(dw^1 + dw^2), \tag{3.105}$$

whence, if f is a solution of Σ, then the 1-forms

$$(\phi \circ j^0 f)^* dx^a = j^0 f^* \phi^* dx^a, \tag{3.106}$$

are linearly independent iff $j^0 f^* dw^1$ and $j^0 f^* dw^2$ are linearly independent. Relations (3.75) and (3.77) show that this will be the case provided that

$$\det \begin{bmatrix} A_1^1 \circ j^0 f & A_2^1 \circ j^0 f \\ A_1^2 \circ j^0 f & A_2^2 \circ j^0 f \end{bmatrix} \neq 0, \tag{3.107}$$

that is, provided that $J \neq 0$.

Thus, to summarize, it has been shown that the magnetogasdynamic equations (3.73)–(3.75) are invariant under the transformations defined by (3.80), (3.81) with the specializations (3.83) and subject to the constraints (3.89) and (3.103).

In the following section, reciprocal-type transformations are constructed in another context, namely, nonlinear heat conduction. A specific boundary value problem is solved using a reduction property associated with the mappings.

3.5 A BÄCKLUND TRANSFORMATION IN NONLINEAR HEAT CONDUCTION

The nonlinear heat equation

$$u_t - [k(u)u_x]_x = 0, \tag{3.108}$$

with variable conductivity $k(u)$ has diverse areas of application, notably in plasma physics [423, 424], boundary layer theory [425], and Darcian filtration [426–428].

Those exact solutions of (3.108) that are known have been incorporated in the class of all similarity solutions of such equations that arise out of invariance under a Lie group of point transformations [429].

In 1979, Rosen [430] derived a specialization of (3.108) of the form

$$u_t - D[u^{-2}u_x]_x = 0 \tag{3.109}$$

in connection with the analysis of the temperature distribution in solid crystalline hydrogen. A novel transformation was introduced which reduced the *nonlinear* heat equation (3.109) to the classical *linear* heat conduction equation

$$u'_t - Du'_{xx} = 0, \tag{3.110}$$

and initial-value problems of importance were thereby solved. Subsequently, in an independent development, Bluman and Kumei [431] rederived the result, but in the context, of Noether transformations.

Here a more general result is presented. Thus, a reciprocal-type transformation is constructed that leaves the nonlinear heat conduction equation invariant. Appropriate specialization leads to the Rosen–Bluman–Kumei transformation.

The General Invariance Property

New independent variables x', t' are introduced, defined by

$$dx' = u\, dx + k(u)u_x\, dt, \tag{3.111}$$

$$dt' = dt, \tag{3.112}$$

3.5 A BÄCKLUND TRANSFORMATION IN NONLINEAR HEAT CONDUCTION

whence, if u' is the new dependent variable and $k'(u')$ is the new thermal conductivity,

$$u'\, dx' + k'(u')u'_{x'}\, dt' = u'u\left[dx + \left(\frac{k(u)u_x}{u} + \frac{k'(u')u'_x}{u'u^2}\right)dt\right] = dx$$

if we set

$$u' = u^{-1}, \tag{3.113}$$

$$k' = u^2 k. \tag{3.114}$$

Thus,

$$u'_{t'} - [k'(u')u'_{x'}]_{x'} = 0, \tag{3.115}$$

and we obtain the interesting result that the nonlinear heat conduction equation (3.108) is invariant under the transformations (3.111)–(3.114).

The Rosen–Bluman–Kumei Transformation

In solid crystalline molecular hydrogen, the temperature distribution $T = T(\mathbf{x}, t)$ is governed by the Fourier equation

$$\rho c_p\, \partial T/\partial t = \nabla(k\nabla T), \tag{3.116}$$

where the specific heat c_p and thermal conductivity k are such that

$$c_p \approx \alpha T^3, \tag{3.117}$$

$$k \approx \beta(T/T_c)^3[1 + (T/T_c)^4]^{-2}, \tag{3.118}$$

where α, β, and T_c are appropriate constants [430]. Substitution of the empirical expressions (3.117) and (3.118) into (3.116) leads to the nonlinear heat conduction equation

$$\partial \theta/\partial t = D\theta^2 \nabla^2 \theta \tag{3.119}$$

for the dimensionless thermal variable

$$\theta(\mathbf{x}, t) = [1 + (T/T_c)^4]^{-1}, \tag{3.120}$$

where D is a diffusion constant.

In one-dimension, (3.119) reduces to

$$\bar{\theta}_t - D[\bar{\theta}^{-2}\bar{\theta}_x]_x = 0, \tag{3.121}$$

where $\bar{\theta} = \theta^{-1}$. Thus, from (3.111)–(3.114), the transformation

$$\begin{aligned}\theta' &= \bar{\theta}^{-1} = \theta,\\ dx' &= \bar{\theta}\, dx + D\bar{\theta}^{-2}\bar{\theta}_x\, dt = \theta^{-1}\, dx - D\theta_x\, dt,\\ dt' &= dt,\end{aligned} \tag{3.122}$$

takes the nonlinear heat equation (3.121) to the associated linear equation

$$\theta'_t - D\theta'_{x'x'} = 0. \tag{3.123}$$

This is the result obtained by Rosen [430], who used it to solve the following initial-value problem for the temperature distribution in a block of solid crystalline hydrogen:

Semi-Infinite Solid Subject to Surface Cooling or Warming

In this case, the boundary-value problem consists of solving

$$\begin{aligned}\theta_t - D\theta_{x'x'} &= 0, \\ \theta &= \theta_0, \quad x = 0, \quad t > 0, \\ \theta &= \theta_i, \quad t = 0, \quad x > 0,\end{aligned} \tag{3.124}$$

where x' is as given in (3.122). Rosen presented the solution

$$\theta = \theta_i + (\theta_0 - \theta_i)(1 - \operatorname{erf}\eta_0)^{-1}(1 - \operatorname{erf}\eta), \quad \eta_0 \le \eta \le \infty, \tag{3.125}$$

in which, from (3.122),

$$\begin{aligned}\eta = x'/2(Dt)^{1/2} &= [\theta_i + (\theta_0 - \theta_i)(1 - \operatorname{erf}\eta_0)^{-1}]\eta \\ &\quad + (\theta_i - \theta_0)(1 - \operatorname{erf}\eta_0)^{-1} \\ &\quad \times [\eta\operatorname{erf}\eta + \pi^{-1/2}\exp(-\eta^2)].\end{aligned} \tag{3.126}$$

On setting $x' = 0$ and $\eta = \eta_0$ in Eq. (3.126), Rosen obtained

$$1 - \theta_0^{-1}\theta_i = \pi^{1/2}\eta_0(1 - \operatorname{erf}\eta_0)\exp\eta_0^2$$

$$= \begin{cases} 1 - \frac{1}{2}\eta_0^{-2} + O(\eta_0^{-4}), & \eta_0 \gtrsim 1, \\ \pi^{1/2}\eta_0 - 2\eta_0^2 + O(\eta_0^3), & |\eta_0| \ll 1, \\ 2\pi^{1/2}\eta_0[\exp\eta_0^2 + O(|\eta_0|^{-1})], & \eta_0 \lesssim -1,\end{cases} \tag{3.127}$$

whence it is seen that there is a unique positive value of η_0 corresponding to surface cooling ($\theta_0 > \theta_i$) and a unique negative value of η_0 corresponding to surface warming ($\theta_0 < \theta_i$). In either case, in the neighborhood of the boundary, the solution (3.125)–(3.127) shows that

$$\theta = \theta_0 - \eta_0 x/(Dt)^{1/2} + O(x^2/Dt).$$

The Rosen–Bluman–Kumei transformation may also be used to solve boundary value problems involving thermally insulated finite slabs [430]. In this connection, it was noted in [431] that

$$\left[\frac{\partial\bar\theta}{\partial x}(x,t)\right]_{x=s(t)} = 0 \Leftrightarrow \left[\frac{\partial\theta'}{\partial x'}(x',t')\right]_{x'=s'(t')} = 0,$$

whence $x = s(t)$ is an insulated boundary in the nonlinear problem governed by (3.121) if and only if the corresponding boundary $x' = s'(t')$ is an insulated boundary in the associated linear problem governed by (3.123).

In general, (3.122) shows that a noninsulating boundary condition for the *nonlinear* heat equation (3.121) on a *fixed* boundary $x = \text{const} = c$ is mapped into a noninsulating boundary condition for the *linear* heat equation (3.123) but at a corresponding *moving* boundary $x' = s'(t')$ traveling with speed

$$\frac{ds'}{dt'} = \left[D\bar{\theta}(x,t)^{-2} \frac{\partial \bar{\theta}}{\partial x}(x,t) \right]_{x=c,\, t=t'}.$$

If, however, the boundary in the nonlinear problem is insulated, then the preceding relation shows that the associated insulated boundary in the canonical linear problem is also fixed.

In conclusion, note that Berryman [432] has recently extended the Rosen–Bluman–Kumei transformation to obtain a correspondence between a *moving* boundary problem for the nonlinear heat equation

$$u_t + [(1 + \delta)u^\delta u_x]_x = 0,$$

and a *fixed* boundary problem for the associated nonlinear heat equation

$$u'_{t'} + [(1 + \delta)(u')^{-\delta-2} u'_{x'}]_{x'} = 0.$$

This result emerges out of the specialization of the transformations (3.111)–(3.114) with $k(u) = -(1 + \delta)u^\delta$; the case $\delta = -2$ gives the Rosen–Bluman–Kumei mapping. Details of the analytic and computational advantages of dealing with the new fixed boundary problem rather than the original moving boundary problem are given in [432].

3.6 BÄCKLUND TRANSFORMATIONS OF THE LOEWNER TYPE

In 1950, Loewner [64] introduced a generalization of the classical Bäcklund transformations (1.9). Thus, transformations of the form

$$\mathbb{B}_i(s^a, u, u_a, v, v_a; s'^a, u', u'_a, v', v'_a) = 0; \quad i = 1, \ldots, 6, \quad a = 1, 2, \quad (3.128)$$

were developed which relate surface elements $\{s^a, u, u_a\}$, $\{s^a, v, v_a\}$ and $\{s'^a, u', u'_a\}$, $\{s'^a, v', v'_a\}$ associated with *pairs* of surfaces $u = u(s^a)$, $v = v(s^a)$ and $u' = u'(s'^a)$, $v' = v'(s'^a)$. It was asserted that the choice of six relations in (3.128) is natural, in that, if the s^a are regarded as independent variables and s'^a, u, v, u', v' as dependent variables, then the number of equations coincides with the number of unknowns.

Loewner's investigation was specifically concerned with the reduction of hodograph systems to canonical form in subsonic, transsonic, and supersonic flow. Accordingly, the analysis in [64] was entirely devoted to the *linear* subclass of Bäcklund transformations of the type (3.128) that can be written in the matrix form[†]

$$\Lambda'_1 = \mathbf{A}\Lambda_1 + \mathbf{B}\Lambda_2 + \mathbf{C}\Lambda + \mathbf{D}\Lambda' + \mathbf{E},$$
$$\Lambda'_2 = \tilde{\mathbf{A}}\Lambda_1 + \tilde{\mathbf{B}}\Lambda_2 + \tilde{\mathbf{C}}\Lambda + \tilde{\mathbf{D}}\Lambda' + \tilde{\mathbf{E}}, \qquad (3.129)$$
$$s'^a = s^a; \quad a = 1, 2,$$

where

$$\Lambda = \begin{bmatrix} u \\ v \end{bmatrix}, \quad \Lambda_a = \begin{bmatrix} u_a \\ v_a \end{bmatrix},$$
$$\Lambda' = \begin{bmatrix} u' \\ v' \end{bmatrix}, \quad \Lambda'_a = \begin{bmatrix} u'_a \\ v'_a \end{bmatrix}, \qquad (3.130)$$

and $\mathbf{A}, \mathbf{B}, \mathbf{C}, \mathbf{D}, \mathbf{E}$ together with $\tilde{\mathbf{A}}, \tilde{\mathbf{B}}, \tilde{\mathbf{C}}, \tilde{\mathbf{D}}$, and $\tilde{\mathbf{E}}$ are 2×2 matrices with entries dependent, in general, on the s^a, $a = 1, 2$.

The hodograph system (3.62) adopts the form

$$\Lambda_1 = \mathbf{H}\Lambda_2, \qquad (3.131)$$

where

$$\mathbf{H} = \begin{bmatrix} 0 & h^1_2(s^1) \\ h^2_1(s^1) & 0 \end{bmatrix}, \qquad (3.132)$$

and the nature of the h^i_j depends on the prevailing flow regime. Loewner sought Bäcklund transformations of the type (3.129) which reduce hodograph systems (3.131) to associated systems

$$\Lambda'_1 = \mathbf{H}'\Lambda'_2, \qquad (3.133)$$

where $\mathbf{H}'$ adopts one of the canonical forms

$$\begin{bmatrix} 0 & -1 \\ 1 & 0 \end{bmatrix}, \quad \begin{bmatrix} 0 & -s \\ 1 & 0 \end{bmatrix}, \quad \begin{bmatrix} 0 & 1 \\ 1 & 0 \end{bmatrix}, \qquad (3.134)$$

corresponding, in turn, to subsonic, transsonic, and supersonic flow. This method is now described in detail.

Thus, application of the integrability requirement

$$\Omega' = \Lambda'_{12} - \Lambda'_{21} = \mathbf{0} \qquad (3.135)$$

[†] The original gasdynamic equations are, of course, nonlinear, but are linearized by the hodograph transformation.

3.6 BÄCKLUND TRANSFORMATIONS OF THE LOEWNER TYPE

to $(3.129)_{1,2}$ yields

$$-\tilde{A}\Lambda_{11}+(A-\tilde{B})\Lambda_{12}+B\Lambda_{22}$$
$$+(A_2-\tilde{A}_1+D\tilde{A}-\tilde{D}A-\tilde{C})\Lambda_1+(B_2-\tilde{B}_1+D\tilde{B}-\tilde{D}B+C)\Lambda_2$$
$$+(C_2-\tilde{C}_1+D\tilde{C}-\tilde{D}C)\Lambda+(D_2-\tilde{D}_1+D\tilde{D}-\tilde{D}D)\Lambda'$$
$$+E_2-\tilde{E}_1+D\tilde{E}-\tilde{D}E=0 \qquad (3.136)$$

on use of the companion integrability condition on Λ, namely,

$$\Omega = \Lambda_{12} - \Lambda_{21} = 0. \qquad (3.137)$$

In general, if

$$|D_2 - \tilde{D}_1 + D\tilde{D} - \tilde{D}D| \neq 0, \qquad (3.138)$$

then (3.136) may be solved for Λ' and substitution into $(3.129)_{1,2}$ produces a pair of matrix equations involving third-order partial derivatives of Λ. However, if we impose the conditions

$$\tilde{A} = B = 0, \qquad A = \tilde{B},$$
$$C_2 - \tilde{C}_1 + D\tilde{C} - \tilde{D}C = 0,$$
$$D_2 - \tilde{D}_1 + D\tilde{D} - \tilde{D}D = 0, \qquad (3.139)$$
$$E_2 - \tilde{E}_1 + D\tilde{E} - \tilde{D}E = 0,$$

then provided that

$$|A_2 - \tilde{C} - \tilde{D}A| \neq 0, \qquad (3.140)$$

a matrix equation of the form (3.131) results.

Similarly, if $|A| \neq 0$, so that the roles of Λ and Λ' may be reversed in (3.129), and if we then write

$$\Lambda_1 = A'\Lambda'_1 + C'\Lambda' + D'\Lambda + E',$$
$$\Lambda_2 = A'\Lambda'_2 + \tilde{C}'\Lambda' + \tilde{D}'\Lambda + \tilde{E}', \qquad (3.141)$$

where

$$A' = A^{-1},$$
$$C' = -A^{-1}D, \qquad D' = -A^{-1}C, \qquad E' = -A^{-1}E, \qquad (3.142)$$
$$\tilde{C}' = -A^{-1}\tilde{D}, \qquad \tilde{D}' = -A^{-1}\tilde{C}, \qquad \tilde{E}' = -A^{-1}\tilde{E},$$

then the integrability condition (3.137) leads to a matrix equation of the form (3.133), if the primed counterparts of conditions $(3.139)_{3-5}$ and (3.140) obtain, namely,

$$C'_2 - \tilde{C}'_1 + D'\tilde{C}' - \tilde{D}'C' = 0,$$
$$D'_2 - \tilde{D}'_1 + D'\tilde{D}' - \tilde{D}'D' = 0, \qquad (3.143)$$
$$E'_2 - \tilde{E}'_1 + D'\tilde{E}' - \tilde{D}'E' = 0,$$

together with

$$|A_2' - \tilde{C}' - \tilde{D}'A'| \neq 0. \tag{3.144}$$

Now, from (3.129) and (3.139)$_{1,2}$,

$$\Lambda_1' - H'\Lambda_2' = A\Lambda_1 + C\Lambda + D\Lambda' + E - H'[A\Lambda_2 + \tilde{C}\Lambda + \tilde{D}\Lambda' + \tilde{E}],$$

whence, the matrix equation (3.131) is mapped by the Bäcklund transformation (3.129) to the associated matrix equation (3.133), and conversely, if the set of conditions (3.139) and (3.143) is augmented by the restrictions

$$\begin{aligned} A^{-1}H'A &= H, & C - H'\tilde{C} &= 0, \\ D - H'\tilde{D} &= 0, & E - H'\tilde{E} &= 0, \end{aligned} \tag{3.145}$$

where (3.136) and its primed counterpart show that

$$\begin{aligned} H &= (A_2 - \tilde{C} - \tilde{D}A)^{-1}(A_1 - C - DA), \\ H' &= (A_2' - \tilde{C}' - \tilde{D}'A')^{-1}(A_1' - C' - D'A'). \end{aligned} \tag{3.146}$$

However,

$$\begin{aligned} C_2' - \tilde{C}_1' + D'\tilde{C}' - \tilde{D}'C' &= -A_2'D + A_1'\tilde{D} + A'(\tilde{D}_1 - D_2) + A'CA'\tilde{D} - A'\tilde{C}A'D \\ &= -(A_2' + A'\tilde{D} + A'\tilde{C}A')D + (A_1' + A'D + A'CA')\tilde{D} \\ &= -(A_2' - \tilde{C}' - \tilde{D}'A')D + (A_1' - C' - D'A')\tilde{D} \\ &= 0, \end{aligned}$$

by virtue of (3.139)$_4$, (3.145)$_3$, and (3.146)$_2$. Hence, condition (3.143)$_1$ is seen to be a consequence of relations (3.139), (3.145), and (3.146). These same conditions may likewise be shown to imply relations (3.143)$_{2,3}$. Moreover, elimination of the primed quantities in (3.146)$_2$ shows that it is equivalent to the relation (3.146)$_1$ while similarly, condition (3.144) is implied by (3.140).

Thus, to summarize, it has been shown that the Bäcklund transformations

$$\begin{aligned} \Lambda_1' &= A\Lambda_1 + H'\tilde{C}\Lambda + H'\tilde{D}\Lambda' + H'\tilde{E}, \\ \Lambda_2' &= A\Lambda_2 + \tilde{C}\Lambda + \tilde{D}\Lambda' + \tilde{E}, \\ s'^a &= s^a, \quad a = 1, 2, \end{aligned} \tag{3.147}$$

subject to the conditions

$$\begin{aligned} \tilde{C}_1 - (H'\tilde{C})_2 + \tilde{D}H'\tilde{C} - H'\tilde{D}\tilde{C} &= 0, \\ \tilde{D}_1 - (H'\tilde{D})_2 + \tilde{D}H'\tilde{D} - H'(\tilde{D})^2 &= 0, \\ \tilde{E}_1 - (H'\tilde{E})_2 + \tilde{D}H'\tilde{E} - H'\tilde{D}\tilde{E} &= 0, \\ A_1 - H'\tilde{C} - H'\tilde{D}A - (A_2 - \tilde{C} - \tilde{D}A)H &= 0 \\ |A_2 - \tilde{C} - \tilde{D}A| &\neq 0, \end{aligned} \tag{3.148}$$

3.6 BÄCKLUND TRANSFORMATIONS OF THE LOEWNER TYPE

transform the matrix equation

$$\Lambda_1 = H\Lambda_2 = A^{-1}H'A\Lambda_2, \qquad (3.149)$$

to the associated matrix equation (3.133).

The preceding Loewner transformations lie within a class of ordinary Bäcklund maps $\psi: J^1(M, N) \times N' \to J^1(M, N')$ described in Chapter 2 (see Example 2.9). Such linear transformations will be shown in this and subsequent chapters to have important applications not only to gasdynamics and nonlinear elasticity, but also to punch, crack, and torsion boundary value problems in linear elasticity.

In conclusion, note that linear auto-Bäcklund transformations have also been obtained by McCarthy [446] for the class of equations

$$\mathbb{D}\phi := (\partial_1^{p_1} \partial_2^{p_2} \cdots \partial_n^{p_n} + \partial_1^{q_1} \partial_2^{q_2} \cdots \partial_n^{q_n} \cdots + \partial_1^{v_1} \partial_2^{v_2} \cdots \partial_n^{v_n})\phi = 0, \qquad (3.150)$$

where $p_1, p_2, \ldots, p_n, q_1, q_2, \ldots, q_n, \ldots, v_1, v_2, \ldots, v_n$ are nonnegative integers. In view of their intrinsic interest, we record three examples given in [446]. In each, the Bäcklund transformation implies that the constituent ϕ_i are solutions of the given equation $\mathbb{D}\phi = 0$.

(i) **The 1 + 1 Linear Diffusion Equation**

$$\mathbb{D}\phi = \phi_t - \phi_{xx} = 0, \qquad (3.151)$$

$$\begin{bmatrix} 1 & -\partial_x \\ -\partial_x & \partial_t \end{bmatrix} \begin{bmatrix} \phi_1 \\ \phi_2 \end{bmatrix} = \begin{bmatrix} 0 \\ 0 \end{bmatrix}. \qquad (3.152)$$

(ii) **The Linearized Korteweg–deVries Equation**

$$\mathbb{D}\phi = \phi_t + \phi_{xxx} = 0, \qquad (3.153)$$

$$\begin{bmatrix} 1 & \partial_x & 0 \\ 0 & \omega & \partial_x \\ \partial_x & 0 & \omega^2 \partial_t \end{bmatrix} \begin{bmatrix} \phi_1 \\ \phi_2 \\ \phi_3 \end{bmatrix} = \begin{bmatrix} 0 \\ 0 \\ 0 \end{bmatrix}. \qquad (3.154)$$

$$(\omega = e^{2\pi i/3}).$$

(iii) **The 2 + 1 Linear Diffusion Equation**

$$\mathbb{D}\phi = \phi_t + \phi_{xx} + \phi_{yy} = 0, \qquad (3.155)$$

$$\begin{bmatrix} \partial_x & \partial_y & \partial_t & 0 \\ \partial_y & -\partial_x & 0 & \partial_t \\ 1 & 0 & -\partial_x & -\partial_y \\ 0 & 1 & -\partial_y & \partial_x \end{bmatrix} \begin{bmatrix} \phi_1 \\ \phi_2 \\ \phi_3 \\ \phi_4 \end{bmatrix} = \begin{bmatrix} 0 \\ 0 \\ 0 \\ 0 \end{bmatrix}. \qquad (3.156)$$

176 3. GASDYNAMICS, HEAT CONDUCTION, AND MAGNETOGASDYNAMICS

3.7 REDUCTION OF THE HODOGRAPH EQUATIONS TO CANONICAL FORM IN SUBSONIC, TRANSSONIC, AND SUPERSONIC GASDYNAMICS

In the context of gasdynamics, the matrix equation (3.136) is here identified with a hodograph system wherein $\mathbf{H}$, whether in subsonic, transsonic, or supersonic flow, has the two salient properties of being independent of the variable s^2 and also of having zero principal diagonal elements. Bäcklund transformations of the type (3.147) are sought that preserve these properties in $\mathbf{H}'$ and that, more particularly, reduce $\mathbf{H}'$ to one of the canonical forms (3.134).

Henceforth, each of the matrices $\mathbf{A}$, $\tilde{\mathbf{C}}$, $\tilde{\mathbf{D}}$, and $\tilde{\mathbf{E}}$ in (3.147) is assumed to be independent of s^2, so that (3.148) reduces to the system of coupled *ordinary* differential equations

$$\tilde{\mathbf{C}}_1 + \tilde{\mathbf{D}}\mathbf{H}'\tilde{\mathbf{C}} - \mathbf{H}'\tilde{\mathbf{D}}\tilde{\mathbf{C}} = 0,$$
$$\tilde{\mathbf{D}}_1 + \tilde{\mathbf{D}}\mathbf{H}'\tilde{\mathbf{D}} - \mathbf{H}'(\tilde{\mathbf{D}})^2 = 0,$$
$$\tilde{\mathbf{E}}_1 + \tilde{\mathbf{D}}\mathbf{H}'\tilde{\mathbf{E}} - \mathbf{H}'\tilde{\mathbf{D}}\tilde{\mathbf{E}} = 0, \qquad (3.157)$$
$$\mathbf{A}_1 - \mathbf{H}'\tilde{\mathbf{C}} - \mathbf{H}'\tilde{\mathbf{D}}\mathbf{A} + (\tilde{\mathbf{C}} + \tilde{\mathbf{D}}\mathbf{A})\mathbf{H} = 0,$$
$$|\tilde{\mathbf{C}} + \mathbf{D}\mathbf{A}| \neq 0.$$

Certain properties of the system (3.157) are immediate. Thus, $(3.157)_2$ shows that

$$\operatorname{tr}\{\tilde{\mathbf{D}}_1\} = [\operatorname{tr}\{\tilde{\mathbf{D}}\}]_1 = \operatorname{tr}\{\mathbf{H}'(\tilde{\mathbf{D}})^2\} - \operatorname{tr}\{\tilde{\mathbf{D}}\mathbf{H}'\tilde{\mathbf{D}}\}$$
$$= \operatorname{tr}\{\mathbf{H}'\tilde{\mathbf{D}} \cdot \tilde{\mathbf{D}}\} - \operatorname{tr}\{\tilde{\mathbf{D}} \cdot \mathbf{H}'\tilde{\mathbf{D}}\}$$
$$= 0,$$

so that the trace of $\tilde{\mathbf{D}}$ is constant. Further, if $\tilde{\mathbf{D}}^*$ denotes the adjoint of $\tilde{\mathbf{D}}$, then

$$\operatorname{tr}\{\tilde{\mathbf{D}}_1\tilde{\mathbf{D}}^*\} = \operatorname{tr}\{\mathbf{H}'\tilde{\mathbf{D}}(\tilde{\mathbf{D}}\tilde{\mathbf{D}}^*)\} - \operatorname{tr}\{\tilde{\mathbf{D}}\mathbf{H}'(\tilde{\mathbf{D}}\tilde{\mathbf{D}}^*)\}$$
$$= \det\tilde{\mathbf{D}}[\operatorname{tr}\{\mathbf{H}'\tilde{\mathbf{D}}\} - \operatorname{tr}\{\tilde{\mathbf{D}}\mathbf{H}'\}]$$
$$= 0, \qquad (3.158)$$

since $\tilde{\mathbf{D}}\tilde{\mathbf{D}}^* = (\det \tilde{\mathbf{D}})\mathbf{I}$. But if $\tilde{\mathbf{D}} = [d^i_j]$, then

$$\operatorname{tr}\{\tilde{\mathbf{D}}_1\tilde{\mathbf{D}}^*\} = (d^1_1)_1 d^2_2 - (d^1_2)_1 d^2_1 - (d^2_1)_1 d^1_2 + (d^2_2)_1 d^1_1 = (\det \tilde{\mathbf{D}})_1,$$

whence, from (3.158), it follows that $\det \tilde{\mathbf{D}}$ is also constant.

In a similar fashion, postmultiplication in turn of $(3.157)_1$ and $(3.157)_{3,4}$ by the adjoint matrices $\tilde{\mathbf{C}}^*$, $\tilde{\mathbf{E}}^*$, and $\tilde{\mathbf{A}}^*$ and computation of traces demonstrates that $\det \tilde{\mathbf{C}}$, $\det \tilde{\mathbf{E}}$, and $\det \tilde{\mathbf{A}}$ are each constant.

3.7 SUBSONIC, TRANSSONIC, AND SUPERSONIC GASDYNAMICS

In the subsequent discussion, the matrices $\mathbf{A}$, $\tilde{\mathbf{C}}$, $\tilde{\mathbf{D}}$ are specialized to be of the forms

$$\mathbf{A} = \begin{bmatrix} a_1^1 & 0 \\ 0 & a_2^2 \end{bmatrix}, \quad \tilde{\mathbf{C}} = \begin{bmatrix} 0 & c_2^1 \\ c_1^2 & 0 \end{bmatrix}, \quad \tilde{\mathbf{D}} = \begin{bmatrix} 0 & d_2^1 \\ d_1^2 & 0 \end{bmatrix}, \quad (3.159)$$

while $\tilde{\mathbf{E}}$ is taken as the 2×2 null matrix. Thus, in particular, if $\mathbf{H}$ adopts the form (3.132), then

$$\mathbf{H}' = [h_j^{i'}] = \begin{bmatrix} 0 & a_1^1 h_2^1 / a_2^2 \\ a_2^2 h_1^2 / a_1^1 & 0 \end{bmatrix}, \quad (3.160)$$

so that the required property of zero principal diagonal elements is preserved.

Further, *in extenso*, equations (3.157) yield

$$\begin{aligned} c_{2,1}^1 + c_2^1 [d_1^1 h_1^{2'} - h_2^1 d_1^2] &= 0, \\ c_{1,1}^2 + c_1^2 [d_2^2 h_2^{1'} - h_1^2 d_2^1] &= 0, \\ d_{2,1}^1 + d_2^1 [d_1^1 h_1^{2'} - h_2^1 d_1^2] &= 0, \\ d_{1,1}^2 + d_1^2 [d_2^2 h_2^{1'} - h_1^2 d_2^1] &= 0, \\ a_{1,1}^1 - h_2^{1'} [c_1^2 + d_1^2 a_1^1] + h_1^2 [c_2^1 + d_2^1 a_2^2] &= 0, \\ a_{2,1}^2 - h_1^{2'} [c_2^1 + d_2^1 a_2^2] + h_2^1 [c_1^2 + d_1^2 a_1^1] &= 0. \end{aligned} \quad (3.161)$$

Since

$$\det \mathbf{A} = a_1^1 a_2^2 = a, \quad \det \tilde{\mathbf{D}} = -d_2^1 d_1^2 = d,$$

where a, d are constants, $(3.161)_{3-6}$ reduce to consideration of a *pair* of Riccati equations for d_2^1 and a_1^1, namely,

$$\begin{aligned} d_{2,1}^1 + h_1^{2'} (d_2^1)^2 + h_2^1 d &= 0, \\ a_{1,1}^1 + a^{-1} h_1^{2'} c_2^1 (a_1^1)^2 + [h_1^{2'} d_2^1 - h_2^{1'} d_1^2] a_1^1 - h_2^{1'} c_1^2 &= 0, \end{aligned} \quad (3.162)$$

while c_2^1, c_1^2 are given by $(3.161)_{1,2}$, once d_2^1 and hence d_1^2 have been obtained. The further restrictions $d_2^1 = d_1^2 = 0$ are now made and accordingly, (3.162) reduces to the single Riccati equation

$$a_{1,1}^1 + \bar{\alpha}(a_1^1)^2 + \bar{\beta} = 0, \quad (3.163)$$

where

$$\bar{\alpha} = a^{-1} h_1^{2'} c_2^1, \quad (3.164)$$
$$\bar{\beta} = -h_2^{1'} c_1^2, \quad (3.165)$$

and c_2^1, c_1^2 are constants by virtue of $(3.161)_{1,2}$.

The discussion is next concerned with reduction to specific canonical forms associated with subsonic, transsonic, and supersonic flow regimes in steady gasdynamics. Subsequently, applications in nonsteady gasdynamics, magnetogasdynamics, elasticity, and nonlinear dielectrics will be developed in detail.

Subsonic Flow

If $M < 1$, a reduction is sought wherein $\mathbf{H}'$ adopts the canonical form $(3.134)_1$, so that $h_2^{1\prime} = -1$, $h_1^{2\prime} = +1$. Thus,

(a) if $\bar{\alpha} = 0$,
$$a_1^1 = -\bar{\beta}s^1 + \delta; \tag{3.166}$$

(b) if $\bar{\beta} = 0$,
$$a_1^1 = [\bar{\alpha}s^1 + \varepsilon]^{-1}; \tag{3.167}$$

(c) if $\bar{\beta}/\bar{\alpha} > 0$,
$$a_1^1 = (\bar{\beta}/\bar{\alpha})^{1/2} \tan\{(\bar{\beta}/\bar{\alpha})^{1/2}(-\bar{\alpha}s^1 + \zeta)\}; \tag{3.168}$$

(d) if $\bar{\beta}/\bar{\alpha} < 0$,
$$a_1^1 = (-\bar{\beta}/\bar{\alpha})^{1/2} \coth\{(-\bar{\beta}/\bar{\alpha})^{1/2}(\bar{\alpha}s^1 + \eta)\}; \tag{3.169}$$

where δ, ε, ζ, η are arbitrary real constants.

Under the present class of Bäcklund transformations, $\mathbf{H}$ is given by

$$\mathbf{H} = \mathbf{A}^{-1}\mathbf{H}'\mathbf{A} = \begin{bmatrix} 0 & h_2^{1\prime}a/(a_1^1)^2 \\ h_1^{2\prime}(a_1^1)^2/a & 0 \end{bmatrix}, \tag{3.170}$$

and in subsonic flow, it follows that reduction of the hodograph system (3.64) to the Cauchy–Riemann equations may be achieved if the real gas function K given by (3.65) is approximated by an expression adopting one of the following forms corresponding, in turn, to (3.166)–(3.169):

(a) $a^2/[-\bar{\beta}s^1 + \delta]^4$, (3.171)

(b) $a^2[\bar{\alpha}s^1 + \varepsilon]^4$, (3.172)

(c) $(a\bar{\alpha}/\bar{\beta})^2 \cot^4\{(\bar{\beta}/\bar{\alpha})^{1/2}(-\bar{\alpha}s^1 + \zeta)\}$, (3.173)

(d) $(a\bar{\alpha}/\bar{\beta})^2 \tanh^4\{(-\bar{\beta}/\bar{\alpha})^{1/2}(\bar{\alpha}s^1 + \eta)\}$. (3.174)

The approximations (a)–(d) to $K(s^1)$ each lead, via (3.22), (3.63), and (3.65), to associated multiparameter approximations K^*, p^*, ρ^*, q^* to the real gas variables K, p, ρ, q. Thus, corresponding to the adiabatic gas law (3.56), if the gas density and local speed of sound are taken to be unity at $q = 0$, then

3.7 SUBSONIC, TRANSSONIC, AND SUPERSONIC GASDYNAMICS

$C = \gamma^{-1}$, and we obtain

$$K = [1 - \tfrac{1}{2}(\gamma + 1)q^2]/[1 - \tfrac{1}{2}(\gamma - 1)q^2]^{(\gamma+1)/(\gamma-1)}, \quad (3.175)$$

$$p = \gamma^{-1}[1 - \tfrac{1}{2}(\gamma - 1)q^2]^{\gamma/(\gamma-1)}, \quad (3.176)$$

$$\rho = [1 - \tfrac{1}{2}(\gamma - 1)q^2]^{1/(\gamma-1)}, \quad (3.177)$$

where

$$M^2 = q^2/[1 - \tfrac{1}{2}(\gamma - 1)q^2], \quad (3.178)$$

so that, over the subsonic range,

$$0 \leq q < [2/(\gamma + 1)]^{1/2}. \quad (3.179)$$

The available parameters in the approximations K^*, p^*, ρ^*, q^* to the corresponding real gas variables may now be chosen so as to achieve best alignment in some specified manner over the range (3.179).

A variety of transformations previously established by diverse means in the literature are now retrieved in a unified manner. Thus, Müller [127] and Sauer [128] investigated approximations of the type (3.171) and (3.172), respectively, while the Kármán–Tsien approximation corresponds simply to the choice $\bar{\alpha} = 0$ in (3.172). On the other hand, approximations of the type (3.173) and (3.174) have been applied extensively in a monograph by Dombrovskii [71] to solve various boundary value problems. In particular, it is noted therein that the approximation (3.174) leads to the following multiparameter $p^*(s^1), \rho^*(s^1),$ and $q^*(s^1)$ relations:

$$p^* = \frac{(a\bar{\alpha}/\bar{\beta})(d_3)^2\{d_1 \tanh(d_1 s^*) + 1\}}{2d_2[(d_1)^2 - 1][d_2 e^{2s^*}\{d_1 \tanh(d_1 s^*) - 1\} + \{d_1 \tanh(d_1 s^*) + 1\}]} + d_4, \quad (3.180)$$

$$\rho^* = \frac{d_2 d^{2s^*}\{d_1 \tanh(d_1 s^*) - 1\} + \{d_1 \tanh(d_1 s^*) + 1\}}{(a\bar{\alpha}/\bar{\beta}) \tanh(d_1 s^*)[d_2 e^{2s^*}\{\tanh(d_1 s^*) - d_1\} + \{\tanh(d_1 s^*) + d_1\}]}, \quad (3.181)$$

$$q^* = \frac{(a\bar{\alpha}/\bar{\beta})d_3 \tanh(d_1 s^*)}{[d_2 e^{2s^*}\{d_1 \tanh(d_1 s^*) - 1\} + \{d_1 \tanh(d_1 s^*) + 1\}]}, \quad (3.182)$$

where $d_1 = (-\bar{\alpha}\bar{\beta})^{1/2}$, $s^* = s^1 + \eta\bar{\alpha}^{-1}$, and the remaining d_i are arbitrary real constants.

In Fig. 3.1, the variation of $p^*, \rho^*,$ and q^* is shown against that of the real adiabatic gas variables $p, \rho,$ and q with the arbitrary parameters in (3.174) and (3.180)–(3.182) assigned so as to give best alignment in the neighborhood of the stagnation point $q = 0$. The corresponding numerical comparison is set out in detail by Power *et al.* [66].

180   3. GASDYNAMICS, HEAT CONDUCTION, AND MAGNETOGASDYNAMICS

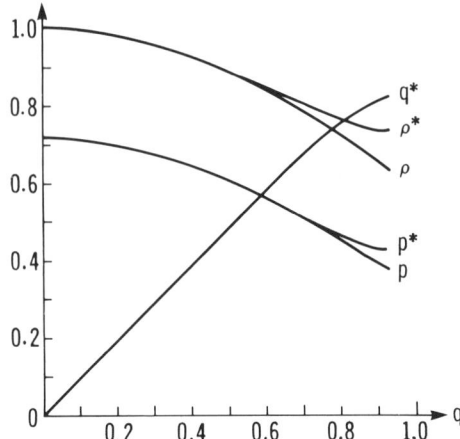

Fig. 3.1 The Dombrovskii transformation. Comparison of real and approximated gas variables over the subsonic range $0 \leq q < (2/\gamma + 1)^{1/2}$: alignment at stagnation conditions.

It is clear that (a)–(d) and the corresponding p^*, ρ^*, q^* relations may be used to approximate the associated real gasdynamic variables in other ways dependent on the part of the subsonic region in which optimal alignment is required. Reduction to the Cauchy–Riemann equations achieved, established complex variable methods of classical hydrodynamics may be adduced to treat a wide range of boundary value problems in subsonic gasdynamics. Reference may be made to the monograph of Dombrovskii [71] and to the extensive bibliography contained therein for a description of such applications. Analogous transformations for hyperbolic systems will be exploited later in this chapter to analyze the reflection of a centered wave in a shock tube. In the next chapter, they will be used extensively in connection with wave propagation in bounded nonlinear elastic and dielectric media. Application to elastic-plastic wave propagation is described in [87, 88].

Transsonic Flow

In transsonic regions $(M \sim 1)$, the hodograph system (3.62) is mixed and in this case, instead of (3.63), the change of variable

$$s^1 = \int_{q_0}^{q} \rho(\sigma) q^{-1}(\sigma) \, d\sigma, \tag{3.183}$$

is introduced, so that the hodograph system reduces to

$$\begin{bmatrix} \phi \\ \psi \end{bmatrix}_{s^1} = \begin{bmatrix} 0 & -K(s^1) \\ 1 & 0 \end{bmatrix} \begin{bmatrix} \phi \\ \psi \end{bmatrix}_{s^2}. \tag{3.184}$$

3.7 SUBSONIC, TRANSSONIC, AND SUPERSONIC GASDYNAMICS

Reduction is sought to the canonical form (3.133) wherein

$$\mathbf{H}' = \begin{bmatrix} 0 & -s^1 \\ 1 & 0 \end{bmatrix}, \quad (3.185)$$

whence, the specialization $h_2^{1\prime} = -s^1$, $h_1^{2\prime} = +1$ in (3.164)–(3.165) reduces the Riccati equation (3.163) to the form

$$a_{1,1}^1 + a^{-1}c_2^1(a_1^1)^2 + c_1^2 s^1 = 0. \quad (3.186)$$

Accordingly,

(a) if $c_2^1 = 0$,

$$a_1^1 = -\tfrac{1}{2}c_1^2(s^1)^2 + \delta; \quad (3.187)$$

(b) if $c_1^2 = 0$,

$$a_1^1 = [a^{-1}c_2^1 s^1 + \varepsilon]^{-1}; \quad (3.188)$$

where δ, ε are again arbitrary real constants.

By virtue of (3.170), it follows that the transsonic system

$$\begin{bmatrix} \phi \\ \psi \end{bmatrix}_{s^1} = \begin{bmatrix} 0 & -s^1 a/(a_1^1)^2 \\ (a_1^1)^2/a & 0 \end{bmatrix} \begin{bmatrix} \phi \\ \psi \end{bmatrix}_{s^2}, \quad (3.189)$$

is reducible to the Tricomi canonical form when a_1^1 adopts either of the forms (3.187) or (3.188). Moreover, (3.189) may be rewritten in the standard form

$$\begin{bmatrix} \phi \\ \psi \end{bmatrix}_{s^{*1}} = \begin{bmatrix} 0 & -K^*(s^{*1}) \\ 1 & 0 \end{bmatrix} \begin{bmatrix} \phi \\ \psi \end{bmatrix}_{s^{*2}}, \quad (3.190)$$

where

$$K^*(s^{*1}) = a^2 s^1/(a_1^1)^4, \quad (3.191)$$

and

$$s^{*1} = a^{-1} \int_{s^0}^{s^1} a_1^1(\sigma)^2 \, d\sigma, \quad s^{*2} = s^2. \quad (3.192)$$

Thus, corresponding to (3.187) and (3.188), we obtain:

(a)
$$K^* = a^2 s^1 [-\tfrac{1}{2}c_1^2(s^1)^2 + \delta]^{-4}, \quad (3.193)$$
$$s^{*1} = a^{-1}[\tfrac{1}{20}(c_1^2)^2(s^1)^5 - \tfrac{1}{3}\delta c_1^2(s^1)^3 + \delta s^1], \quad (3.194)$$

(b)
$$K^* = a^2 s^1 [a^{-1} c_2^1 s^1 + \varepsilon]^4, \quad (3.195)$$
$$s^{*1} = -(c_2^1)^{-1}[a^{-1} c_2^1 s^1 + \varepsilon]^{-1}, \quad (3.196)$$

where, in the preceding, the reference constant s^0 has been taken to be zero. In particular, the specialization $c_1^2 = 0$ in (3.193) and (3.194) produces the well-known transsonic approximation due to Frankl' [129] and subsequently extensively developed by Tomotika and Tamada [130].

Supersonic Flow

In supersonic regions ($M > 1$) the hodograph system (3.62) is hyperbolic and in this case, the change of variable

$$s^1 = i \int_{q_0}^{q} [1 - M(\sigma)^2]^{1/2} \sigma^{-1} \, d\sigma \qquad (3.197)$$

is introduced, so that (3.62) reduces to

$$\begin{bmatrix} \phi \\ \psi \end{bmatrix}_{s^1} = \begin{bmatrix} 0 & iK(s^1)^{1/2} \\ -iK(s^1)^{-1/2} & 0 \end{bmatrix} \begin{bmatrix} \phi \\ \psi \end{bmatrix}_{s^2}, \qquad (3.198)$$

where now $K < 0$.

Reduction to the canonical form associated with the classical wave equation is now sought so that $h_2^{1'} = h_1^{2'} = 1$ in (3.164)–(3.165). Thus, subject only to the change $\bar{\beta} = -c_1^2$ instead of the value $\bar{\beta} = c_1^2$, if a_1^1 adopts one of the forms (3.166)–(3.169) with s^1 now given by (3.197), then such a reduction may be achieved. The corresponding K^* are, from (3.170) and (3.198), given by

$$K^* = -a^2/(a_1^1)^4. \qquad (3.199)$$

In particular, the approximation corresponding to (3.168) was exploited by Khristianovich [131] in supersonic flow. Also, as noted previously, the analog of the Kármán–Tsien approximation in supersonic régimes wherein K is replaced by an appropriate constant, has been investigated by Coburn [68].

It has been seen that the linear Bäcklund transformations introduced by Loewner provide a convenient unifying framework for a wide class of approximations in subsonic, transsonic, and supersonic gasdynamics. In fact, the reciprocal and adjoint transformations introduced earlier in this chapter may also be generated readily via the Loewner formulation. Thus, the Bäcklund transformations (3.147) of the type

$$\Lambda_1' = \mathbf{A}\Lambda_1, \qquad \Lambda_2' = \mathbf{A}\Lambda_2, \qquad s'^a = s^a, \qquad a = 1, 2, \qquad (3.200)$$

with $\mathbf{A}$ a constant matrix that adopts, in turn, the forms

$$\begin{bmatrix} a_1^1 & 0 \\ 0 & a_2^2 \end{bmatrix}, \qquad (3.201)$$

$$\begin{bmatrix} 0 & a_2^1 \\ a_1^2 & 0 \end{bmatrix}, \qquad (3.202)$$

may be shown to lead to the reciprocal and adjoint relations, respectively, when applied to the hodograph system. The details have been set out by Rogers [132] and are not repeated here. The important point is that, seen in this light, both such transformations may be regarded as of *linear* Bäcklund type.

3.8 BÄCKLUND TRANSFORMATIONS IN ALIGNED NONDISSIPATIVE MAGNETOGASDYNAMICS

Invariant transformations of a reciprocal type have been constructed in Section 3.4 for steady, two-dimensional magnetogasdynamics. Here, Bäcklund transformations of the Loewner type are introduced under the additional assumption that $u^j(x^a)$, $H^j(x^a)$ are aligned, so that

$$H^j = k\rho u^j, \tag{3.203}$$

where, in general,

$$u^j \, \partial k/\partial x^j = 0 \tag{3.204}$$

while here, k is assumed constant.

The alignment condition (3.203) together with the equations of continuity and motion may be readily combined to produce a Bernoulli integral (Seebass [134]). With the assumption of uniform stagnation ethalpy, this serves, under homentropic conditions, to determine $p(q)$ and $\rho(q)$ relations once the gas law is specified.

On introduction of new variables u^{*j}, p^*, ρ^*, according to[†]

$$u^{*j} = \{1 - \mu k^2 \rho\} u^j, \tag{3.205}$$

$$p^* = \Pi_m, \tag{3.206}$$

$$\rho^* = \{1 - \mu k^2 \rho\}^{-1} \rho, \quad 1 - \mu k^2 \rho \neq 0, \tag{3.207}$$

the governing magnetogasdynamic equations reduce to the associated gasdynamic system

$$u^{*j} \frac{\partial \rho^*}{\partial x^j} + \rho^* \frac{\partial u^{*j}}{\partial x^j} = 0, \quad j = 1, 2, \tag{3.208}$$

$$\rho^* u^{*j} \frac{\partial u^{*i}}{\partial x^j} + \frac{\partial p^*}{\partial x^i} = 0, \quad i, j = 1, 2, \tag{3.209}$$

together with

$$\frac{\partial u^{*2}}{\partial x^1} - \frac{\partial u^{*1}}{\partial x^2} = 0. \tag{3.210}$$

[†] See Grad [72] and Iurév [133].

184   3. GASDYNAMICS, HEAT CONDUCTION, AND MAGNETOGASDYNAMICS

Equations (3.208) and (3.210) allow the introduction, in turn, of $\psi^*(x^a)$, $\phi^*(x^a)$, according to

$$(\rho^* q^*)^{-1} d\psi^* = -\sin\theta^* dx^1 + \cos\theta^* dx^2, \tag{3.211}$$

$$q^{*-1} d\phi^* = \cos\theta^* dx^1 + \sin\theta^* dx^2, \tag{3.212}$$

where

$$q^* e^{i\theta^*} = u^{*1} + iu^{*2}. \tag{3.213}$$

In analogy with the gas dynamics procedure, the hodograph system

$$\begin{bmatrix} \phi^* \\ \psi^* \end{bmatrix}_{q^*} = \begin{bmatrix} 0 & \dfrac{(M^2-1)(1-\mu k^2 \rho)}{[1-\mu k^2 \rho(1-M^2)]\rho q} \\ \dfrac{\rho}{(1-\mu k^2 \rho)^2 q} & 0 \end{bmatrix} \begin{bmatrix} \phi^* \\ \psi^* \end{bmatrix}_{\theta^*}, \tag{3.214}$$

may now be established, subject to the requirement that

$$0 < |J(x^1, x^2; q^*, \theta^*)| < \infty. \tag{3.215}$$

In *sub-Alfvénic regions* ($\mu k^2 \rho > 1$), it is seen that if $M > 1$, the hodograph system (3.214) is elliptic. On the other hand, if $M < 1$, since $(\mu k^2 \rho)^{-1} = q^2/b^2 = A^2$, where $b = (\mu/\rho)^{1/2} H$ is the *Alfvén speed* and A is the *Alfvén number*, the system is elliptic in the region

$$\frac{q^2}{b^2} + \frac{q^2}{c^2} < 1,$$

and hyperbolic in the region

$$\frac{q^2}{b^2} + \frac{q^2}{c^2} > 1.$$

In *super-Alfvénic regions* ($\mu k^2 \rho < 1$), as in conventional gasdynamics, the hodograph system is elliptic for $M < 1$ and hyperbolic for $M > 1$. The various régimes in the (M, A) plane are illustrated in Fig. 3.2. Transcritical regions have been investigated by Seebass [134] and Tamada [135].

Bäcklund transformations of the type described in Section 3.7 may now be employed in a similar manner to reduce the hodograph system (3.214) to the appropriate canonical forms in elliptic or hyperbolic régimes. Thus, if attention is restricted to subsonic super-Alfvénic regions, under the change of variable

$$s^1 = \int_{q_0}^{q} \left[\frac{(1-M^2)(1-\mu k^2 \rho(1-M^2))}{(1-\mu k^2 \rho)} \right]^{1/2} \frac{dq}{q}, \qquad s^2 = \theta^*, \tag{3.216}$$

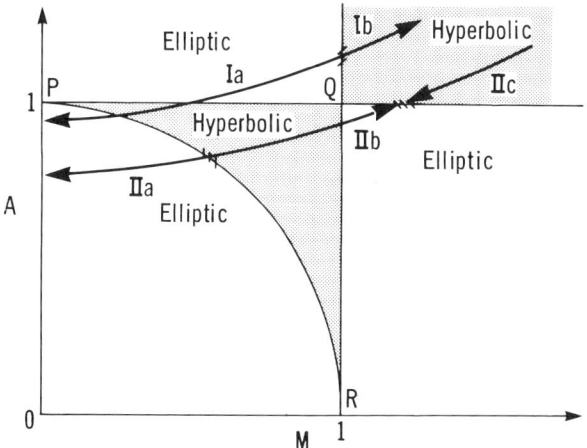

Fig. 3.2 The sub-Alfvénic and super-Alfvénic regions in the (M, A) plane (Seebass [134]).

the system (3.214) reduces to the form

$$\begin{bmatrix} \phi^* \\ \psi^* \end{bmatrix}_{s^1} = \begin{bmatrix} 0 & -K^{1/2}(s^1) \\ K^{-1/2}(s^1) & 0 \end{bmatrix} \begin{bmatrix} \phi^* \\ \psi^* \end{bmatrix}_{s^2}, \qquad (3.217)$$

where

$$K = (1 - M^2)(1 - \mu k^2 \rho)^3/\rho^2[1 - \mu k^2 \rho(1 - M^2)]. \qquad (3.218)$$

Bäcklund transformations may be now employed to reduce the system (3.217) to the Cauchy–Riemann system when K is approximated by forms analogous to (3.171)–(3.174) as set out in the previous section. In particular, the analog of the Kármán–Tsien approximation wherein K is replaced by an appropriate constant may be shown to lead to a hypothetical constitutive law of the form

$$p = A - \frac{B}{\rho} + \frac{\mu k^2 C \rho^2}{[1 - \mu k^2 \rho]^2}, \qquad (3.219)$$

where A, $B > 0$, $C > 0$ are real parameters. The consequences of such an approximation are investigated in [132]. It is noted that the Kármán–Tsien approximation of nonconducting gasdynamics is recovered in the limit $k \to 0$.

3.9 INVARIANT TRANSFORMATIONS IN NONSTEADY GASDYNAMICS

The subject of invariant transformations in nonsteady gasdynamics has been developed in a series of papers by Nikol'skii [75], Tomilov [76], Movsesian [77], and Ustinov [78]. Reference may also be made to the work of Smith [136] on invariant transformations of the substitution principle type. Here, a reciprocal invariant transformation is constructed and is shown to correspond to a Bäcklund transformation of a Monge–Ampère equation. Note that these reciprocal relations have since been extended to nonsteady, quasi one-dimensional, oblique field magnetogasdynamics by Rogers and Kingston [137].

The governing equations of inviscid, one-dimensional nonsteady gasdynamics, neglecting heat conduction and heat radiation are, in the usual notation,

$$\frac{\partial \rho}{\partial t} + \frac{\partial}{\partial x^1}(\rho u^1) = 0, \tag{3.220}$$

$$\rho\left[\frac{\partial u^1}{\partial t} + u^1 \frac{\partial u^1}{\partial x^1}\right] + \frac{\partial p}{\partial x^1} = 0, \tag{3.221}$$

$$\frac{\partial s}{\partial t} + u^1 \frac{\partial s}{\partial x^1} = 0, \tag{3.222}$$

together with an equation of state (3.4).

The equations of continuity and motion (3.220)–(3.221) imply the existence of $\psi(x^1, t)$, $\bar{\xi}(x^1, t)$ such that

$$d\psi = \rho \, dx^1 - \rho u^1 \, dt, \tag{3.223}$$

$$a_1 \, d\bar{\xi} = \rho u^1 \, dx^1 - [p + \rho(u^1)^2 + a_2] \, dt, \tag{3.224}$$

where $a_1 \neq 0$ and a_2 are real constants.

Introduction of new variables u'^1, ρ', x'^1, t', according to

$$u'^1 = \frac{-a_1 u^1}{(p + a_2)}, \quad a_1 \neq 0, \quad p + a_2 \neq 0, \tag{3.225}$$

$$\rho' = \frac{a_3 \rho (p + a_2)}{(p + \rho(u^1)^2 + a_2)}, \quad a_3 \neq 0, \tag{3.226}$$

$$x'^1 = x^1, \tag{3.227}$$

$$t' = \bar{\xi}, \tag{3.228}$$

provided that

$$0 < |(p + \rho(u^1)^2 + a_2)| < \infty, \tag{3.229}$$

shows that

$$\rho' \, dx'^1 - \rho' u'^1 \, dt' = a_3[\rho \, dx^1 - \rho u^1 \, dt],$$

whence,

$$\frac{\partial \rho'}{\partial t'} + \frac{\partial}{\partial x'^1}(\rho' u'^1) = 0, \tag{3.230}$$

3.9 INVARIANT TRANSFORMATIONS IN NONSTEADY GASDYNAMICS

where in the above, a_3 is a nonzero real constant. Hence, the equation of continuity is preserved in the (x'^1, t') space.

Moreover, since

$$\frac{\partial}{\partial x^1} = \frac{\partial}{\partial x'^1} + \frac{\rho u^1}{a_1} \frac{\partial}{\partial t'}, \qquad \frac{\partial}{\partial t} = -\frac{[p + \rho(u^1)^2 + a_2]}{a_1} \frac{\partial}{\partial t'},$$

the equation of motion (3.221) transforms into

$$\rho' \left[\frac{\partial u'^1}{\partial t'} + u'^1 \frac{\partial u'^1}{\partial x'^1} \right] + \left[\frac{a_1^2 a_3}{(p + a_2)^2} \right] \frac{\partial p}{\partial x'^1} = 0. \tag{3.231}$$

Hence, if we set

$$p' = a_4 - a_1^2 a_3 / (p + a_2), \tag{3.232}$$

where a_4 is an arbitrary real constant, then (3.231) becomes

$$\rho' \left[\frac{\partial u'^1}{\partial t'} + u'^1 \frac{\partial u'^1}{\partial x'^1} \right] + \frac{\partial p'}{\partial x'^1} = 0, \tag{3.233}$$

so that the equation of motion is likewise preserved in the (x'^1, t') space.

Finally, (3.222) yields

$$\frac{\partial s}{\partial t} + u^1 \frac{\partial s}{\partial x^1} = -\left(\frac{p + a_2}{a_1} \right) \left[\frac{\partial s'}{\partial t'} + u'^1 \frac{\partial s'}{\partial x'^1} \right], \tag{3.234}$$

so that, since $p + a_2 \neq 0$,

$$\frac{\partial s'}{\partial t'} + u'^1 \frac{\partial s'}{\partial x'^1} = 0, \tag{3.235}$$

where

$$s' = \Phi(s) \tag{3.236}$$

and Φ is assumed to be an invertible function of the specific entropy s.

Thus, the system of nonsteady gasdynamics equations (3.220)–(3.222) is invariant under the reciprocal-type transformations

$$\begin{aligned}
u'^1 &= -a_1 u^1 / (p + a_2), & a_1 &\neq 0, \quad p + a_2 \neq 0, \\
p' &= a_4 - a_1^2 a_3 / (p + a_2), & a_3 &\neq 0, \\
\rho' &= a_3 \rho (p + a_2) / (p + \rho(u^1)^2 + a_2), & & \\
s' &= \Phi(s), \quad x' = x, & t' &= \bar{\xi},
\end{aligned} \tag{3.237}$$

subject to the condition (3.229).

The reciprocal relations $(3.237)_{1-4}$ may be inverted to yield

$$u^1 = -a_1 a_3 u'^1/(a_4 - p'), \quad p = [a_1^2 a_3/(a_4 - p')] - a_2,$$
$$\rho = \rho'/a_3[1 - \rho'(u'^1)^2/(a_4 - p')], \quad s = \Phi^{-1}(s'). \quad (3.238)$$

This establishes the following result (Rogers [81]):

THEOREM 3.5 If $\{u^1(x^1, t), p(x^1, t), \rho(x^1, t), s(x^1, t)\}$ constitutes a solution set of equations (3.220)–(3.222), then so also does the four-parameter system

$$\left\{ \frac{-a_1 a_3 u^1(x^1, t')}{[a_4 - p(x^1, t')]}, \frac{a_1^2 a_3}{[a_4 - p(x^1, t')]} - a_2, \right.$$
$$\left. \frac{\rho(x^1, t')}{a_3[1 - \rho(x^1, t')(u^1(x^1, t'))^2/(a_4 - p(x^1, t'))]}, \Phi^{-1}(s(x^1, t')) \right\}$$

where t' is defined by (3.228), which expression is integrable by virtue of (3.220) and (3.221).

As in the case of the analogous transformations in steady gasdynamics, the invariance does not, in general, extend to the equation of state. However, the parameters a_i, $i = 1, \ldots, 4$, are available for the approximation of the real gas law.

That the preceding reciprocal relations constitute a Bäcklund transformation may readily be seen in terms of Martin's formulation of the system (3.220)–(3.221), wherein the particle trajectories and isobars are adopted as curvilinear coordinates in the (x^1, t) plane. Thus, with p, ψ as independent variables, Martin [138] noted that (3.220)–(3.221) reduce to consideration of the Monge–Ampère equation

$$\xi_{pp}\xi_{\psi\psi} - (\xi_{\psi p})^2 = \tau_p(p, \psi), \quad (3.239)$$

where $\xi = a_1 \bar{\xi} + (p + a_2)t$, and u^1, ρ, x^1, t are given by

$$u^1 = \xi_\psi, \quad \rho = \tau^{-1}(p, \psi),$$
$$x^1_p = \xi_\psi \xi_{pp}, \quad x^1_\psi = \xi_\psi \xi_{p\psi} + \tau, \quad (3.240)$$
$$t = \xi_p,$$

and (3.239) represents the integrability condition on $(3.240)_{3,4}$.

The reciprocal relations (3.237) correspond to the Bäcklund transformation of (3.239) given by

$$\xi'_{p'} = a_1^{-1}[-(p + a_2)\xi_p + \xi],$$
$$\xi'_{\psi'} = -a_1 \xi_\psi/(p + a_2),$$
$$p' = a_4 - a_1^2 a_3/(p + a_2), \quad (3.241)$$
$$\psi' = a_3 \psi,$$

3.10 BÄCKLUND TRANSFORMATIONS IN LAGRANGIAN GASDYNAMICS

with $p + a_2 \neq 0$ and

$$a_3 \tau' = \frac{(p + a_2)\tau + (\xi_\psi)^2}{(p + a_2)}. \tag{3.242}$$

On the other hand, Ustinov's transformation as set out in [78] corresponds to the simple Bäcklund transformation of (3.239) given by

$$\xi'_{p'} = \psi, \quad \xi'_{\psi'} = \xi_\psi, \quad p' = \tau, \quad \psi' = -\xi_p, \tag{3.243}$$

together with

$$\tau' = p, \tag{3.244}$$

and the requirement that

$$0 < |\xi_{pp}\tau_\psi - \xi_{p\psi}\tau_p| < \infty. \tag{3.245}$$

Ustinov [78] employed this transformation to generate a new solution of the system (3.220)–(3.222) associated with the flow between a piston and a nonuniform shock wave.

In conclusion, we remark that the above Bäcklund transformations also have application to the analysis of plane, rotational, hypersonic flow past thin aerofoils in view of a general similitude developed by Hayes [139] and Goldsworthy [140]. Therein, the governing equations of steady hypersonic flow may be shown, under certain circumstances, to reduce to the system (3.220)–(3.222). High Mach number flows of importance may thereby be shown to be directly analogous to associated nonsteady piston-driven motions.

3.10 BÄCKLUND TRANSFORMATIONS IN LAGRANGIAN GASDYNAMICS. REFLECTION OF A CENTERED WAVE IN A SHOCK TUBE

The Lagrangian description of one-dimensional nonsteady inviscid gasdynamics consists of the pair of equations

$$-\partial p/\partial X = \rho_0 \, \partial v/\partial t, \tag{3.246}$$

$$\partial v/\partial X = \partial e/\partial t, \tag{3.247}$$

augmented by an appropriate state law which, in homentropic conditions, adopts the form

$$p = -\sum(e). \tag{3.248}$$

Here p, ρ are the gas pressure and material density, $e = (\rho_0/\rho) - 1$ is the stretch, while $v(X, t)$ represents the material velocity; ρ_0 is the density of the gas in its reference state.

Substitution of the constitutive law (3.248) into (3.246) yields

$$A^2(e)\, \partial e/\partial X = \partial v/\partial t, \tag{3.249}$$

where

$$A(e) = (-\rho_0^{-1}\, dp/de)^{1/2} \tag{3.250}$$

is the Lagrangian signal speed. The governing equations now become

$$\begin{bmatrix} e \\ v \end{bmatrix}_X = \begin{bmatrix} 0 & 1/A^2(e) \\ 1 & 0 \end{bmatrix} \begin{bmatrix} e \\ v \end{bmatrix}_t, \tag{3.251}$$

and under the hodograph transformation wherein X, t are taken as the new dependent variables and v, e are the new independent variables, then subject to the requirement

$$0 < |J(v, e; X, t)| < \infty, \tag{3.252}$$

(3.251) reduces to the *linear* matrix system

$$\begin{bmatrix} X \\ t \end{bmatrix}_c = \begin{bmatrix} 0 & A(c) \\ A(c)^{-1} & 0 \end{bmatrix} \begin{bmatrix} X \\ t \end{bmatrix}_v, \tag{3.253}$$

where

$$c = \int_0^e A(\tau)\, d\tau. \tag{3.254}$$

The Loewner-type Bäcklund transformations may now be introduced as in Section 3.7 to achieve the reduction of (3.253) to the hyperbolic canonical form

$$\begin{bmatrix} X' \\ t' \end{bmatrix}_c = \begin{bmatrix} 0 & 1 \\ 1 & 0 \end{bmatrix} \begin{bmatrix} X' \\ t' \end{bmatrix}_v, \tag{3.255}$$

for certain multiparameter gas laws. The latter may be used for both local and global approximation to the adiabatic state relation

$$p = \rho_0 A_0^2 \gamma^{-1}[(1 + e)^{-\gamma} - 1]. \tag{3.256}$$

It turns out (Cekirge and Varley [85]) that the most appropriate such model gas law for the approximation of (3.256) in the range

$$0.1\rho_0 \leq \rho \leq \rho_0 \tag{3.257}$$

is that in which the Lagrangian sound speed adopts the form

$$A/A_0 = \theta_2 \tan^2\{\theta_0 + \theta_1 c/A_0\}, \tag{3.258}$$

where θ_i, $i = 1, 2, 3$, are arbitrary constants. Use of relation (3.250) shows

3.10 BÄCKLUND TRANSFORMATIONS IN LAGRANGIAN GASDYNAMICS

that the model (p, e) laws associated with (3.258) are given parametrically by

$$\frac{p}{\rho_0 A_0^2} = p_0 + \theta_1^{-1}\theta_2 \left[\frac{\theta_1 c}{A_0} - \tan\left(\theta_0 + \frac{\theta_1 c}{A_0}\right) \right], \quad (3.259)$$

$$e = e_0 - (\theta_1 \theta_2)^{-1} \left[\cot\left(\theta_0 + \frac{\theta_1 c}{A_0}\right) + \frac{\theta_1 c}{A_0} \right], \quad (3.260)$$

where p_0, e_0 are further arbitrary constants. In Fig. 3.3, the variation of $T = -p$ with e is compared for the real adiabatic gas law (3.256) and the model laws (3.259)–(3.260) with appropriate choice of the available parameters to give alignment over the range $0.1\rho_0 \leq \rho \leq \rho_0$. A least squares approximation is employed and detailed comparison tables are given by Cekirge and Varley [85].

We now present a summary of the work by Cekirge and Varley on the application of the model laws (3.259)–(3.260) to the analysis of the pressure variation at the closed end of a shock tube during the reflection of a centered simple wave.

The situation envisaged is one in which an adiabatic gas with constitutive law of type (3.256) is contained at high pressure between the closed end of a shock tube and a diaphragm at $X = D$. This membrane is suddenly removed, whereupon the initial discontinuity at the interface between the high pressure region and the external atmosphere divides into two waves. Thus, a constant strength shock wave moves away from the closed end of the tube into the quiet atmosphere while a centered rarefaction wave moves backward over the high pressure gas. Once this disturbance reaches the rigid end

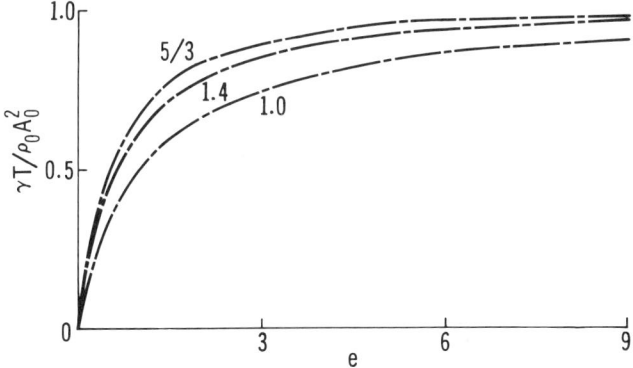

Fig. 3.3 Comparison of the exact stress–strain relations for gases with isentropic exponents $\gamma = 1$, 1.4, and $\tfrac{5}{3}$ with model stress–strain relations. The strain e varies over the range $0 \leq e \leq 9$, which corresponds to a density variation $0.1\rho_0 \leq \rho \leq \rho_0$. —— denotes model equation; – – denotes Law $p/p_0 = (\rho/\rho_0)^\gamma$ (Cekirge and Varley [85]).

$X = 0$, it undergoes total reflection as an expansion wave and proceeds to catch up with the contact discontinuity represented by $x = x(D, t)$ in the t-x plane. At this interface, the rarefaction wave is partially reflected and partially transmitted. The transmitted part subsequently moves as a simple wave until it, in turn, catches up with the shock, whereupon again partial reflection and transmission take place. The Cekirge–Varley analysis presented here is valid only until such time as shock formation occurs in the region $0 \leq X \leq D$. It deals specifically with the initial centered simple wave region and the subsequent interaction régime that occurs on the reflection of the disturbance at $X = 0$.

It is convenient to proceed in terms of the Riemann formulation. Thus, on return to the canonical form (3.255), it is seen that

$$X' = M(f) + N(g), \tag{3.261}$$

$$t' = -M(f) + N(g), \tag{3.262}$$

where M, N are arbitrary functions of the Riemann invariants

$$f = (c - v)/2, \tag{3.263}$$

$$g = (c + v)/2. \tag{3.264}$$

From (3.251), it follows that

$$f_t + A(c) f_X = 0, \tag{3.265}$$

and

$$g_t - A(c) g_X = 0, \tag{3.266}$$

whence, f is constant on the α-characteristic wavelets given by

$$\left. \frac{dX}{dt} \right|_\alpha = A(c), \tag{3.267}$$

while g is constant on the β-characteristic wavelets given by

$$\left. \frac{dX}{dt} \right|_\beta = -A(c). \tag{3.268}$$

Accordingly,

$$f = F(\alpha), \tag{3.269}$$

$$g = G(\beta), \tag{3.270}$$

where, for convenience, the characteristic wavelets of the α- and β-waves are subsequently labeled according to

$$\alpha = t \quad \text{at} \quad X = 0, \tag{3.271}$$

$$\beta = t \quad \text{at} \quad X = D. \tag{3.272}$$

3.10 BÄCKLUND TRANSFORMATIONS IN LAGRANGIAN GASDYNAMICS

Note that relations (3.263)–(3.264) together with (3.269)–(3.270) allow c, v to be expressed in terms of $F(\alpha)$ and $G(\beta)$.

Insertion of expressions (3.261) and (3.262) into the original Bäcklund transformation allows the complete integration of the hodograph system (3.253). An important class of particular solutions is that in which

$$X = 1 - \phi(\alpha)A^{1/2} - \bar{M}\tau(\alpha), \qquad (3.273)$$

$$t = \phi(\alpha)A^{-1/2} + \tau(\alpha), \qquad (3.274)$$

where, in the notation of Section 3.7,

$$\bar{M} = -(\bar{\alpha}/\bar{\beta})\det \mathbf{A}, \qquad (3.275)$$

and $\phi(\alpha)$, $\tau(\alpha)$ are subject to the compatibility condition

$$d\tau/d\alpha = 2\bar{\beta}(\det \mathbf{A})^{-1/2}\phi(\alpha)\,dF/d\alpha. \qquad (3.276)$$

Kazakia and Varley [141] show that relations (3.273)–(3.274) are appropriate for the analysis of the early stages of evolution of a centered fan in a wide range of nonlinear continua. The procedure is outlined here for the shock tube problem alluded to above. A more detailed discussion of the method is given in the next chapter in the context of both nonlinear elasticity and nonlinear dielectrics.

When the membrane in the shock tube bursts, the centered wave so generated which travels toward $X = 0$ crosses, in turn, two distinct regions, I and II say. In region I there is a simple wave, with $F \equiv 0$, while in region II, the centered wave interacts with the wave reflected from $X = 0$, so that F is no longer identically zero. We now proceed to the analysis of the state variables in these two regions. In the sequel, as in [141], distance, time, and velocity are measured in units of D, D/A_0, and A_0, respectively.

In the reference state ahead of the advancing simple wave, $A = 1$ and the hodograph equations (3.253) integrate, subject to conditions (3.271), (3.272), to give

$$X = \tfrac{1}{2}\{1 - \alpha + \beta\}, \qquad (3.277)$$

$$t = \tfrac{1}{2}\{1 + \alpha + \beta\}. \qquad (3.278)$$

At the front of the centered wave, $\beta = 0$, whence,

$$\begin{aligned} X = \tfrac{1}{2}(1 - \alpha), &\quad t = \tfrac{1}{2}(1 + \alpha), \quad -1 \leq \alpha \leq 1, \\ A = 1, &\quad F = 0. \end{aligned} \qquad (3.279)$$

Insertion of this information into (3.273) and (3.274) shows that, in the incident simple wave region,

$$\phi = \tfrac{1}{2}(1 + \alpha), \quad \tau = 0, \quad -1 \leq \alpha \leq 1, \qquad (3.280)$$

so that
$$A = (1 - X)/t, \tag{3.281}$$
$$\alpha = 2[(1 - X)t]^{1/2} - 1, \quad -1 \le \alpha \le 1. \tag{3.282}$$
Further, since $F \equiv 0$ in region I, it follows from (3.263) and (3.269) that
$$v = c(A) = c[(1 - X)/t]. \tag{3.283}$$
The β-characteristics are given by
$$c(A) - F(\alpha) = G = \text{const}, \tag{3.284}$$
so that, in region I, they are the rays
$$(1 - X)/t = \text{const}. \tag{3.285}$$

The incident centered wave remains a simple wave until it begins to interact with the wave reflected from $X = 0$. Since $X = 0$ is a perfectly rigid boundary,
$$v = 0, \quad F = G = \frac{c}{2} \quad \text{at } X = 0, \tag{3.286}$$
while (3.273) and (3.274) show that
$$\phi = m(1 - \bar{M}\alpha)/(m^2 - \bar{M}), \tag{3.287}$$
$$\tau = (m^2\alpha - 1)/(m^2 - \bar{M}), \tag{3.288}$$
where
$$A^{1/2} = m(\alpha) \quad \text{at} \quad X = 0. \tag{3.289}$$

From (3.286) and (3.289), together with the model constitutive law (3.258), it is seen that
$$dm/d\alpha = A^{-1/2}(dA/dc)(dF/d\alpha) = [\mu + vA](dF/d\alpha), \tag{3.290}$$
where
$$A_0^{1/2}\mu = 2\theta_1\theta_2^{1/2}, \tag{3.291}$$
$$A_0^{3/2}v = 2\theta_1\theta_2^{-1/2}. \tag{3.292}$$

Substitution of the relations (3.287) and (3.288) into the compatibility condition (3.276) shows that $m(\alpha)$ is the solution of the initial-value problem
$$(1 - \bar{M}\alpha) \, dm/d\alpha + m(m^2 - \bar{M}) = 0, \quad m = 1 \quad \text{at} \quad \alpha = 1, \tag{3.293}$$
so that
$$m = \left[\frac{\bar{M}(\bar{M} - 1)}{(\bar{M}\alpha - 1)^2 + \bar{M} - 1}\right]^{1/2}. \tag{3.294}$$

3.10 BÄCKLUND TRANSFORMATIONS IN LAGRANGIAN GASDYNAMICS

Insertion of the expression (3.294) into (3.287) and (3.288) now generates $\phi(\alpha)$ and $\tau(\alpha)$, whence, from (3.273) and (3.274), the Lagrangian sound speed $A(X, t)$ is obtained explicitly in the form

$$A = \bar{M}(X^2 + \bar{M} - 1)/[(\bar{M}t - 1)^2 + \bar{M} - 1]. \quad (3.295)$$

At $X = 0$, (3.295) together with the constitutive law (3.258) yields

$$t = \bar{M}^{-1}[1 - (1 - \bar{M})^{1/2} \csc\{\theta_0 + \theta_1 c\}], \quad (3.296)$$

and if, as in [85], we set $\bar{M} = -\cot^2 \theta_0$ and revert to the original variables, we retrieve the relation

$$\frac{A_0 t}{D} = 1 + \sec^2 \theta_0 \left\{ \frac{\sin \theta_0}{\sin(\theta_0 + (\theta_1 c / A_0))} - 1 \right\} \quad (3.297)$$

obtained by Cekirge and Varley but in a different manner.

The temporal variation of the pressure p and stretch e at $X = 0$ is now obtained parametrically through relations (3.259), (3.260), and (3.297). In Fig. 3.4, the variation of p/p_0 with $A_0 t/(D - X)$ in the incident centered wave is compared with its variation with $A_0 t/D$ at the end of the shock tube [85]. The available constants in (3.259) and (3.260) have been chosen so that there is alignment at reference conditions, that is,

$$A = A_0, \quad T = e = 0 \quad \text{at} \quad c = 0.$$

Comparison of the pressure variation results with those obtained by Owczarek [156] by means of a graphical procedure are shown. Agreement is seen to be excellent.

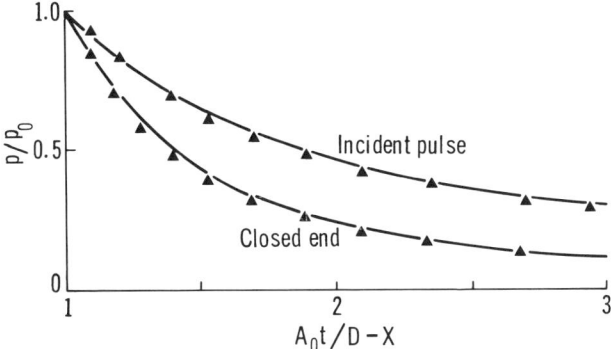

Fig. 3.4 The predicted pressure variation at the closed end of a shock tube during the reflection of a centered simple wave compared with that calculated by Owczarek (1964). ——— denotes theoretical prediction; ▲ denotes numerical values computed by Owczarek (Cekirge and Varley [85]).

In conclusion, we remark that Mortell and Seymour [144] have recently employed the Cekirge–Varley model gas laws to analyze the propagation of disturbances generated by an oscillating piston at one end of a shock tube, the other end of which is closed. The reader is referred to that work for details. In the next chapter, we shall see that the model state laws approach has, in fact, wide applicability outside gasdynamics. Thus, this approach will be used to solve important problems in both nonlinear elasticity and nonlinear dielectrics.

CHAPTER **4**

Bäcklund Transformations and Wave Propagation in Nonlinear Elastic and Nonlinear Dielectric Media

4.1 PROPAGATION OF LARGE AMPLITUDE WAVES IN NONLINEAR ELASTIC MEDIA. REDUCTION TO CANONICAL FORM OF THE RIEMANN REPRESENTATION

In an extensive analysis of the one-dimensional propagation of large amplitude, longitudinal disturbances in nonlinear elastic media of finite extent, Cekirge and Varley [85] observed that notable analytic progress can be made for a distinctive class of model constitutive laws. Thus, for these materials, the Riemann characteristic formulation allows the exact solution of important problems involving the interaction of large amplitude elastic waves. In particular, one such model law was shown to provide a reasonable approximation to the behavior of saturated soil, dry sand, and clay silt under dynamic compression, while another was employed to describe a class of hard elastic materials. The approach developed in [85] has been subsequently applied in a variety of areas by Kazakia and Varley [141, 142],

Cekirge [143], Kazakia and Venkataraman [94], and Mortell and Seymour [144].

In a separate development (Rogers [86]), it was shown that, in fact, Riemann's characteristic equations may be integrated generally for the Cekirge–Varley constitutive laws by means of a Bäcklund transformation. Further, that this approach has more general application was indicated by Rogers et al. [95] in the context of electromagnetic wave propagation through nonlinear dielectric media. Links may, in turn, be established between the Bäcklund transformation approach and not only the Karal–Keller wave front expansion method but also the Bergman integral operator formalism.

We begin with a description of the reduction to canonical form of the characteristic equations for a class of important model constitutive laws. The relevant background material on the dynamic response of nonlinear elastic materials is to be found, for example, in the work of Eringen and Suhubi [145].

In the finite deformation of an elastic body wherein the material points $\mathbf{X}$ move to new points $\mathbf{x}$ according to

$$\mathbf{x} = \mathbf{x}(\mathbf{X}, t), \tag{4.1}$$

the extension e is defined by

$$e := \{ds^* - dS^*\}/dS^*, \tag{4.2}$$

where

$$dS^{*2} = dX_k dX_k = c_{kl} dx_k dx_l, \tag{4.3}$$

$$ds^{*2} = dx_k dx_k = C_{kl} dX_k dX_l \tag{4.4}$$

and

$$c_{kl}(\mathbf{x}, t) = \delta_{KL} X_{K,k} X_{L,l}, \tag{4.5}$$

$$C_{KL}(\mathbf{X}, t) = \delta_{kl} x_{k,K} x_{l,L} \tag{4.6}$$

are termed the Cauchy and Green deformation tensors, respectively. In (4.5) and (4.6), the usual notation

$$X_{K,k} = \frac{\partial X_K}{\partial x_k}, \qquad x_{k,K} = \frac{\partial x_k}{\partial X_K},$$

is adopted.

The Lagrangian and Eulerian strain tensors E_{KL} and e_{kl} are introduced in turn through the relations

$$E_{KL} = \tfrac{1}{2}\{C_{KL} - \delta_{KL}\}, \tag{4.7}$$

$$e_{kl} = \tfrac{1}{2}\{\delta_{kl} - c_{kl}\}, \tag{4.8}$$

while the particle displacement $\mathbf{u}(\mathbf{X}, t)$ is given by

$$\mathbf{x} = \mathbf{X} + \mathbf{u}(\mathbf{X}, t). \tag{4.9}$$

4.1 PROPAGATION OF LARGE AMPLITUDE WAVES

In the Lagrangian formulation, the equations of motion become

$$\frac{\partial T_{ij}}{\partial X_j} = \rho_0 \frac{\partial^2 x_i}{\partial t^2}, \quad i,j = 1, 2, 3, \tag{4.10}$$

where T_{ij} denote the components of the stress tensor **T**, while ρ_0 determines the density of the medium in its undeformed state.

If the dynamic response of the elastic material is both isotropic and homogeneous with respect to the undeformed state, **T** is given in terms of the deformation gradients by a constitutive law

$$\mathbf{T} = \mathbf{T}\{x_{k,K}(\mathbf{X}, t)\}. \tag{4.11}$$

In the case of uniaxial deformation,

$$x = x(X, t), \tag{4.12}$$

$$T = T(\partial x/\partial X, t), \tag{4.13}$$

where the normal traction T per unit undeformed area on the plane $X = $ constant at time t is, by virtue of (4.9) and (4.10), related to its displacement $u(X, t)$ normal to the plane by

$$\partial T/\partial X = \rho_0 \partial^2 u/\partial t^2. \tag{4.14}$$

The material velocity $v(X, t)$ of the plane $X = $ const relative to the bounding plane $X = 0$ is given by

$$v = \partial x/\partial t, \tag{4.15}$$

whence, from (4.14),

$$\partial T/\partial X = \rho_0 \partial v/\partial t. \tag{4.16}$$

Furthermore, for the uniaxial deformation under consideration,

$$e = (\partial x/\partial X) - 1, \tag{4.17}$$

which relation, together with (4.15), provides the compatibility condition

$$\partial v/\partial X = \partial e/\partial t. \tag{4.18}$$

Thus, (4.16) and (4.18) represent the Lagrangian description of the deformation. According to (4.13) these equations are to be augmented by an appropriate constitutive law

$$T = \sum(e), \tag{4.19}$$

whence, on substitution into (4.16),

$$A^2(e) \partial e/\partial X = \partial v/\partial t, \tag{4.20}$$

where

$$A(e) = (\rho_0^{-1} d\Sigma/de)^{1/2} \qquad (4.21)$$

is the signal speed. Equations (4.18) and (4.20) describe the variations of the kinematic variables $e(X, t)$ and $v(X, t)$ which, once determined, lead through (4.15) or (4.17) and the specified side conditions to $x(X, t)$.

On introduction of the hodograph transformation, wherein X, t are taken as the new dependent variables and v, e are the new independent variables, we see that

$$\begin{aligned} v_X &= Jt_e, & v_t &= -JX_e, \\ e_X &= -Jt_v, & e_t &= JX_v, \end{aligned} \qquad (4.22)$$

where $J = J(v, e; X, t)$ and we require that

$$0 < |J| < \infty. \qquad (4.23)$$

Use of relations (4.22) in (4.18) and (4.20) produces the hodograph system

$$\Lambda_c = \mathbf{H}\Lambda_v, \qquad (4.24)$$

where

$$\Lambda = \begin{bmatrix} X \\ t \end{bmatrix}, \qquad (4.25)$$

$$\mathbf{H} = \begin{bmatrix} 0 & A(c) \\ A(c)^{-1} & 0 \end{bmatrix}, \qquad (4.26)$$

and c is the new strain measure given by

$$c = \int_0^e A(s)\,ds. \qquad (4.27)$$

The results of Section 3.7 may now be applied, *mutatis mutandis*, to system (4.24). Thus, the Bäcklund transformation

$$\begin{aligned} \Lambda'_c &= \mathbf{A}\Lambda_c + \mathbf{H}'\tilde{\mathbf{C}}\Lambda, \\ \Lambda'_v &= \mathbf{A}\Lambda_v + \tilde{\mathbf{C}}\Lambda, \\ c' &= c, \qquad v' = v, \end{aligned} \qquad (4.28)$$

with

$$\mathbf{A} = \begin{bmatrix} a_1^1 & 0 \\ 0 & a_2^2 \end{bmatrix}, \qquad (4.29)$$

$$\tilde{\mathbf{C}} = \begin{bmatrix} 0 & c_2^1 \\ c_1^2 & 0 \end{bmatrix}, \qquad (4.30)$$

$$\mathbf{H}' = \begin{bmatrix} 0 & 1 \\ 1 & 0 \end{bmatrix}, \qquad (4.31)$$

4.1 PROPAGATION OF LARGE AMPLITUDE WAVES

reduces (4.24) to a canonical form associated with the classical wave equation provided

$$\mathbf{H} = \mathbf{A}^{-1}\mathbf{H}'\mathbf{A}, \tag{4.32}$$

where

$$a^1_{1,c} + \bar{\alpha}(a^1_1)^2 + \bar{\beta} = 0, \qquad a^1_1 = [(\det \mathbf{A})A(c)^{-1}]^{1/2}, \tag{4.33}$$

and $\bar{\alpha}$, $\bar{\beta}$ are the constants introduced in (3.164)–(3.165). It is recalled that $\det \mathbf{A}$ and the c^i_j are constants.

Consequently, reduction to canonical form via the Bäcklund transformation (4.28) is available when

(a) $\bar{\alpha} = (\det \mathbf{A})^{-1} c^1_2 = 0$,

$$A(c) = (\det \mathbf{A})/[-\bar{\beta}c + \delta]^2, \tag{4.34}$$

(b) $\bar{\beta} = -c_1^2 = 0$,

$$A(c) = (\det \mathbf{A})[\bar{\alpha}c + \varepsilon]^2, \tag{4.35}$$

(c) $\bar{\beta}/\bar{\alpha} > 0$,

$$A(c) = (\bar{\alpha}/\bar{\beta}) \det \mathbf{A} \cot^2\{(\bar{\beta}/\bar{\alpha})^{1/2}(-\bar{\alpha}c + \zeta)\}, \tag{4.36}$$

(d) $\bar{\beta}/\bar{\alpha} < 0$,

$$A(c) = (-\bar{\alpha}/\bar{\beta}) \det \mathbf{A} \tanh^2\{(-\bar{\beta}/\bar{\alpha})^{1/2}(\bar{\alpha}c + \eta)\}. \tag{4.37}$$

Here, δ, ε, ζ, η are arbitrary real constants.

The canonical form

$$\mathbf{\Lambda}'_c = \mathbf{H}'\mathbf{\Lambda}'_v, \qquad \mathbf{H}' = \begin{bmatrix} 0 & 1 \\ 1 & 0 \end{bmatrix} \tag{4.38}$$

yields

$$X'_c = t'_v, \qquad t'_c = X'_v, \tag{4.39}$$

so that

$$X' = N(r) + M(s), \qquad t' = N(r) - M(s), \tag{4.40}$$

where

$$r = \tfrac{1}{2}(c + v), \qquad s = \tfrac{1}{2}(c - v), \tag{4.41}$$

are the Rieman invariants.

The application of the preceding Bäcklund transformations to the analysis of specific initial-boundary-value problems is taken up in Section 4.3 following a discussion of the model state laws implicit in the Riccati equation (4.33).

4.2 THE MODEL CONSTITUTIVE LAWS

In the preceding section, it was shown that the characteristic equations are reducible to the canonical form (4.38) by a single Bäcklund transformation subject to the requirement (4.33), that is,

$$A_c = \mu A^{1/2} + \nu A^{3/2}, \tag{4.42}$$

where

$$\mu = 2\bar{\alpha}(\det \mathbf{A})^{1/2}, \quad \nu = 2\bar{\beta}(\det \mathbf{A})^{-1/2}. \tag{4.43}$$

Nonlinear elastic media characterized by the relation (4.42) have been investigated by Cekirge and Varley [85]. In particular, it was shown that the associated model constitutive laws can be used to approximate locally any given stress–strain relation in some vicinity of $(t, e) = (0,0)$ to within an error of $O(e^4)$. It was further shown that Bell's parabolic law may be approximated by one of the model laws to an error of less than 1% over the range of strain in which Bell's law is well supported by empirical data. These and other aspects of the model laws are summarized in the sequel. The interested reader should consult the work of Kazakia and Varley [141] for application to other materials.

Reduction to the same canonical form is available for more general constitutive laws by iteration of the Bäcklund transformations. However, it will be seen that the model laws based simply on (4.42) have a remarkable versatility although they can only approximate the response of materials over ranges in which the Lagrangian sound speed A has either a monotonically increasing or decreasing dependence on e. Thus, the laws cannot be used to approximate stress–strain relations with a point of inflexion. Such laws occur, for instance, in the axial compression of polycrystalline materials.

The model constitutive laws determined by the relation (4.42) are now described in detail based on the subdivision (a)–(d) of the preceding section.

(a) Relations (4.19), (4.21), and (4.27) together show that materials with a Lagrangian sound speed A given by (4.34) have a nonlinear stress–strain law of the type

$$T = \bar{\lambda}[e + \bar{\mu}]^{-1/3} + \bar{\nu}, \tag{4.44}$$

where $\bar{\lambda}$, $\bar{\mu}$, and $\bar{\nu}$ are material parameters available for approximation purposes.

In the case of an *ideally hard material*, the Lagrangian sound speed increases monotonically without bound. If the material is ideally hard in tension, the strain e can never exceed a limiting value e_1, however large T. Likewise, if the material is ideally hard in compression, however large

4.2 THE MODEL CONSTITUTIVE LAWS

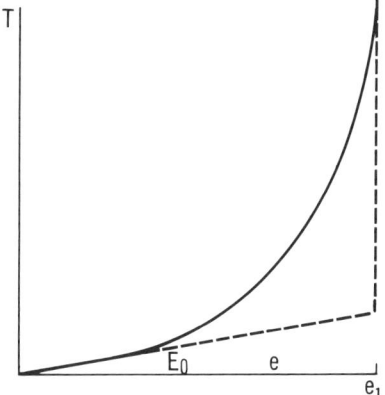

Fig. 4.1 A typical stress–strain relation for an ideally hard elastic material.

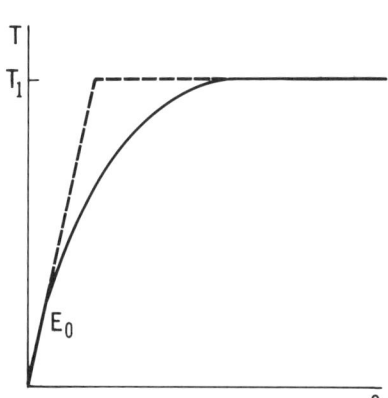

Fig. 4.2 A typical stress–strain relation for an ideally soft elastic material.

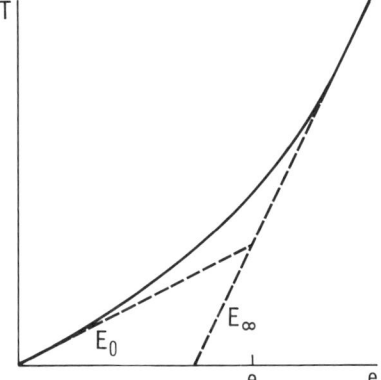

Fig. 4.3 A typical stress–strain relation for a hard elastic material.

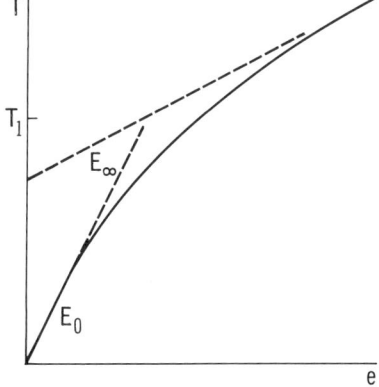

Fig. 4.4 A typical stress–strain relation for a soft elastic material.[†]

$-T$, the quantity $-e$ can never exceed some limiting value $-e_1$. Typical behavior of an ideally hard material is illustrated in Fig. 4.1.

During dynamic compression, fully saturated soil, dry sand, and clay silt all harden in that the local Young's modulus increases with the loading until the material becomes almost rigid. Dry sand and clay silt do not, in general, adhere to the same stress–strain law in unloading as they do in loading. On the other hand, in the case of saturated soils, according to Critescu [146], provided there is no seepage in unloading so that the soil remains fully saturated, hysteresis can be neglected.

[†] Figs. 4.1 through 4.4 are from Cekirge and Varley [85].

In [85], stress–strain laws of the type (4.44) were used to model the behavior of hard materials under dynamic compression. To this end, the parameters $\bar{\lambda}$, $\bar{\mu}$, and $\bar{v}$ in (4.44) were set by the conditions

$$T|_{e=0} = 0, \quad dT/de|_{e=0} = E_0 = \rho_0 A_0^2, \quad dT/de \to \infty \quad \text{as} \quad e \to e_1. \quad (4.45)$$

Here, A_0 is the value of the Lagrangian sound speed in the medium at $e = 0$, E_0 is the local Young's modulus at $e = 0$, while e_1 is the locking strain which e asymptotically approaches with increasing stress. Accordingly, (4.44) adopts the form

$$T/(\rho_0 A_0^2 e_1) = 3[(1 - e/e_1)^{-1/3} - 1]. \quad (4.46)$$

Comparison of this model law with the experimental stress–strain laws corresponding to dynamic compression of saturated soil, dry sand, and clay silt is illustrated in Fig. 4.5. It is observed that the model represents a decided improvement on the linear elastic-perfectly hard model commonly adopted for locking materials. Cekirge and Varley [85] undertook the analysis of the decay of a pulse in a layer of fully saturated soil bounded by sea water above and by rock below. The pulse was assumed to originate at a depth which was large compared with this layer so that the pulse was regarded as essentially plane on crossing the soil–rock boundary. The dynamic response of the soil to the disturbance generated by the transmitted

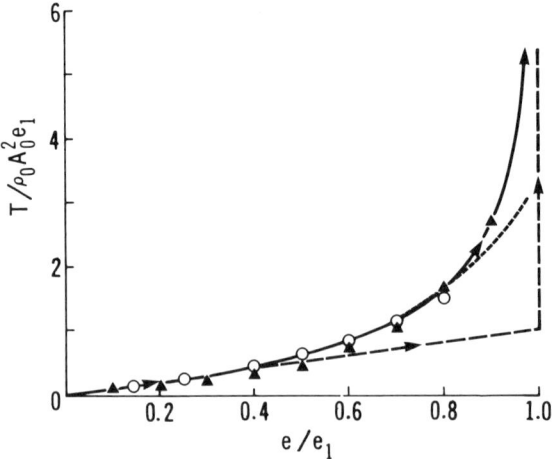

Fig. 4.5 Comparison of the experimental stress–strain relations during the dynamic compression of saturated soil, dry sand, and clay silt with the theoretical law $T/(\rho_0 A_0^2 e_1) = 3[(1 - e/e_1)^{-1/3} - 1]$. ——, Theoretical law; ▲, dry sand (Allen et al., 1957); ○, saturated soil (Liahov, 1964); . . ., clay silt (Ginsburg 1964); --, linear elastic, perfectly hard material (Cekirge and Varley [85]).

4.2 THE MODEL CONSTITUTIVE LAWS

energy at this interface was modeled by a constitutive relation of the type (4.46).

(b) If the Lagrangian sound speed is given by the expression (4.35), then the corresponding model stress–strain law is of the form

$$T = \bar{\lambda}[e + \bar{\mu}]^{-3} + \bar{v}, \tag{4.47}$$

where again, $\bar{\lambda}$, $\bar{\mu}$, and $\bar{v}$ denote parameters available for the approximation of empirical constitutive laws. In particular, the imposition of the conditions

$$T|_{e=0} = 0, \quad dT/de|_{e=0} = E_0 = \rho_0 A_0^2, \quad dT/de \to 0 \text{ as } T \to T_1, \tag{4.48}$$

leads to the model law

$$T/T_1 = 1 - (1 + \rho_0 A_0^2 e/3T_1)^{-3}. \tag{4.49}$$

This stress–strain relation may be used to model *ideally soft materials*, that is, media in which, for all practical purposes, the Lagrangian sound speed decreases monotonically to zero as e increases without bound. For those such materials which soften in extension, no matter how large e becomes, the traction T never exceeds the limiting value T_1. Likewise, for such materials which soften in compression, however large $-e$, the quantity $-T$ can never exceed $-T_1$. A typical stress–strain law for an ideally soft elastic material is shown in Fig. 4.2. Recall that during expansion a gas exhibits ideally soft behavior. In that case, $T_1 = p_0$, the pressure in the reference state.

(c) By virtue of (4.36), or equivalently, on appropriate choice of $\bar{\zeta}$,

$$A(c) = (\bar{\alpha}/\bar{\beta}) \det \mathbf{A} \tan^2[(\bar{\beta}/\bar{\alpha})^{1/2}(\bar{\alpha}c + \bar{\zeta})], \tag{4.50}$$

if we introduce θ_i, $i = 0, 1, 2$, according to

$$\theta_0 = (\bar{\beta}/\bar{\alpha})^{1/2}\bar{\zeta}, \quad \theta_1 = (\bar{\alpha}\bar{\beta})^{1/2} A_0, \quad \theta_2 = A_0^{-1}(\bar{\alpha}/\bar{\beta}) \det \mathbf{A}, \tag{4.51}$$

then relations (4.19), (4.21), and (4.27) together lead to the model stress–strain laws in the parametric form presented in [85], namely,

$$T/(\rho_0 A_0^2) = t_0 + \theta_1^{-1}\theta_2[\tan(\theta_0 + \theta_1 c/A_0) - \theta_1 c/A_0], \tag{4.52}$$

$$e = e_0 - (\theta_1\theta_2)^{-1}[\cot(\theta_0 + \theta_1 c/A_0) + \theta_1 c/A_0], \tag{4.53}$$

where t_0, e_0 are arbitrary constants of integration.

If it is required that there be alignment at the reference state $c = 0$, so that

$$A|_{c=0} = A_0, \quad T|_{c=0} = 0, \quad e|_{c=0} = 0, \tag{4.54}$$

then

$$\theta_2 = \cot^2\theta_0, \quad t_0 = -\theta_1^{-1}\cot\theta_0, \quad e_0 = \theta_1^{-1}\tan\theta_0. \tag{4.55}$$

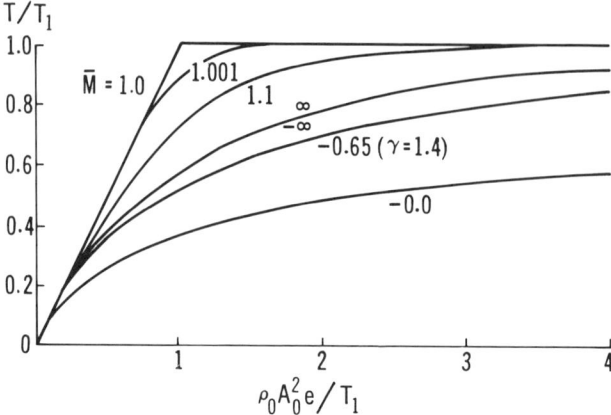

Fig. 4.6 Typical variations of T/T_1 with $\rho_0 A_0^2 e/T_1$ for ideally soft elastic materials. The case $\bar{M} = -0.065$ corresponds to a gas with isentropic exponent $\gamma = 1.4$. (Cekirge and Varley [85]).

The constitutive laws given by (4.52) and (4.53) may be employed to model a class of ideally soft materials parametrized in terms of

$$\bar{M} = -\mu/(A_0 v) = -\cot^2 \theta_0, \quad 0 < \theta_0 \le \pi/2, \quad (4.56)$$

with variation in the range $-\infty < \bar{M} \le 0$. In the limit as $\bar{M} \to -\infty$, the law (4.49) is retrieved. This limiting case together with the stress–strain laws corresponding to $\bar{M} = -0.65$ and $\bar{M} \to 0$ are illustrated in Fig. 4.6.

Finally, as $\bar{\theta}_0 = \theta_0 - \pi/2$ and $\bar{c} = \theta_1 c/A_0$ vary in the ranges

$$0 \le \bar{\theta}_0 \le \pi/2, \quad -\bar{\theta}_0 \le \bar{c} \le 0, \quad (4.57)$$

the parameter $\bar{M} = -\mu/(A_0 v) = -\tan^2 \bar{\theta}_0$ varies in the range $-\infty \le \bar{M} \le 0$ and a class of ideally hard materials is obtained. The stress–strain laws corresponding to $\bar{M} = -10$ and $\bar{M} \to 0, \bar{M} \to -\infty$ are shown in Fig. 4.7. The limiting case $\bar{M} \to 0$ produces the locking material stress–strain law (4.46).

(d) The Lagrangian sound speed (4.37) may be rewritten in accordance with the notation of [85] as

$$\frac{A}{A_0} = \bar{M}\left(\frac{1 + \bar{M}^{1/2} \tanh \bar{c}}{\bar{M}^{1/2} + \tanh \bar{c}}\right)^2, \quad (4.58)$$

where

$$\mu = 2A_\infty A_0^{-1/2}/[(A_0 + A_\infty)e_1], \quad v = -2A_0^{-1/2}/[(A_0 + A_\infty)e_1],$$
$$\bar{c} = \bar{M}^{1/2} c/[(1 + \bar{M})e_1 A_0], \quad T_1 = E_0 e_1 = \rho_0 A_0^2 e_1. \quad (4.59)$$

Here, $e = e_1$ is the strain corresponding to the point at which the tangents to the stress–strain curve at $e = 0$ and $|e| = \infty$ intersect (see Figs. 4.1–4.4).

4.2 THE MODEL CONSTITUTIVE LAWS

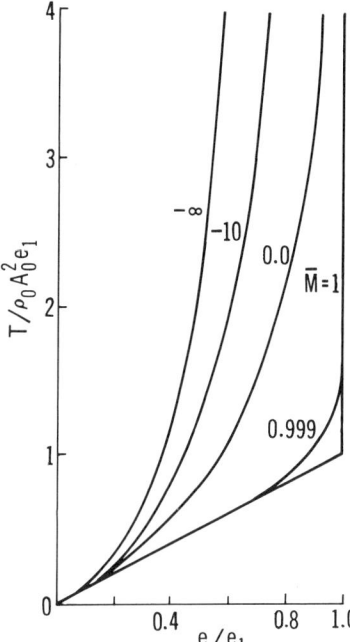

Fig. 4.7 Typical variations of $T/\rho_0 A_0^2 e_1$ with e/e_1 for ideally hard elastic materials (Cekirge and Varley [85]).

The model constitutive laws corresponding to (4.58) are given parametrically by

$$\frac{T}{T_1} = (1 + \bar{M})\left[1 + \bar{M}^{1/2}\left(\bar{c} - \frac{1 + \bar{M}^{1/2}\tanh\bar{c}}{\bar{M}^{1/2} + \tanh\bar{c}}\right)\right], \quad (4.60)$$

$$\frac{e}{e_1} = (1 + \bar{M}^{-1})\left[1 + \bar{M}^{-1/2}\left(\bar{c} - \frac{\bar{M}^{1/2} + \tanh\bar{c}}{1 + \bar{M}^{1/2}\tanh\bar{c}}\right)\right]. \quad (4.61)$$

Bell [147] has shown that during dynamic uniaxial compression many polycrystalline solids adhere to a stress–strain law which can be approximated by

$$T/T_M = (e_E/e_M)^{1/2}, \quad (4.62)$$

outside the neighborhood of $e_E = 0$. Here, the subscripts E and M refer, respectively, to experimental and model state variables, while $T_M(<0)$ and $e_M(<0)$ designate any two values of stress and strain that occur simultaneously. To be specific, $-T_M$ may be taken as the maximum compressive stress and $-e_M$ as the maximum compressive strain that occur during the deformation.

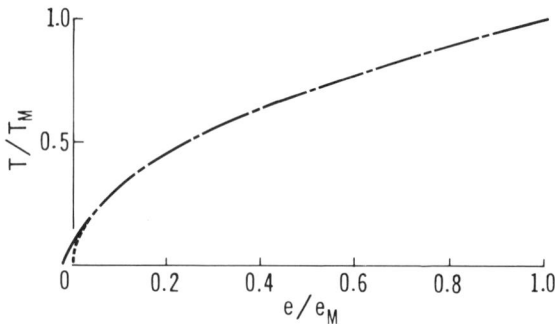

Fig. 4.8 Comparison of Bell's parabolic stress–strain relation with a model stress–strain relation. ——, Model equation; - - -, Bell's parabolic law (Cekirge and Varley [85]).

Bell's law may be rewritten in the form

$$-T = 2\rho_0\bar{\gamma}^2(-e)^{1/2}, \tag{4.63}$$

where

$$T_M = -2\rho_0\bar{\gamma}^2|e_M|^{1/2}, \qquad A_M = \bar{\gamma}|e_M|^{-1/4}, \qquad c_M = -\tfrac{4}{3}\bar{\gamma}|e_M|^{3/4}. \tag{4.64}$$

In [85], the model laws associated with relations (4.60)–(4.61) were used to approximate the law (4.63) over the range $0.1c_M \leq c \leq c_M$. The results are displayed in Fig. 4.8. The maximum error in T and e under the approximation is in the neighborhood of $0.\dot{3}\%$.

The stress–strain laws given by (4.60) and (4.61) model nonideal elastic response as $\bar{c}$ varies in the range $0 \leq \bar{c} < \infty$. In Figs. 4.9 and 4.10, the $(T/T_1, e/e_1)$-relations implicit in these laws are illustrated for certain values of the parameter $\bar{M}$. Thus, Fig. 4.9 shows curves descriptive of the behavior of soft elastic materials. This class of medium is bounded by the type (a) law

$$T/T_1 = 1 - (1 + 3e/e_1)^{-1/3}, \tag{4.65}$$

corresponding to the specialization $\bar{M} = 0$ and by Hooke's law

$$T/T_1 = e/e_1, \tag{4.66}$$

associated with the case $\bar{M} = 1$. The curve $M = 0.341$ corresponds to the approximation to Bell's parabolic law shown previously in Fig. 4.8. By contrast, the curves shown in Fig. 4.10 correspond to cases when (4.60)–(4.61) model nonideal hard elastic behavior. This class of materials is bounded by those governed by Hooke's law (4.66) on the one hand and those with a type (b) stress–strain relation

$$T/T_1 = \tfrac{1}{3}[(1 - e/e_1)^{-3} - 1], \tag{4.67}$$

4.2 THE MODEL CONSTITUTIVE LAWS

on the other; the law (4.67) corresponds to the limiting case $\bar{M} \to +\infty$. As $\bar{M}$ and $\bar{c}$ vary in the ranges

$$1 \leq \bar{M} < \infty, \quad -\eta \leq \bar{c} < 0, \tag{4.68}$$

the laws generated by (4.60) and (4.61) describe ideally soft elastic behavior. Certain such stress–strain laws are presented in Fig. 4.6 along with other

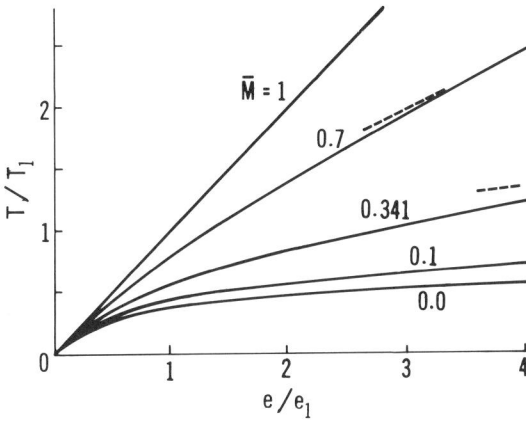

Fig. 4.9 Typical variations of T/T_1 with e/e_1 for soft elastic materials. The parameter $\bar{M} = A_\infty/A_0$. The case $\bar{M} = 0.341$ corresponds to materials which satisfy Bell's parabolic law (Cekirge and Varley [85]).

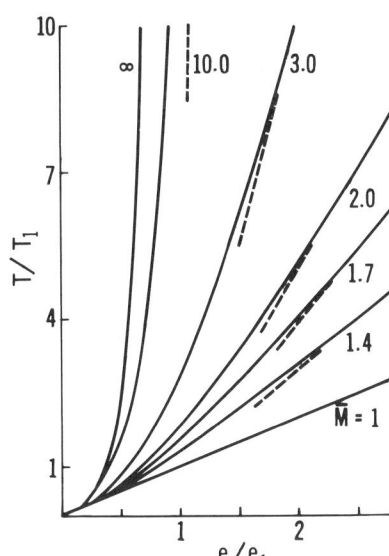

Fig. 4.10 Typical variations of T/T_1 with e/e_1 for hard elastic materials. The parameter $\bar{M} = A_\infty/A_0$ (Cekirge and Varley [85]).

ideally soft materials modeled by the relations (4.52)–(4.53). The limiting case $\bar{M} \to \infty$ produces the explicit stress–strain law (4.49).

If $\bar{M}$ and $\bar{c}$ vary in the ranges

$$0 \leq \bar{M} \leq 1, \quad -\eta_0 \leq \bar{c} \leq 0, \tag{4.69}$$

where

$$\tanh \eta_0 = \bar{M}^{1/2}, \tag{4.70}$$

we obtain a class of ideally hard materials, various examples of which are shown in Fig. 4.7 along with certain ideally hard elastic materials modeled by the relations (4.52)–(4.53). In the limit as $\bar{M} \to 1$, the linear elastic–perfectly hard response is obtained. On the other hand, as $\bar{M} \to 0$, the explicit stress–strain law (4.46) is retrieved, namely that used previously to approximate the response of clays and saturated soils.

The preceding discussion has been directed, in the main, at the application of the model laws (4.42) to the *global* matching of experimental constitutive laws. Should it be required to obtain a *local* fit at $e = 0$ to a stress–strain relation that in some neighborhood of $e = 0$ adopts the form

$$T = \rho_0 A_0^2 [e + pe^2 + qe^3 + O(e^4)], \tag{4.71}$$

then the prescriptions

$$T\big|_{e=0} = 0, \quad \frac{dT}{de}\bigg|_{e=0} = \rho_0 A_0^2,$$

$$\frac{d^2 T}{de^2}\bigg|_{e=0} = 2\rho_0 [\mu A_0^{5/2} + \nu A_0^{7/2}], \tag{4.72}$$

$$\frac{d^3 T}{de^3}\bigg|_{e=0} = \rho_0 [5\mu A_0^{5/2} + 7\nu A_0^{7/2}][\mu A_0^{1/2} + \nu A_0^{3/2}],$$

show that the constitutive law obtained by the integration of (4.42) subject to

$$\mu = \left(\frac{7p^2 - 6q}{2p}\right) A_0^{-1/2}, \quad \nu = \left(\frac{6q - 5p^2}{2p}\right) A_0^{-3/2}, \tag{4.73}$$

$$A(0) = A_0,$$

has Taylor series representation that, except when $p = 0$, agrees with the prescribed relation (4.71) up to terms $O(e^4)$. Accordingly, with the stated exception of the case with $p = 0$, the model constitutive laws implicit in (4.42) can be used to approximate locally, to within an error $O(e^4)$, any specified constitutive law at points where the stress can be expanded as a Taylor series in the strain. It is suggested in [85] that this local approach is appropriate in the investigation of small amplitude deformations involving weak shocks, resonant vibrations of crystals, or other small amplitude deformations where the effect of small local nonlinearities can lead to aggregate first order effects.

4.3 REFLECTION AND TRANSMISSION OF A LARGE AMPLITUDE SHOCKLESS PULSE AT A BOUNDARY

Here, we outline the application of the model laws derived in the preceding section to the analysis of the reflection and transmission of large amplitude longitudinal waves in a nonlinear elastic medium bounded by the material planes $X = 0$ and $X = D$ normal to the axis of stretch. These boundaries separate the enclosed nonlinear elastic medium from other elastic media stretched along the same axis. The surrounding elastic media, which are in general nonlinear, are assumed to be effectively semi-infinite in extent, inasmuch as their deformation at the interfaces with the enclosed medium is not significantly affected by any energy which may be reflected from their outer boundaries.

It is appropriate for the problem in hand to proceed in terms of a characteristic formulation. Thus, Equations (4.18) and (4.20) together with (4.27) show that

$$c_t - A(c)v_X = 0, \tag{4.74}$$
$$v_t - A(c)c_X = 0, \tag{4.75}$$

whence, in terms of the Riemann invariants r s,

$$r_t - A(c)r_X = 0, \tag{4.76}$$
$$s_t + A(c)s_X = 0. \tag{4.77}$$

Hence, s is constant along the right propagating characteristic wavelets $\alpha(X, t) = $ const defined by

$$dX/dt|_\alpha = +A(c), \tag{4.78}$$

while r is constant along the left propagating characteristic wavelets $\beta(X, t) = $ const defined by

$$dX/dt|_\beta = -A(c). \tag{4.79}$$

At the $\alpha = $ const characteristic wavelet,

$$\frac{dx}{dt} = \frac{\partial x}{\partial t}\bigg|_X + \frac{\partial x}{\partial X}\bigg|_t \frac{dX}{dt} = v + (e + 1)A(c) = v + a(c), \tag{4.80}$$

where $a(c)$ is the local speed of sound given by

$$a(c) = (1 + e)A. \tag{4.81}$$

Likewise, at the $\beta = \text{const}$ characteristic wavelet,

$$dx/dt = v - a. \tag{4.82}$$

Relations (4.76)–(4.79) show further that, in the notation of [85],

$$f = s = F(\alpha), \qquad g = r = G(\beta), \tag{4.83}$$

so that c, v may be expressed in terms of the *signal functions* F, G through the relations

$$c = G(\beta) + F(\alpha), \qquad v = G(\beta) - F(\alpha). \tag{4.84}$$

In what follows, it is convenient to label the α, β characteristic wavelets so that

$$\alpha = t \quad \text{at the material boundary} \quad X = 0, \tag{4.85}$$

$$\beta = t \quad \text{at the material boundary} \quad X = D. \tag{4.86}$$

Accordingly, from (4.83), this corresponds to the prescription

$$f = F(t) \quad \text{at} \quad X = 0, \tag{4.87}$$

$$g = G(t) \quad \text{at} \quad X = D. \tag{4.88}$$

The analysis of any specific deformation essentially consists of two parts. Thus, on the one hand, it is required to determine the form of the signal functions $F(\alpha)$ and $G(\beta)$ carried by the α-wave and β-wave components of the disturbance from prescribed initial and boundary conditions. On the other hand, in order to determine the state variables as functions of (X, t), it is necessary to determine $\alpha(X, t)$, $\beta(X, t)$.

For a medium which obeys Hooke's law,

$$T = T_0 + E_0 e, \tag{4.89}$$

where $E_0 > 0$ is Young's modulus, the Lagrangian sound speed reduces to

$$A = (E_0/\rho_0)^{1/2} = A_0, \tag{4.90}$$

while the characteristic equations (4.78), (4.79) yield

$$\alpha = \alpha(X - A_0 t), \qquad \beta = \beta(X + A_0 t). \tag{4.91}$$

In particular, use of the conditions (4.85)–(4.86) yields

$$\alpha = t - X/A_0, \qquad \beta = t + (X - D)/A_0. \tag{4.92}$$

4.3 REFLECTION AND TRANSMISSION OF A LARGE AMPLITUDE PULSE

Hence, in the present Lagrangian formulation, the relations (4.84) imply that any one-dimensional longitudinal deformation of the linear elastic medium (4.89) can be represented as the linear superposition of a nondistorting, nonattenuated progressing α-wave moving to the right and carrying a signal $F(\alpha)$ and a progressing β-wave moving to the left carrying the signal $G(\beta)$. The deformation analysis reduces to the derivation of the signal functions $F(\alpha)$, $G(\beta)$ carried by these waves from specified boundary and initial data. By virtue of (4.84) and (4.92), if v and c are prescribed at the boundaries $X = 0$ and $X = D$, the resolution of $F(\alpha)$ and $G(\beta)$ depends on the solution of simultaneous functional equations.

It was shown by Cekirge and Varley that it is possible to determine the current values of the state variables at the boundary $X = 0$ from the current value of g there and that, similarly, the values of the state variables at $X = D$ can be determined from the current values of f at that boundary. As a consequence, the temporal variation of the state variables at both boundaries can be determined from this analysis if the temporal variation of g at $X = 0$ and f at $X = D$ can be obtained. In general, such problems are not soluble analytically save for very special media undergoing very special deformations. However, for the class of nonlinear elastic media described in the preceding section, important boundary-value problems of this type can, in fact, be solved completely. The remainder of this section is devoted to the description of such a class of problems involving the reflection and transmission of large amplitude shockless pulses at an interface.

A large amplitude shockless pulse is assumed incident at the interface $X = 0$ having traversed a nonlinear elastic medium bounded by $X = 0$ and $X = D$. The extent of the slab is taken to be large compared to the width of the pulse. It is further assumed that until the front reaches $X = 0$, it is moving into an equilibrium configuration, so that the pulse acts as a simple wave with

$$f = F(\alpha) = 0, \qquad g = G(\beta) = v = c. \tag{4.93}$$

Once the pulse front reaches the boundary $X = 0$, it generates not only a reflected α-wave which moves toward the boundary $X = D$ but also a transmitted β-wave which moves into the adjoining medium (Fig. 4.11). The incident pulse, following arrival at $X = 0$, will begin to interact with the reflected α-wave. Accordingly, during this period of interaction, it is no longer a simple wave.

The disturbance which is transmitted into the surrounding medium M_L is, on the other hand, a simple wave. Thus, if the subscript L designates the value of state variables in M_L so that its constitutive law is

$$T_L = T_L(c_L), \tag{4.94}$$

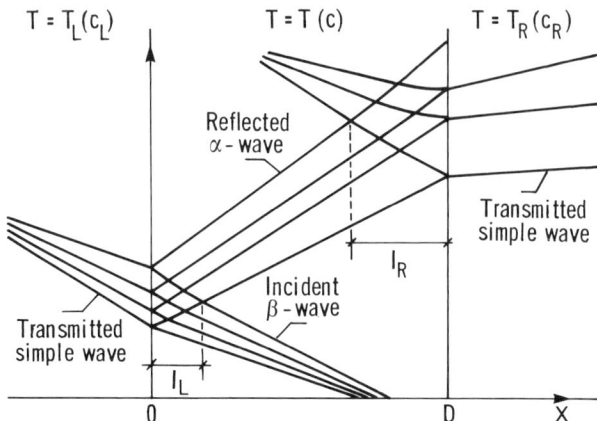

Fig. 4.11 The wave system that is set up when an incident pulse is partly reflected and partly transmitted at an interface between two elastic materials (Cekirge and Varley [85]).

it is seen that, in the transmitted wave, whatever the transmitted signal $G_L(\beta_L)$,

$$g_L = v_L = c_L. \tag{4.95}$$

In particular, relations (4.95) hold at the boundary $X = 0$; moreover, since the boundary is a material plane,

$$v = v_L, \quad T = T_L \quad \text{at} \quad X = 0. \tag{4.96}$$

Relations (4.94)–(4.96) imply that

$$T(c) = T_L(c_L) \quad \text{at} \quad X = 0, \tag{4.97}$$

where

$$\begin{aligned} c &= g + f, \\ c_L &= v_L = v = g - f. \end{aligned} \tag{4.98}$$

Combination of (4.97) and (4.98) provides an implicit relation for the *reflection function*

$$f = L(g) \quad \text{at} \quad X = 0, \tag{4.99}$$

where the *local reflection coefficient* $l(g)$ is given by

$$l(g) := L'(g) = [i(g) - 1]/[i(g) + 1], \tag{4.100}$$

and $i(g)$ is the *local impedance* of the boundary $X = 0$ given by

$$i(g) := \frac{dT_L}{dc_L} \bigg/ \frac{dT}{dc}. \tag{4.101}$$

4.3 REFLECTION AND TRANSMISSION OF A LARGE AMPLITUDE PULSE

Certain particular cases are of interest. Thus, if the interface is perfectly free, then regardless of the amplitude of the incident pulse,

$$c = 0 \quad \text{at} \quad X = 0, \tag{4.102}$$

so that, in view of (4.98)$_1$ and (4.99),

$$f = L = -g \quad \text{at} \quad X = 0, \tag{4.103}$$

corresponding to vanishing local impedance $i(g)$. On the other hand, if $X = 0$ represents a rigid boundary,

$$v = 0 \quad \text{at} \quad X = 0, \tag{4.104}$$

whence,

$$f = L = g \quad \text{at} \quad X = 0, \tag{4.105}$$

corresponding to infinite impedance $i(g)$. If $X = 0$ separates two media with a Hooke constitutive law, then both of the Lagrangian sound speeds A and A_L are constant so that the impedance is independent of g.

Given the constitutive laws of M_L and the bounded medium M, the variables f, c, v, and g_L are determined at $X = 0$ in terms of g by the relations

$$f = L(g), \quad c = g + L(g), \quad v = g - L(g) = g_L. \tag{4.106}$$

Once the reflected α-wave has traveled through the interaction region $X < I_L$ (see Fig. 4.11), it traverses a region where, prior to its arrival, the medium is in equilibrium. In this region, the reflected pulse is a simple wave in which

$$g = 0, \quad c = -v = f. \tag{4.107}$$

If the subscript R denotes the value of the state variables in the medium M_R occupying the region $X > D$ and if the constitutive law of the material is

$$T_R = T_R(c_R), \tag{4.108}$$

then

$$T(c) = T_R(c_R) \quad \text{at} \quad X = D, \tag{4.109}$$

where

$$c = g + f, \quad c_R = -v_R = -v = f - g. \tag{4.110}$$

Combination of (4.109) and (4.110) shows that

$$g = R(f) \quad \text{at} \quad X = D, \tag{4.111}$$

where the reflection coefficient $r(f)$ is defined by

$$r(f) := R'(f) = [j(f) - 1]/[j(f) + 1], \tag{4.112}$$

and the impedance $j(f)$ is given by

$$j(f) := \frac{dT_R}{dc_R} \bigg/ \frac{dT}{dc}. \tag{4.113}$$

The matter of the change of amplitude of the pulse following multiple reflections from the boundaries may now readily be handled after the manner of Cekirge and Varley.

We now turn to the application of the model constitutive laws. Thus, the Bäcklund transformation (4.28), together with the relations (4.83), yields

$$\begin{bmatrix} X' \\ t' \end{bmatrix}_\alpha = \begin{bmatrix} a_1^1 & 0 \\ 0 & a_2^2 \end{bmatrix} \begin{bmatrix} X \\ t \end{bmatrix}_\alpha + \dot{F}(\alpha) \begin{bmatrix} c_1^2 & -c_2^1 \\ -c_1^2 & c_2^1 \end{bmatrix} \begin{bmatrix} X \\ t \end{bmatrix}, \tag{4.114}$$

$$\begin{bmatrix} X' \\ t' \end{bmatrix}_\beta = \begin{bmatrix} a_1^1 & 0 \\ 0 & a_2^2 \end{bmatrix} \begin{bmatrix} X \\ t \end{bmatrix}_\beta + \dot{G}(\beta) \begin{bmatrix} c_1^2 & c_2^1 \\ c_1^2 & c_2^1 \end{bmatrix} \begin{bmatrix} X \\ t \end{bmatrix}. \tag{4.115}$$

In particular,

$$t'_\alpha = a_2^2 t_\alpha + \dot{F}(\alpha)[-c_1^2 X + c_2^1 t],$$
$$t'_\beta = a_2^2 t_\beta + \dot{G}(\beta)[c_1^2 X + c_2^1 t],$$

or in view of (4.40), (4.43), and the relation $a_2^2 = [A(c) \det \mathbf{A}]^{1/2}$,

$$2A^{1/2} t_\alpha + [\mu(t - \alpha) + vX]\dot{F}(\alpha) = m(\alpha), \tag{4.116}$$

$$2A^{1/2} t_\beta + [\mu(t - \beta) + v(D - X)]\dot{G}(\beta) = n(\beta), \tag{4.117}$$

where

$$m(\alpha) := \dot{F}(\alpha)[-2\dot{M}(s)(\det \mathbf{A})^{-1/2} - \mu\alpha], \tag{4.118}$$

$$n(\beta) := \dot{G}(\beta)[2\dot{N}(r)(\det \mathbf{A})^{-1/2} - \mu\beta + vD]. \tag{4.119}$$

Relations (4.116) and (4.117) are in accord with those derived by Cekirge and Varley [85]. Moreover, since

$$dt/d\alpha = \partial t/\partial\alpha + (d\beta/d\alpha)(\partial t/\partial\beta)$$
$$= \partial t/\partial\alpha + (d\beta/d\alpha)A^{-1}(\partial X/\partial\beta)$$
$$= \partial t/\partial\alpha - A^{-1} \partial X/\partial\alpha = 2\partial t/\partial\alpha \quad \text{at fixed } X,$$

and similarly,

$$dt/d\beta = 2\partial t/\partial\beta \quad \text{at fixed } X,$$

it follows that, at any constant X, (4.116) and (4.117) reduce to the pair of *ordinary* nonlinear differential equations

$$A^{1/2} dt/d\alpha + [\mu(t - \alpha) + vX]\dot{F}(\alpha) = m(\alpha), \tag{4.120}$$

$$A^{1/2} dt/d\beta + [\mu(t - \beta) + v(D - X)]\dot{G}(\beta) = n(\beta). \tag{4.121}$$

4.3 REFLECTION AND TRANSMISSION OF A LARGE AMPLITUDE PULSE

It is observed that, in view of the labeling in (4.85) and (4.86),

$$m(t) = A^{1/2} \quad \text{at} \quad X = 0, \tag{4.122}$$

$$n(t) = A^{1/2} \quad \text{at} \quad X = D. \tag{4.123}$$

Once the functions F, G, m, and n have been determined from the prescribed initial and boundary conditions, the variation with t of α and β, and thereby the variation of any of the state variables, may be calculated at any fixed station X.

In what follows, the initial deformation of an impulsively loaded slab of elastic material that softens on loading is now analyzed by a procedure due to Kazakia and Varley [141, 142]. The load is assumed compressive if the material softens on compression and tensile if the material softens on tension.

It is supposed that at $t = 0$, the slab is in equilibrium in a reference state wherein all the state variables $\{v, c, T, e\}$ are zero and the Lagrangian sound speed A has the value A_0. At $t = 0$, the traction at $X = D$ is subject to a discontinuous change to T_a. A centered wave at $X = D$ is thereby generated which travels with speed A_0 toward the interface $X = 0$ where it is partially reflected and partially transmitted. As the centered wave crosses toward $X = 0$, it traverses two distinct regions. In region I, it is a centered simple wave with $F = 0$ and during its passage, at any fixed station X, the traction T increases monotonically in time to T_a. The latter value is then retained until the arrival of the wave that is reflected from $X = 0$.

As T changes from zero to T_a, the state variables v, c, and G, which are equal in region I, change from zero to c_a while e changes to e_a and the Lagrangian speed A decreases from A_0 to A_a. If the applied load is tensile, $T_a > 0$ and $c_a > 0$, while if the applied load is compressive, $T_a < 0$ and $c_a < 0$.

In region II, the centered wave interacts with the wave reflected from $X = 0$, and the signal function F is no longer identically zero, but rather is to be calculated by means of the relation (4.97).

It was shown in [85] that as the centered wave crosses both regions I and II, Eq. (4.117) reduces to[†]

$$2A^{1/2} t_\beta + [\mu t + v(1 - X)] \dot{G}(\beta) = 0. \tag{4.124}$$

But,

$$\dot{G}(\beta) = c_\beta = A^{-1/2}(\mu + vA)^{-1} A_\beta, \tag{4.125}$$

$$t_\beta = (\mu - vA)^{-1}[\mu t + v(1 - X)]_\beta, \tag{4.126}$$

[†] Following Kazakia and Varley, distance, time, and velocity are henceforth measured in units of D, D/A_0, and A_0, respectively. Stress is measured in units of $\rho_0 A_0^2$ and the material constants μ, v in units of $A_0^{-1/2}$ and $A_0^{-3/2}$, respectively.

whence, on substitution in (4.124) and integration, we obtain
$$\mu t + v(1 - X) = \phi(\alpha)A^{-1/2}(\mu + vA), \tag{4.127}$$
and insertion of the latter expression into (4.126) yields
$$t_\beta = \phi(\alpha)[A^{-1/2}]_\beta. \tag{4.128}$$
Equations (4.127) and (4.128) show that in the centered wave, X and t can be expressed in terms of A and α by relations of the type
$$X = 1 - \phi(\alpha)A^{1/2} - \bar{M}\tau(\alpha), \tag{4.129}$$
$$t = \phi(\alpha)A^{-1/2} + \tau(\alpha). \tag{4.130}$$

Once $\phi(\alpha)$ and $\tau(\alpha)$ have been determined, relations (4.129) and (4.130) together determine $A(X,t)$. At specified X, Eq. (4.129) gives A explicitly in terms of the characteristic parameter α. Insertion of this expression into (4.130) determines t as an explicit function of α. The variation of A with t is thereby obtained parametrically through α. In a similar manner, (4.129) and (4.130) together determine the variation of A and X with α at fixed time t.

Once $A(X,t)$ has been obtained via (4.129) and (4.130), the strain measure $c(X,t)$ is determined by the prevailing A-c relation. The traction $T(X,t)$ and strain $e(X,t)$ are given by the relations
$$T = \rho_0 \int_0^c A(s)\,ds, \tag{4.131}$$
$$e = \int_0^c A^{-1}(s)\,ds, \tag{4.132}$$
respectively, together with $c(X,t)$. The trajectories of constant stress and strain are generated by holding A fixed in (4.129) and (4.130) and allowing α to vary. The same equations also give the trajectories of the α-characteristics. Thus, on elimination of A, they yield
$$(t - \tau)(1 - X - \bar{M}\tau) = \phi^2, \tag{4.133}$$
and since τ, ϕ are constant on a given α-characteristic, it follows that (4.133) gives the X-t relation at any characteristic, which is accordingly a hyperbola in the X-t plane.

An immediate consequence of the hodograph system (4.25)–(4.26) is that the Lagrangian distance measure $X(\alpha,\beta)$ and time measure $t(\alpha,\beta)$ satisfy the equations
$$X_\alpha = -At_\alpha, \tag{4.134}$$
$$X_\beta = At_\beta. \tag{4.135}$$

4.3 REFLECTION AND TRANSMISSION OF A LARGE AMPLITUDE PULSE

Insertion of the expressions (4.129) and (4.130) for X and t into (4.134) shows that $F(\alpha)$, $\phi(\alpha)$, and $\tau(\alpha)$ are not independent but rather, that they are related through the compatibility condition

$$d\tau/d\alpha = v\phi \, dF/d\alpha. \tag{4.136}$$

The material velocity v can now be expressed in terms of A and α via the relation

$$v = c(A) - 2F(\alpha), \tag{4.137}$$

where it is understood that the appropriate A-c relationship is to be used in (4.137) as determined by the material response considerations outlined in the preceding section.

The trajectories of the β-characteristics are given by relation (4.84)$_1$, since on such a characteristic,

$$c(A) - F(\alpha) = G = \text{const.} \tag{4.138}$$

In region I, it is observed that in the reference state where $A = 1$, the hodograph equations (4.134), (4.135) integrate subject, to conditions (4.85) and (4.86), to provide

$$\alpha = t - X, \quad \beta = t + X - 1. \tag{4.139}$$

At the front of the centered wave, $\beta = 0$, so that

$$X = \tfrac{1}{2}(1 - \alpha), \quad t = \tfrac{1}{2}(1 + \alpha), \quad -1 \leq \alpha \leq 1,$$
$$A = 1, \quad F = 0. \tag{4.140}$$

Insertion of these data into (4.129) and (4.130) shows that in region I,

$$\phi = \tfrac{1}{2}(1 + \alpha), \quad -1 \leq \alpha \leq 1, \quad \tau = 0. \tag{4.141}$$

Hence, in that simple wave regime,

$$A = (1 - X)/t, \quad A_a \leq A \leq 1, \tag{4.142}$$

$$\alpha = 2[(1 - X)t]^{1/2} - 1, \quad -1 \leq \alpha \leq 1. \tag{4.143}$$

Moreover, since from (4.137), $F = 0$ in region I, it follows that

$$v = c(A) = c[(1 - X)/t], \tag{4.144}$$

while (4.138) shows that the β-characteristics are the rays

$$(1 - X)/t = \text{const.} \tag{4.145}$$

Relations (4.142)–(4.145) correspond to the well-known centered simple wave solutions of the hodograph equations (Courant and Friedrichs [148]).

The incident centered wave remains a simple wave until it enters region II where it interacts with the wave reflected from $X = 0$. The trajectory of the front of this reflected wave is obtained by setting $\alpha = 1$ in (4.143) and so is given by

$$(1 - X)t = 1. \tag{4.146}$$

At this front, (4.142) and (4.146) together show that

$$A = (1 - X)^2,$$

and since interaction region II is completely crossed by the wave front when $A = A_a$, in (4.146),

$$0 \leq X \leq 1 - A_a^{1/2}, \qquad 1 \leq t \leq A_a^{-1/2}. \tag{4.147}$$

In the first instance, the case in which the boundary $X = 0$ is perfectly free is considered. Thus,

$$c = T = 0, \qquad A = 1 \quad \text{at} \quad X = 0, \tag{4.148}$$

and (4.129), (4.130) show that

$$\phi = (1 - \bar{M}\alpha)/(1 - \bar{M}), \qquad \tau = (\alpha - 1)/(1 - \bar{M}). \tag{4.149}$$

Substitution of these expressions into the compatibility condition (4.136) and integration of the latter, subject to the condition $F = 0$ at $\alpha = 1$, shows that

$$F = \mu^{-1} \ln \phi. \tag{4.150}$$

The back of the centered wave on which $G = c_a$ reaches $X = 0$ when $F = -G = -c_a$; whence (4.150) shows that relations (4.149) prevail when $1 \leq \alpha \leq \alpha_a$, where

$$\alpha_a = 1 + \left(\frac{1 - \bar{M}}{\bar{M}}\right)[1 - \exp(-\mu c_a)]. \tag{4.151}$$

Insertion of expressions (4.149) into (4.129) and (4.130) and subsequent elimination of the characteristic parameter α provide an explicit expression for $A(X, t)$, namely,

$$A = \tfrac{1}{4}\{(1 - \theta) + [(1 - \theta)^2 + 4\bar{M}\theta]^{1/2}\}^2, \tag{4.152}$$

where

$$\theta = X/(1 - \bar{M}t). \tag{4.153}$$

The material velocity v is now determined by (4.137), together with (4.150), so that

$$v = c(A) - 2\mu^{-1} \ln \phi, \tag{4.154}$$

4.3 REFLECTION AND TRANSMISSION OF A LARGE AMPLITUDE PULSE

where $A(X, t)$ is given by (4.152) and (4.153) while $\phi(X, t)$ is then determined by

$$\phi = X/(1 - A^{1/2}) = (1 - \bar{M}t)A^{1/2}/(A^{1/2} - \bar{M}). \tag{4.155}$$

Relation $c(A)$ is to be obtained from the prevailing A-c law.

Since ϕ is constant at any α-characteristic, the trajectories of the latter may be generated by setting $\phi = $ const in (4.155). On the other hand, the trajectories of the β-characteristics as they are refracted by the reflected α-waves are given by (4.138) together with (4.150), the A-c law and the $A(X, t)$, $\phi(X, t)$ relations. Thus, at constant β,

$$c(A) - \mu^{-1} \ln \phi = G = \text{const},$$

whence

$$\phi \exp[-\mu c(A)] = \frac{X}{1 - A^{1/2}} \exp[-\mu c(A)]$$

$$= \frac{(1 - \bar{M}t)A^{1/2}}{A^{1/2} - \bar{M}} \exp[-\mu c(A)] = \text{const}$$

$$= \frac{1 - \bar{M}\bar{\beta}}{1 - \bar{M}},$$

where $\bar{\beta}$ is the arrival time of the β-characteristic at $X = 0$. In particular, the trajectory of the back of the centered wave corresponds to $\bar{\beta} = \alpha_a$ and so is parametrized in terms of A according to

$$X = (1 - A^{1/2})\exp[\mu\{c(A) - c_a\}],$$

$$t = \bar{M}^{-1}\left[1 - \frac{(A^{1/2} - \bar{M})}{A^{1/2}}\exp[\mu\{c(A) - c_a\}]\right], \quad A_a \leq A \leq 1. \tag{4.156}$$

Relation (4.153) implies that the constant levels of θ and hence of the state variables A, T, c, and e (but not, in general, v) propagate with constant speed $|\bar{M}\theta|$. The θ-wave is centered at $(X, t) = (0, \bar{M}^{-1})$.

At a constant station X, from (4.152) and (4.153),

$$dA/dt|_{X=\text{const}} = \{\bar{M}\theta/(1 - \bar{M}t)\}\dot{A}(\theta), \tag{4.157}$$

and $\dot{A}(\theta)$ becomes unbounded so that a shock forms when $\theta = \theta_s$, $A = A_s$, where

$$(1 - \theta_s)^2 + 4\bar{M}\theta_s = 0, \quad A_s = \tfrac{1}{4}(1 - \theta_s)^2 = -\bar{M}\theta_s. \tag{4.158}$$

An examination of (4.158) shows that A_s is imaginary when $0 < \bar{M} < 1$ so that no shock can form for nonideal materials in region II. On the other hand, if T_1 designates the limiting value of T as $A \to 0$, then provided that

T_a/T_1 is large enough for A to attain the value A_s, a shock can form in an ideal material. Since $A_s = A_s(\bar{M})$, the criterion for shock formation depends only on the material parameter $\bar{M}$. If the condition is met, the shock appears at the front of the reflected wave. Certain aspects of this shock phenomenon are discussed further in [142].

In the case of a centered wave incident at a perfectly rigid interface, the reflected wave continues to soften the material and accordingly, no shocks form. Indeed, Kazakia and Varley point out that in an ideally soft material, if T_a/T_1 exceeds a certain critical value (dependent only on $\bar{M}$), the incident wave is refracted so strongly by the reflected wave that part of it never reaches the boundary $X = 0$.

When $X = 0$ is a perfectly rigid interface, the analysis parallels that of the shock tube problem in Section 3.10. Thus,

$$v = 0, \quad F = G = \tfrac{1}{2}c \quad \text{at} \quad X = 0, \tag{4.159}$$

while (4.122), (4.129), and (4.130) combine to show that

$$\phi = \frac{m}{m^2 - \bar{M}}(1 - \bar{M}\alpha), \quad \tau = \frac{m^2\alpha - 1}{m^2 - \bar{M}}, \tag{4.160}$$

and it remains to determine $m(\alpha)$ at the rigid boundary. In this connection, the constitutive law (4.42) together with conditions (4.122) and (4.159) at $X = 0$ show that

$$dF/d\alpha = (\mu + vm^2)^{-1}\, dm/d\alpha. \tag{4.161}$$

On appropriate substitution of relations (4.160) into the compatibility condition (4.136) and use of (4.161), it is seen that $m(\alpha)$ is given by

$$(1 - \bar{M}\alpha)\, dm/d\alpha + m(m^2 - \bar{M}) = 0, \quad m = 1 \quad \text{at} \quad \alpha = 1, \tag{4.162}$$

whence

$$m = \left[\frac{\bar{M}(\bar{M} - 1)}{(\bar{M}\alpha - 1)^2 + \bar{M} - 1}\right]^{1/2}. \tag{4.163}$$

Thus, from (4.160),

$$\phi = \frac{1 - \bar{M}}{1 - \bar{M}\alpha}\left[\frac{(\bar{M}\alpha - 1)^2 + \bar{M} - 1}{\bar{M}(\bar{M} - 1)}\right]^{1/2}, \quad \tau = \frac{1 - \alpha}{1 - \bar{M}\alpha}, \tag{4.164}$$

while

$$F = \tfrac{1}{2}c(m^2), \tag{4.165}$$

where the relation $c(m^2)$ is determined by the prevailing A-c law. Insertion of expressions (4.160) for $\phi(\alpha)$ and $\tau(\alpha)$ into (4.129) and (4.130) now delivers

4.3 REFLECTION AND TRANSMISSION OF A LARGE AMPLITUDE PULSE

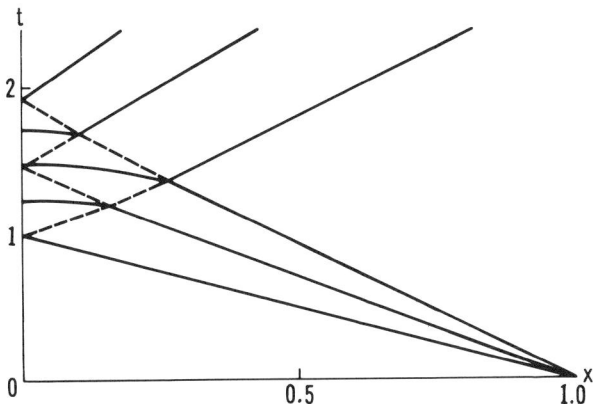

Fig. 4.12 A typical reflection of a centered wave from a rigid boundary in a nonideal material ($M = 0.36$). The centered wave is always completely reflected in a finite time. ——, Trajectories of constant levels of stress; ---, the characteristic curves (Kazakia and Varley [141]).

$A(X, t)$ and $\alpha(X, t)$ in the explicit forms

$$A = \bar{M}(X^2 + \bar{M} - 1)/[(\bar{M}t - 1)^2 + \bar{M} - 1], \quad (4.166)$$

$$\alpha = [(\bar{M} - 1)t + (t - 1)X]/[\bar{M} - 1 + (\bar{M}t - 1)X], \quad (4.167)$$

while the material velocity $v(X, t)$ is then given by (4.137) together with (4.165).

The trajectories of the β-characteristics may be obtained directly from (4.79) and (4.166) and are the curves

$$[(\bar{M} - 1)t + (1 - t)X]/[(\bar{M} - 1) + (1 - \bar{M}t)X] = \text{const.} \quad (4.168)$$

In Figs. 4.12 and 4.13, the trajectories of the characteristics in region II are shown for a typical nonideal material ($\bar{M} = 0.357$) and a typical ideal material ($\bar{M} = 1.1$) respectively. If T_a/T_1 is sufficiently large, the reflected wave produces yield in the material. The circumstances in which this occurs are discussed in [141].

The above two cases in which the interface $X = 0$ is perfectly free or perfectly rigid correspond, in turn, to $i \equiv 0$ and $i \equiv \infty$, where i is the impedance of the boundary. For other values of i when the material M_L is nonlinear, the forms of $\phi(\alpha)$, $\tau(\alpha)$, and $F(\alpha)$ in (4.129), (4.130), and (4.137) cannot, in general, be obtained analytically. The method of computation to be adopted is as follows:

Condition (4.99) shows that

$$F(\alpha) = L(\bar{G}), \quad (4.169)$$

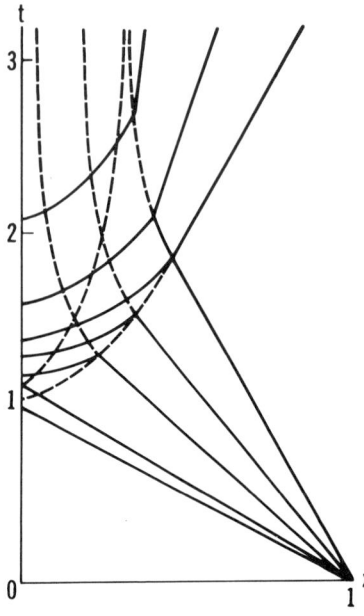

Fig. 4.13 A typical reflection of a centered wave from a rigid boundary in an ideal material ($M = 1.1$). If the applied traction is sufficiently large, the material yields at the rigid boundary and the centered wave is not reflected in a finite time. ———, Trajectories of constant levels of stress; ——— the characteristic curves (Kazakia and Varley [141]).

where

$$G = \bar{G}(\alpha) \quad \text{at} \quad X = 0. \tag{4.170}$$

On the other hand, (4.106) and (4.122) combine to show that, in terms of $\bar{G}$,

$$m(\alpha) = A^{1/2}[\bar{G} + L(\bar{G})]. \tag{4.171}$$

Relations (4.160), which are valid for any interface, rigid or otherwise, then give ϕ and τ in terms of $m(\alpha)$ and hence $\bar{G}$. Accordingly, it remains only to determine $\bar{G}(\alpha)$. This may be achieved by substitution of the expressions (4.169) and (4.171) for $F(\alpha)$ and $m(\alpha)$ together with those of (4.160) for ϕ and τ into the compatibility condition (4.136). This results in the equation

$$\bar{M}A^{1/2}[\bar{G} + L(\bar{G})]\,d\alpha/d\bar{G} + \mu(\bar{M}\alpha - 1) = 0, \tag{4.172}$$

for $\bar{G}(\alpha)$. Integration of (4.172) subject to the condition

$$\bar{G} = 0 \quad \text{at} \quad \alpha = 1, \tag{4.173}$$

yields

$$\alpha = 1 + \frac{1 - \bar{M}}{\bar{M}}\left(1 - \exp\left\{-\mu \int_0^{\bar{G}} A^{-1/2}[s + L(s)]^{-1}\,ds\right\}\right). \tag{4.174}$$

Condition (4.173) is a consequence of the fact that the front of the centered wave at which $G = 0$ arrives at $X = 0$ at $t = 1$. Relation (4.174) gives $\bar{G}(\alpha)$

implicitly. Once determined, $F(\alpha)$ and $m(\alpha)$ follow from (4.169) and (4.171) while ϕ and τ are given by (4.160). The variations of the state variables A and v in the interaction region II are then given by relations (4.129), (4.130), and (4.137). If the material does not experience yield at the interface, since $\bar{G} = c_a$ at the back of the incident centered wave, it follows that $\bar{G}$ varies over the interval $(0, c_a)$ in (4.174). If, on the other hand, the material yields at the interface, then $\bar{G}$ varies over the range $(0, \bar{G}^*)$ where

$$\bar{G}^* + L(\bar{G}^*) = c_1 \tag{4.175}$$

(Kazakia and Varley [141]).

In the absence of shock formation, the reflected wave, having crossed interaction region II, becomes a simple wave in which $G \equiv c_a$; where $(4.106)_{1,2}$ show that

$$c = c_a + F(\alpha), \tag{4.176}$$

at the back of the incident centered wave. Equations (4.129) and (4.130) with

$$A = A[c_a + F(\alpha)] = A_B(\alpha), \tag{4.177}$$

provide the parametric representation

$$X = X_B(\alpha), \qquad t = t_B(\alpha), \tag{4.178}$$

for the trajectory of the back of the incident centered wave. The curve (4.178) separates interaction region II and the simple wave region.

Relations (4.147) imply that the wave front emerges from the interaction region II at $X = 1 - A_a^{1/2}$ when $t = A_a^{-1/2}$. Thereafter, it moves with constant speed A_a toward the boundary $X = 1$ to arrive at time $t = 2A_a^{-1/2}$.

Conditions under which the wave reflected from $X = 0$ focuses to form a shock are set out by Kazakia and Varley [141, 142]. A detailed analysis of subsequent reflection at the loaded boundary $X = 1$ is also presented there. Rather than pursue these matters, we now turn to a recent development of the method with application in electromagnetic theory.

4.4 ELECTROMAGNETIC WAVE PROPAGATION IN NONLINEAR DIELECTRIC MEDIA. REDUCTION TO CANONICAL FORM VIA BÄCKLUND TRANSFORMATIONS

Wave propagation in nonlinear isotropic dielectrics has been studied by Jeffrey [149], Broer [150], Jeffrey and Korobeinikov [151], and Venkataraman and Rivlin [152] with a view to understanding the mechanism underlying the formation and propagation of discontinuities. Certain physical consequences of nonlinearity in the refractive index have been explored by

De Martini *et al.* in [153] where, in particular, they discuss distortion effects in nonlinear optical experiments in the propagation of light pulses.

Kazakia and Venkataraman [94] investigated a class of nonlinear boundary value problems for Maxwell's equations, involving the propagation of a linearly polarized plane electromagnetic wave in a nonlinear dielectric slab of finite extent. The nonlinear medium was assumed to be embedded in a linear dielectric. A variant of the Cekirge–Varley procedure was adopted wherein the Riemann characteristic equations were shown to be integrable for a certain class of **B-H** and **D-E** constitutive laws. An exact representation of a centered fan which describes the interaction of this wave with an oncoming signal was presented. Here the work of Kazakia and Venkataraman [94] is set in the context of Bäcklund transformation theory.

In the absence of charge and current density, Maxwell's equations of electrodynamics reduce to

$$\nabla \times \mathbf{E} + \frac{\partial \mathbf{B}}{\partial t} = \mathbf{0}, \tag{4.179}$$

$$\nabla \cdot \mathbf{D} = 0, \tag{4.180}$$

$$\nabla \times \mathbf{H} - \frac{\partial \mathbf{D}}{\partial t} = \mathbf{0}, \tag{4.181}$$

$$\nabla \cdot \mathbf{B} = 0, \tag{4.182}$$

where **E**, **H**, **D**, and **B** denote in turn, the electric field, the magnetic field, the electric displacement field, and the magnetic flux. Maxwell's equations are augmented by appropriate constitutive laws descriptive of the material behavior. Here, certain nonlinear **B-H**, **D-E** state laws will be adopted.

In the case of electromagnetic waves polarized in the (x_2, x_3) plane and propagating along the **i** direction with

$$\begin{aligned} \mathbf{E} &= E_3(x_1, t)\mathbf{k}, & \mathbf{H} &= H_2(x_1, t)\mathbf{j}, \\ \mathbf{D} &= D_3(x_1, t)\mathbf{k}, & \mathbf{B} &= B_2(x_1, t)\mathbf{j}, \end{aligned} \tag{4.183}$$

Maxwell's equations (4.179)–(4.182) yield

$$\partial E_3/\partial x_1 = \partial B_2/\partial t, \qquad \partial H_2/\partial x_1 = \partial D_3/\partial t. \tag{4.184}$$

These dynamic equations are here supplemented by the constitutive relations

$$D_3 = D(E_3), \qquad B_2 = B(H_2), \tag{4.185}$$

insertion of which into (4.184) shows that

$$\frac{\partial E}{\partial x} = B'(H)\frac{\partial H}{\partial t}, \qquad \frac{\partial H}{\partial x} = D'(E)\frac{\partial E}{\partial t}, \tag{4.186}$$

where indices have now been dropped.

4.4 ELECTROMAGNETIC WAVE PROPAGATION

Under the hodograph transformation in which E and H are taken as independent variables and x and t are taken as dependent variables,

$$E_x = Jt_H, \quad E_t = -Jx_H,$$
$$H_x = -Jt_E, \quad H_t = Jx_E, \tag{4.187}$$

where $J = J(E, H; x, t)$ and it is assumed that

$$0 < |J| < \infty. \tag{4.188}$$

Accordingly, the governing equations (4.186), together with (4.187), lead to the hodograph system

$$\begin{bmatrix} x \\ t \end{bmatrix}_h = \begin{bmatrix} 0 & A^*(e, h) \\ A^{*-1}(e, h) & 0 \end{bmatrix} \begin{bmatrix} x \\ t \end{bmatrix}_e, \tag{4.189}$$

where

$$h = \int_{H_0}^{H} [B'(\zeta)]^{1/2} d\zeta, \tag{4.190}$$

$$e = \int_{E_0}^{E} [D'(\eta)]^{1/2} d\eta, \tag{4.191}$$

$$A^*(e, h) = [B'(H)D'(E)]^{-1/2}. \tag{4.192}$$

The initial state of the medium is assumed to be uniform with constant values E_0 and H_0 for the electric and magnetic intensity fields, respectively.

A consequence of Loewner's work as outlined in Chapter 3 is that the Bäcklund transformation

$$\Lambda'_h = \mathbf{A}\Lambda_h + \mathbf{\Omega}'\tilde{\mathbf{C}}\Lambda, \quad \Lambda'_e = \mathbf{A}\Lambda_e + \tilde{\mathbf{C}}\Lambda, \quad h' = h, \quad e' = e \tag{4.193}$$

takes the matrix system

$$\Lambda_h = \mathbf{\Omega}\Lambda_e = \mathbf{A}^{-1}\mathbf{\Omega}'\mathbf{A}\Lambda_e \tag{4.194}$$

to the associated system

$$\Lambda'_h = \mathbf{\Omega}'\Lambda'_e, \tag{4.195}$$

subject to the requirements

$$\mathbf{A}_h - \mathbf{\Omega}'\tilde{\mathbf{C}} - (\mathbf{A}_e - \tilde{\mathbf{C}})\mathbf{\Omega} = \mathbf{0}, \quad |\mathbf{A}_e - \tilde{\mathbf{C}}| \neq 0, \tag{4.196}$$

$$\tilde{\mathbf{C}}_h - (\mathbf{\Omega}'\tilde{\mathbf{C}})_e = \mathbf{0}. \tag{4.197}$$

In the present context, we take

$$\Lambda = \begin{bmatrix} x \\ t \end{bmatrix}, \quad \mathbf{\Omega} = \begin{bmatrix} 0 & A^*(e, h) \\ A^{*-1}(e, h) & 0 \end{bmatrix}, \tag{4.198}$$

$$\Lambda' = \begin{bmatrix} x' \\ t' \end{bmatrix}, \quad \mathbf{\Omega}' = \begin{bmatrix} 0 & 1 \\ 1 & 0 \end{bmatrix}, \tag{4.199}$$

while in (4.193), $\mathbf{A}$ and $\tilde{\mathbf{C}}$ are assumed to be of the form

$$\mathbf{A} = \begin{bmatrix} a_1^1 & 0 \\ 0 & a_2^2 \end{bmatrix}, \qquad (4.200)$$

$$\tilde{\mathbf{C}} = \begin{bmatrix} c_1^1 & c_2^1 \\ c_1^2 & c_2^2 \end{bmatrix}, \qquad (4.201)$$

where the a_j^i and c_j^i are dependent on both h and e and are constrained by conditions (4.196) and (4.197). *In extenso*, if $\Omega = [\omega_j^i]$, then (4.196) yields

$$\begin{aligned} a_{1,h}^1 - c_1^2 + c_2^1 \omega_1^2 &= 0, & a_{1,e}^1 - c_1^1 + c_2^1/\omega_2^1 &= 0, \\ a_{2,h}^2 - c_2^1 + c_1^2 \omega_2^1 &= 0, & a_{2,e}^2 - c_2^2 + c_1^2/\omega_1^2 &= 0; \end{aligned} \qquad (4.202)$$

whence, in particular, det $\mathbf{A} = a_1^1 a_2^2$ is constant. Thus, the system (4.202) may be reduced to consideration of the pair of nonlinear equations

$$\begin{aligned} a_{1,h}^1 - c_1^2 + c_2^1 (a_1^1)^2/\det \mathbf{A} &= 0, \\ a_{1,e}^1 - c_1^1 + c_2^2 (a_1^1)^2/\det \mathbf{A} &= 0. \end{aligned} \qquad (4.203)$$

It is readily seen that (4.203) together with (4.197), admit the particular solution

$$\mathbf{A} = \begin{bmatrix} (ae+b)^{-1}(ch+d)^{-1} & 0 \\ 0 & (ae+b)(ch+d) \end{bmatrix},$$

$$\tilde{\mathbf{C}} = \begin{bmatrix} 0 & c(ae+b) \\ 0 & a(ch+d) \end{bmatrix}, \qquad (4.204)$$

with det $\mathbf{A} = 1$. Thus, the Bäcklund transformation (4.193) with the specializations (4.204) reduces the hodograph system

$$\begin{bmatrix} x \\ t \end{bmatrix}_h = \begin{bmatrix} 0 & (ae+b)^2(ch+d)^2 \\ (ae+b)^{-2}(ch+d)^{-2} & 0 \end{bmatrix} \begin{bmatrix} x \\ t \end{bmatrix}_e, \qquad (4.205)$$

to the hyperbolic canonical form given by (4.195) and (4.199). This shows that

$$x' = N(r) + M(s), \qquad t' = N(r) - M(s), \qquad (4.206)$$

where

$$r = (h+e)/2, \qquad s = (h-e)/2 \qquad (4.207)$$

represent the Riemann invariants. Insertion of expressions (4.206) into the Bäcklund relations (4.193) and subsequent integration yield

$$x = -[N(r) - M(s)]A^{*1/2} - 2J(r)M(s) + 2\int_{r_0}^{r} J'(r)N(r)\,dr - 2\int_{s_0}^{s} K(s)M'(s)\,ds \qquad (4.208)$$

4.4 ELECTROMAGNETIC WAVE PROPAGATION

and
$$t = [N(r) - M(s)]A^{*-1/2}, \qquad (4.209)$$
where
$$A^* = (ae + b)^2(ch + d)^2, \qquad (4.210)$$
$$= J(r) + K(s). \qquad (4.211)$$

A model constitutive law of the type (4.210) was introduced by Kazakia and Venkataraman [94] and adopted in a subsequent analysis of the evolution of a centered fan in a nonlinear dielectric slab. On use of (4.190)–(4.192) it is seen that the relation (4.210) corresponds to model D-E, B-H laws of the type

$$D_M = D_M(E) = D_0 + (ab)^{-1}[1 - \{1 + 3ab^{-3}(E - E_0)\}^{-1/3}], \qquad (4.212)$$

and

$$B_M = B_M(H) = B_0 + (cd)^{-1}[1 - \{1 + 3cd^{-3}(H - H_0)\}^{-1/3}]. \qquad (4.213)$$

Following Kazakia and Venkataraman, nondimensional variables may be introduced according to

$$\bar{D} = (D - D_0)/D_1, \qquad \bar{E} = (E - E_0)/E_1,$$
$$\bar{B} = (B - B_0)/B_1, \qquad \bar{H} = (H - H_0)/H_1, \qquad (4.214)$$
$$\bar{A}^* = A^*/A_0,$$

where

$$D_1 = (ab)^{-1}, \qquad E_1 = \tfrac{1}{3}a^{-1}b^3,$$
$$B_1 = (cd)^{-1}, \qquad H_1 = \tfrac{1}{3}c^{-1}d^3, \qquad (4.215)$$
$$A_0 = b^2 d^2,$$

and A_0 represents the electromagnetic wave speed in the equilibrium state wherein $E = E_0$ and $H = H_0$. Hence, on introduction of the relations (4.214) into (4.212) and (4.213), these state laws become

$$\bar{D} = 1 - (1 + \bar{E})^{-1/3}, \qquad \bar{B} = 1 - (1 + \bar{H})^{-1/3}, \qquad (4.216)$$

corresponding to

$$\bar{A}^* = (1 + \bar{E})^{2/3}(1 + \bar{H})^{2/3}. \qquad (4.217)$$

In the case of small field intensities wherein $D(E)$ and $B(H)$ admit the Taylor series expansions

$$D = \varepsilon_1 E + \varepsilon_2 |E|E + O(E^3), \qquad (4.218)$$
$$B = \mu_1 H + \mu_2 |H|H + O(H^3), \qquad (4.219)$$

the choice of the material constants a, b, c, d in (4.212) and (4.213) according to

$$a = \mp \frac{1}{2} \varepsilon_2 \varepsilon_1^{-7/4}, \qquad b = \varepsilon_1^{-1/4} \left[1 \mp \frac{3}{2} \frac{\varepsilon_2}{\varepsilon_1} E_0 \right]^{1/3},$$
$$c = \mp \frac{1}{2} \mu_2 \mu_1^{-7/4}, \qquad d = \mu_1^{-1/4} \left[1 \mp \frac{3}{2} \frac{\mu_2}{\mu_1} H_0 \right]^{1/3}, \qquad (4.220)$$

allows a local approximation to (4.218) and (4.219) valid to within an error of $O(E^3)$ and $O(H^3)$, respectively. The upper and lower signs are associated with, in turn, positive and negative values of the electric and magnetic fields.

The importance of the proposed state laws (4.212) and (4.213) in the case of large amplitude disturbances depends on how well they may be adapted to empirically or theoretically determined material response over a finite range of E and H.[†] The matter of global fit of the model constitutive law (4.212) to a particular class of D-E laws is taken up in [94].

4.5 EVOLUTION OF A LARGE AMPLITUDE CENTERED FAN IN A NONLINEAR DIELECTRIC SLAB

In this section, we consider the propagation of a large amplitude disturbance in a nonlinear slab with material response modeled by D-E, B-H laws of type (4.212), (4.213). The disturbance is generated by the arrival of a constant electromagnetic shock at the boundary $x = x_0$ (see Fig. 4.14). The analysis is based on that of Kazakia and Venkataraman [94].

It is convenient to proceed in terms of a characteristic formulation wherein

$$r_t - A^*(e, h) r_x = 0, \qquad (4.221)$$

$$s_t + A^*(e, h) s_x = 0, \qquad (4.222)$$

so that the Riemann invariant s is constant on any right propagating characteristic wavelet $\alpha(x, t) = $ const given by

$$dx/dt\big|_\alpha = +A^*(e, h), \qquad (4.223)$$

while the Riemann invariant r is constant on any left propagating characteristic wavelet $\beta(x, t) = $ const given by

$$dx/dt\big|_\beta = -A^*(e, h). \qquad (4.224)$$

[†] Additional multiparameter model D-E and B-H state laws for which the characteristic equations are reducible to canonical form are to be found in [95] (see also Donato and Fusco [154]).

4.5 LARGE AMPLITUDE CENTERED FAN IN A DIELECTRIC SLAB

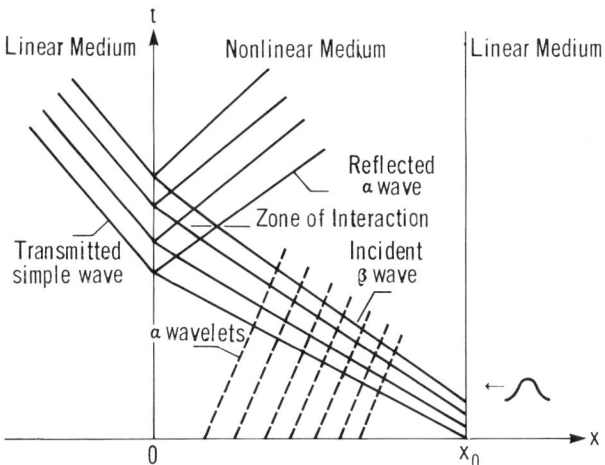

Fig. 4.14 The initial wave pattern produced inside the slab during the reflection and transmission of an arbitrary signal at the interface with a linear dielectric (Kazakia and Venkataraman [94]).

Thus, in the notation of [94],

$$r = \tfrac{1}{2}(h + e) = G(\beta), \qquad s = -\tfrac{1}{2}(h - e) = F(\alpha), \tag{4.225}$$

so that h and e may be expressed in terms of the signal functions $F(\alpha)$, $G(\beta)$ according to

$$e = G(\beta) + F(\alpha), \qquad h = G(\beta) - F(\alpha). \tag{4.226}$$

The dielectric slab is assumed to be bounded by the planes $x = 0$ and $x = x_0$ and to be embedded in dielectric media with linear properties (Fig. 4.14). The α- and β-characteristic wavelets are labeled in accordance with the usual convention

$$\alpha = t \quad \text{at} \quad x = 0, \tag{4.227}$$

$$\beta = t \quad \text{at} \quad x = x_0. \tag{4.228}$$

An arbitrary shockless pulse is assumed to be generated at $x = x_0$ and to traverse the dielectric slab. Until the wave front reaches $x = 0$, it is taken to be moving into an equilibrium region wherein $E = E_0$, $H = H_0$. Consequently, the pulse initially acts like a simple wave with

$$F(\alpha) \equiv 0, \tag{4.229}$$

$$G(\beta) = e = h. \tag{4.230}$$

However, once the pulse front reaches the interface $x = 0$, a reflected α-wave is produced which proceeds toward $x = x_0$. At the same time, a transmitted wave is produced which moves into the adjoining linear dielectric medium. Following its arrival at $x = 0$, the incident pulse begins to interact with the reflected α-wave and as a consequence, in this region of interaction, the disturbance no longer acts like a simple wave.

By contrast, the wave which is transmitted into the adjoining linear dielectric medium M_L is a simple wave. Thus, if the subscript L denotes the value of state variables in M_L so that its constitutive laws are

$$D_L = \varepsilon_L E_L, \tag{4.231}$$

$$B_L = \mu_L H_L, \tag{4.232}$$

then in the transmitted wave, whatever the transmitted signal $G_L(\beta_L)$,

$$e_L = h_L. \tag{4.233}$$

In particular, relation (4.233) holds at $x = 0$, while in addition, we have the boundary conditions

$$E = E_L, \quad H = H_L, \quad \text{at} \quad x = 0 \tag{4.234}$$

(Born and Wolf [155]). As a consequence, from (4.190), (4.191), and (4.231)–(4.234), at $x = 0$,

$$E_L - E_0 = (\mu_L/\varepsilon_L)^{1/2}[H_L - H_0] = E - E_0 = (\mu_L/\varepsilon_L)^{1/2}[H - H_0], \tag{4.235}$$

while the local reflection and transmission coefficients are given by

$$\frac{dF}{dG} = \left[\left(\frac{\mu_L \varepsilon_N}{\mu_N \varepsilon_L}\right)^{1/2} - 1\right] \bigg/ \left[\left(\frac{\mu_L \varepsilon_N}{\mu_N \varepsilon_L}\right)^{1/2} + 1\right], \tag{4.236}$$

together with

$$\frac{dG_L}{dG} = 2\left(\frac{\mu_L}{\mu_N}\right)^{1/2} \bigg/ \left[\left(\frac{\mu_L \varepsilon_N}{\mu_N \varepsilon_L}\right)^{1/2} + 1\right], \tag{4.237}$$

where

$$\mu_N := B'(H), \quad \varepsilon_N := D'(E). \tag{4.238}$$

It was observed by Kazakia and Venkataraman that the specialization

$$x = -\phi^*(\alpha)A^{*1/2} + \tau^*(\alpha), \tag{4.239}$$

$$t = \phi^*(\alpha)A^{*-1/2}, \tag{4.240}$$

of relations (4.208) and (4.209) provides, on appropriate choice of $\phi^*(\alpha)$ and $\tau^*(\alpha)$, a representation of the interaction between the incident centered fan and an arbitrary oncoming disturbance. The functions $\phi^*(\alpha)$, $\tau^*(\alpha)$ are

4.5 LARGE AMPLITUDE CENTERED FAN IN A DIELECTRIC SLAB

related to this oncoming signal and moreover, from (4.224), satisfy a compatibility condition of the form

$$d\tau^*(\alpha)/d\alpha = 2c_1^{-1}\phi^*(\alpha)[c_1 - 1 - 2\bar{F}(\alpha)]\,d\bar{F}(\alpha)/d\alpha, \qquad (4.241)$$

where here and in the sequel we proceed in terms of the state laws (4.216) with associated signal functions $\bar{F}, \bar{G}$ given by

$$\begin{aligned} 2\bar{F} &= [(1+\bar{E})^{1/3} - 1] - c_1[(1+\bar{H})^{1/3} - 1], \\ 2\bar{G} &= c_1^{-1}[(1+\bar{E})^{1/3} - 1] + [(1+\bar{H})^{1/3} - 1], \end{aligned} \qquad (4.242)$$

where $c_1 = ad/bc$. Furthermore, distances are henceforth normalized with respect to x_0 while time, together with the characteristic variables α, β, is normalized with respect to x_0/A_0.

In region I (see Fig. 4.15), wherein the centered wave moves through a reference state with $A^* = 1$, the characteristic equations (4.223), (4.224) integrate, subject to conditions (4.227), (4.228) to give

$$x = \tfrac{1}{2}[1 - \alpha + \beta], \qquad (4.243)$$

$$t = \tfrac{1}{2}[1 + \alpha + \beta]. \qquad (4.244)$$

At the front of the centered pulse, $\beta = 0$, so that matching this solution with (4.239), (4.240) yields, in region I,

$$\phi^*(\alpha) = \tfrac{1}{2}(1 + \alpha), \qquad \tau^*(\alpha) = 1, \qquad -1 \le \alpha \le 1. \qquad (4.245)$$

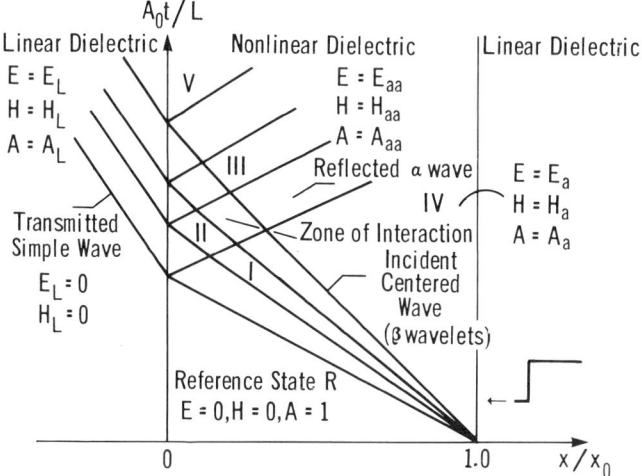

Fig. 4.15 The wave system that is set up when a centered fan is partly reflected and partly transmitted at the interface with a linear dielectric (Kazakia and Venkataraman [94]).

Insertion of these expressions for $\phi^*(\alpha)$ and $\tau^*(\alpha)$ into (4.239)–(4.240) and subsequent elimination of α shows that, in region I,

$$A^* = (1 - x)/t, \qquad A_a^* \leq A^* \leq 1, \tag{4.246}$$

where A_a^* represents the value of A^* at the back of the centered wave. Hence, since $\bar{F} \equiv 0$ in the simple wave region, from (4.217) and (4.242), the field variables $\bar{E}$ and $\bar{H}$ are given implicitly in terms of x, t by

$$(1 - x)/t = \{1 + c_1^{-1}[(1 + \bar{E})^{1/3} - 1]\}(1 + \bar{E})^{2/3} \tag{4.247}$$

$$= \{1 + c_1[(1 + \bar{H})^{1/3} - 1]\}^2(1 + \bar{H})^{2/3}. \tag{4.248}$$

The incident centered wave remains a simple wave until it enters interaction region II. In region I,

$$\alpha = 2[(1 - x)t]^{1/2} - 1, \qquad -1 \leq \alpha \leq 1, \tag{4.249}$$

and the trajectory of the boundary of this simple wave region with the adjoining interaction region corresponds to $\alpha = 1$ and so is given by

$$(1 - x)t = 1. \tag{4.250}$$

At the boundary $x = 0$, (4.235) shows that

$$\bar{E} = c_2 \bar{H}, \tag{4.251}$$

where

$$c_2 = (\mu_L/\varepsilon_L)^{1/2} H_1/E_1 = c_1 i_0, \tag{4.252}$$

and the quantity

$$i_0 = (\mu_L/\varepsilon_L)^{1/2} (d/b)^2 \tag{4.253}$$

represents the ratio of the intrinsic impedances in their equilibrium state.

Use of relation (4.217) together with (4.251) in expression (4.239) at $x = 0$ shows that $\phi^*(\alpha)$ and $\tau^*(\alpha)$ are related by

$$\tau^*(\alpha) = \frac{\phi^*(\alpha) h(\alpha)}{c_2^{1/3}} \left(\frac{c_2 - 1}{h^3(\alpha) - 1} \right)^{2/3}, \tag{4.254}$$

where

$$\bar{H}\big|_{x=0} + 1 = \frac{h^3(\alpha)(c_2 - 1)}{c_2(h^3(\alpha) - 1)}. \tag{4.255}$$

Furthermore, from (4.242) at $x = 0$,

$$2\bar{F}(\alpha) = c_1 - 1 + (1 - c_3 h(\alpha)) \left(\frac{c_2 - 1}{h^3(\alpha) - 1} \right)^{1/3} \equiv 2\hat{F}(h), \tag{4.256}$$

4.5 LARGE AMPLITUDE CENTERED FAN IN A DIELECTRIC SLAB

where

$$c_3 = c_1 c_2^{-1/3}. \tag{4.257}$$

Elimination of $\tau^*(\alpha)$ and $\bar{F}(\alpha)$ in the compatibility condition (4.241) and subsequent integration subject to the condition

$$\phi^* = 1 \quad \text{at} \quad h = c_2^{1/3} \tag{4.258}$$

(a consequence of (4.254) at $\alpha = 1$), yields

$$\phi^* = \left[\frac{(c_2^{2/3} + c_2^{1/3} + 1)(h-1)^2}{(c_2^{1/3} - 1)^2(h^2 + h + 1)} \right]^{1/6(c_3^{-1} + c_3)}$$

$$\times \exp[3^{-1/2}(c_3^{-1} - c_3)\{\tan^{-1}[3^{-1/2}(1+2h)] - \tan^{-1}[3^{-1/2}(1+2c_2^{1/3})]\}]$$

$$\equiv \hat{\phi}(h). \tag{4.259}$$

Introduction of (4.217) into relations (4.239)–(4.240) and elimination of $\bar{H}$ by means of (4.242)$_1$ now give

$$(1+\bar{E})^{1/3} = \frac{1}{2}\left[(1+2\hat{F}(h)-c_1) + \left\{(1+2\hat{F}(h)-c_1)^2 + 4c_1 \frac{\hat{\phi}(h)}{t}\right\}^{1/2}\right] \tag{4.260}$$

$$= \frac{1}{2}\left[(1+2\hat{F}(h)-c_1) + \left\{(1+2\hat{F}(h)-c_1)^2 + 4c_1 \left(\frac{\hat{\tau}(h)-x}{\hat{\phi}(h)}\right)\right\}^{1/2}\right], \tag{4.261}$$

where

$$\hat{\tau}(h) = \frac{c_3}{c_1}\left(\frac{c_2-1}{h^3-1}\right)^{2/3} h\hat{\phi}(h). \tag{4.262}$$

On the other hand, elimination of $\bar{E}$ between (4.242)$_1$ and (4.260)–(4.261) provides

$$(1+\bar{H})^{1/3} = \frac{1}{2c_1}\left[\left\{(1+2\hat{F}(h)-c_1)^2 + 4c_1 \frac{\hat{\phi}(h)}{t}\right\}^{1/2} - (1+2\hat{F}(h)-c_1)\right] \tag{4.263}$$

$$= \frac{1}{2c_1}\left[\left\{(1+2\hat{F}(h)-c_1)^2 + 4c_1 \left(\frac{\hat{\tau}(h)-x}{\hat{\phi}(h)}\right)\right\}^{1/2} - (1+2\hat{F}(h)-c_1)\right]. \tag{4.264}$$

Elimination of the parameter h between (4.260) and (4.261) determines $\bar{E}(x,t)$, while similarly, elimination of h between (4.263) and (4.264) gives $\bar{H}(x,t)$. In particular, the variations of $\bar{E}$ or $\bar{H}$ at fixed station x or specified time t are readily accessible.

If attention is restricted to nonmagnetic media so that the material constant c is zero in (4.210), the state variables in the evolution of the centered

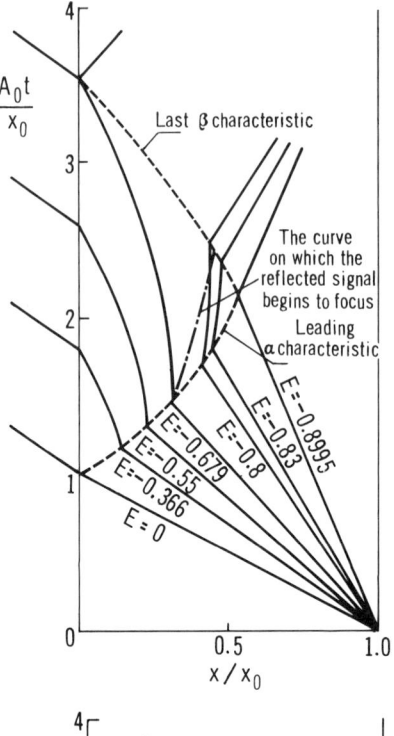

Fig. 4.16 Constant E-level trajectories within the slab (Kazakia and Venkataraman [94]).

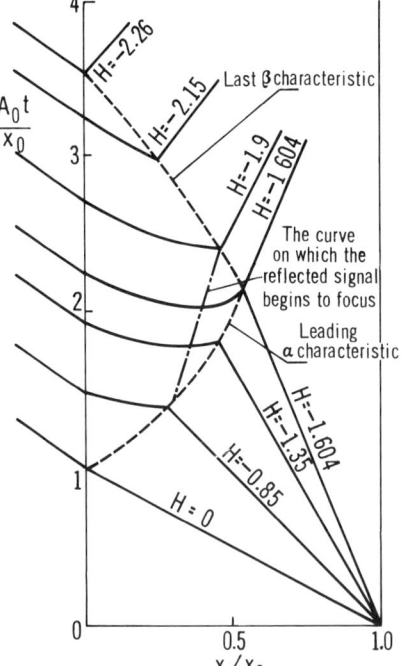

Fig. 4.17 Constant H-level trajectories within the slab (Kazakia and Venkataraman [94]).

4.5 LARGE AMPLITUDE CENTERED FAN IN A DIELECTRIC SLAB

fan were obtained in [94] in terms of

$$\bar{H} = c_1^{-1} H^*,$$

by taking the limit as $c_1^{-1} \to 0$ of the expressions (4.247), (4.248), (4.255), (4.256), and (4.259)–(4.264). The trajectories of constant E and H levels in regions I and II are shown in Figs. 4.16 and 4.17.

Finally, we remark that the preceding results are only valid until such time as the disturbance reflected from the boundary $x = 0$ forms a shock. This occurs at stations $x = x_s$, where the field variables experience unbounded gradients so that

$$dE/dt\big|_{x=x_s} \to \infty, \qquad dH/dt\big|_{x=x_s} \to \infty.$$

Kazakia and Venkataraman show that there is shock formation on the reflected front when

$$t = t_s = 2/(1 - i_0), \qquad h \equiv 1.$$

Accordingly, when $i_0 > 1$, no shock occurs.

CHAPTER 5

Bäcklund Transformations and Stress Distribution Theory in Elastostatics

5.1 WEINSTEIN'S CORRESPONDENCE PRINCIPLE IN THE CONTEXT OF BÄCKLUND TRANSFORMATIONS OF THE GENERALIZED STOKES–BELTRAMI SYSTEM

Stokes–Beltrami systems have their origin in the study of axially symmetric solutions of the n-dimensional Laplace equation

$$\nabla_n^2 f = 0, \tag{5.1}$$

where

$$\nabla_n^2 = \sum_{i=1}^{n} \frac{\partial^2}{\partial x^{i2}}. \tag{5.2}$$

Thus, if solutions ϕ of (5.1) are sought that depend only on the two variables

$$x = x^1, \qquad y = \left[\sum_{i=2}^{n} (x^i)^2\right]^{1/2} \tag{5.3}$$

5.1 WEINSTEIN'S PRINCIPLE: STOKES–BELTRAMI SYSTEMS

then $\phi(x, y)$ satisfies the equation

$$\frac{\partial}{\partial x}\left(y^p \frac{\partial \phi}{\partial x}\right) + \frac{\partial}{\partial y}\left(y^p \frac{\partial \phi}{\partial y}\right) = 0, \tag{5.4}$$

where $p = n - 2$ and $y \geq 0$. Accordingly, ϕ may be regarded as an axially symmetric potential in a *space of* $n = p + 2$ *dimensions*. The conventional notation $\phi\{p\}$ is adopted, on occasion, in order to indicate the dependence of ϕ on the parameter p.

Here, it will prove convenient to study the *generalized axially symmetric potential equation* (5.4) in the context of the *generalized Stokes–Beltrami system*

$$\begin{bmatrix} \phi \\ \psi \end{bmatrix}_y = \begin{bmatrix} 0 & -y^{-p} \\ y^p & 0 \end{bmatrix} \begin{bmatrix} \phi \\ \psi \end{bmatrix}_x, \tag{5.5}$$

where the existence of $\psi\{-p\}$ is guaranteed by (5.4). In the case $p = 0$, the Cauchy–Riemann system is retrieved, while the case $p = 1$ was the subject of a series of important papers by Beltrami [313] that were originally published circa 1880. There, Beltrami founded a generalized theory of analytic functions later to be developed in a systematic manner by Bers and Gelbart [314] and Weinstein [98]. A survey of recent developments in this subject, together with an extensive list of references is given by Bauer and Ruscheweyh [315].

In terms of applications, it was Arndt [316] who first noted the importance of the Stokes–Beltrami system in $n = 5$ dimensions in connection with the theory of torsion of shafts of revolution. On the other hand, Weinstein [317] demonstrated that in the elliptic régime, Tricomi's equation of transsonic gasdynamics (see Chapter 3) may be reduced to a Stokes–Beltrami system with $n = \frac{5}{3}$.

Invariance properties of the Cauchy–Riemann system have had a major impact on classical potential theory. Thus, invariance under conformal mappings has been widely used to transform two-dimensional harmonic boundary value problems to canonical problems with simple boundaries whose solution is readily obtained. It is natural therefore to inquire as to whether the more general Stokes–Beltrami system (5.5) admits invariance or reducibility properties of practical importance. In this connection, Beltrami early demonstrated that the classical case $p = 1$ admits simple invariance properties whereby sequences of axially symmetric harmonic functions may readily be generated (Bers and Gelbart [314]). Subsequently, Weinstein [318] established a correspondence principle of considerable utility in continuum mechanics, while further invariance properties in the case $p = 1$ were noted by Parsons [319].

In later papers, Rogers [320] and Rogers and Kingston [321] demonstrated that both Parson's invariance properties and Weinstein's correspondence principle may be readily retrieved from a study of the invariance and reducibility properties of the Stokes–Beltrami system (5.5) under matrix Bäcklund transformations of the type introduced in Chapter 3. Thus, Bäcklund transformations

$$\Omega'_x = \tilde{\mathbf{A}}\Omega_x + \tilde{\mathbf{B}}\Omega + \tilde{\mathbf{C}}\Omega',$$
$$\Omega'_y = \mathbf{A}\Omega_y + \mathbf{B}\Omega + \mathbf{C}\Omega', \tag{5.6}$$
$$x' = x, \quad y' = y,$$

were introduced that transform the Stokes–Beltrami system

$$\Omega_y = \Lambda\Omega_x, \quad \Lambda = \begin{bmatrix} 0 & -y^{-p} \\ y^p & 0 \end{bmatrix}, \tag{5.7}$$

to an associated system

$$\Omega'_y = \Lambda'\Omega'_x, \quad \Lambda' = \begin{bmatrix} 0 & -y^{-q} \\ y^q & 0 \end{bmatrix}, \tag{5.8}$$

where

$$\Omega = \begin{bmatrix} \phi \\ \psi \end{bmatrix}, \quad \Omega' = \begin{bmatrix} \phi' \\ \psi' \end{bmatrix},$$

and $\tilde{\mathbf{A}}, \tilde{\mathbf{B}}, \tilde{\mathbf{C}}, \mathbf{A}, \mathbf{B}, \mathbf{C}$ are 2×2 matrices with entries dependent on x, y. In particular, linking Bäcklund transformations were established in [321] with $q = p + 2$, $p \neq -1$ and the matrices specialized according to

$$\tilde{\mathbf{A}} = \mathbf{A} = \begin{bmatrix} 2(ax+b) & y^{-p-1}[a(x^2-y^2)+2(bx+d)] \\ -y^{p+1}[a(x^2-y^2)+2(bx+d)] & 2y^2(ax+b) \end{bmatrix},$$
$$\tilde{\mathbf{B}} = \begin{bmatrix} a(p+2) & 2y^{-p-1}(ax+b) \\ 0 & ay^2 + (p+1)[ax^2 + 2(bx+d)] \end{bmatrix},$$
$$\mathbf{B} = \begin{bmatrix} 0 & -ay^{-p} - (p+1)y^{-p-2}[ax^2 + 2(bx+d)] \\ a(p+2)y^{p+2} & 2y(ax+b) \end{bmatrix}, \tag{5.9}$$
$$\mathbf{C} = \begin{bmatrix} c & 0 \\ 0 & c \end{bmatrix}, \quad \tilde{\mathbf{C}} = \begin{bmatrix} 0 & -cy^{-p-2} \\ cy^{p+2} & 0 \end{bmatrix},$$

where a, b, c, d are arbitrary parameters assumed not all to vanish simultaneously. Relations (5.6) and (5.9) together constitute a four-parameter class of correspondence principles linking the Stokes–Beltrami system (5.7) with

5.1 WEINSTEIN'S PRINCIPLE: STOKES–BELTRAMI SYSTEMS

the associated system

$$\begin{bmatrix} \phi' \\ \psi' \end{bmatrix}_y = \begin{bmatrix} 0 & -y^{-(p+2)} \\ y^{(p+2)} & 0 \end{bmatrix} \begin{bmatrix} \phi' \\ \psi' \end{bmatrix}_x.$$

In particular, if we set

$$a = b = c = 0, \quad d \neq 0,$$

the Bäcklund relations yield

$$\phi'_x = 2dy^{-p-1}\psi_x, \quad \phi'_y = 2dy^{-p-1}\psi_y - 2d(p+1)y^{-p-2}\psi,$$

whence,

$$\phi' = 2dy^{-p-1}\psi, \quad p \neq -1,$$

or, in Weinstein's notation,

$$\psi\{p\} = \delta y^{p+1}\phi\{p+2\}, \tag{5.10}$$

where $\delta = 1/(2d)$. The relation (5.10) is known as the *Weinstein Correspondence Principle*. It implies that given a $\psi\{p\}$, an associated $\phi\{p+2\}$ may be determined up to a multiplicative constant, in two more dimensions and vice versa.

If

$$L_k(f) := f_{xx} + f_{yy} + ky^{-1}f_y,$$

we see that

$$L_k(\phi) = 0, \quad L_k(\psi) = 0,$$

where $k = p$ and $k = -p$, respectively. Thus, Weinstein's correspondence principle may be written as

$$f\{-k\} = y^{k+1}f\{k+2\}, \tag{5.11}$$

where, in this context, $f\{k\}$ designates that f is a solution of

$$L_k(f) = 0. \tag{5.12}$$

In an alternative notation adopted by Burns [322], (5.11) is written as

$$f_k^{(1)} \leftrightarrow y^{1-k}f_{2-k}^{(1)}, \tag{5.13}$$

and induction readily establishes a more general result for the iterated equation of generalized axially symmetric potential theory

$$L_k^n(f) = 0, \quad L_k^n := (L_k)^n, \tag{5.14}$$

namely, that

$$f_k^{(n)} \leftrightarrow y^{1-k}f_{2-k}^{(n)}. \tag{5.15}$$

Burns applied this result in the case $n = 2$ to systematize the study of problems involving Stokes flow of a viscous fluid past such bodies as a spindle, lens, or torus. The work may be compared to that of Pell and Payne [323–325].

5.2 APPLICATION OF WEINSTEIN'S CORRESPONDENCE PRINCIPLE IN ELASTOSTATICS. ASSOCIATED AXIALLY SYMMETRIC PUNCH, CRACK, AND TORSION BOUNDARY-VALUE PROBLEMS

The governing equations of linear isotropic elasticity consist of the equilibrium equations

$$g^{jk}\tau_{ij,k} + F_i = 0, \qquad i,j = 1, 2, 3, \tag{5.16}$$

together with the linear stress–strain relations[†]

$$\tau_{ij} = \lambda\bar{\theta}g_{ij} + 2\mu e_{ij}, \qquad i,j = 1, 2, 3, \tag{5.17}$$

where

$$\bar{\theta} = g^{ij}e_{ij}, \tag{5.18}$$

and

$$e_{ij} = \tfrac{1}{2}[u_{i,j} + u_{j,i}]. \tag{5.19}$$

Here τ_{ij} and e_{ij} represent the components of the stress and strain tensors, respectively, while u_i and F_i denote, in turn, the components of the displacement and body force vectors; g_{ij} and g^{ij} are the usual components of the covariant and contravariant metric tensor, while λ, μ are the Lamé elastic parameters. The medium is taken to be homogeneous, so that λ and μ are constants.

The above system of equations holds at every interior point of the elastic body under deformation. Moreover, on the surface Σ of the body, the stresses must fulfill the equilibrium requirements

$$\tau_{ij}v^j = T_i^v, \tag{5.20}$$

where $\mathbf{T}^v$ is the stress vector acting on the surface element of Σ with normal v. In addition, to the system (5.16)–(5.20) must be adjoined the Saint-Venant compatibility conditions

$$e_{ij,kl} + e_{kl,ij} - e_{ik,jl} - e_{jl,ik} = 0, \tag{5.21}$$

required for the continuity of the displacements as defined by (5.19).

[†] These correspond to Hooke's law "*ut tensio sic vis*" annunciated in 1678.

5.2 PUNCH, CRACK, AND TORSION PROBLEMS

In what follows, it is shown that Weinstein's correspondence principle may be used to link axially symmetric punch, crack, and torsion problems [99, 326]. In this context, it is convenient to introduce cylindrical coordinates r, θ, z so that the equilibrium equations (5.16) become, under the assumption $F_i = 0$,

$$\frac{\partial \tau_{rr}}{\partial r} + \frac{1}{r}\frac{\partial \tau_{r\theta}}{\partial \theta} + \frac{\partial \tau_{rz}}{\partial z} + \frac{\tau_{rr} - \tau_{\theta\theta}}{r} = 0,$$

$$\frac{\partial \tau_{r\theta}}{\partial r} + \frac{1}{r}\frac{\partial \tau_{\theta\theta}}{\partial \theta} + \frac{\partial \tau_{\theta z}}{\partial z} + \frac{2\tau_{r\theta}}{r} = 0, \quad (5.22)$$

$$\frac{\partial \tau_{rz}}{\partial r} + \frac{1}{r}\frac{\partial \tau_{\theta z}}{\partial \theta} + \frac{\partial \tau_{zz}}{\partial z} + \frac{\tau_{rz}}{r} = 0,$$

while the strain–displacement relations (5.19) yield

$$e_{rr} = \frac{\partial u_r}{\partial r}, \quad e_{\theta\theta} = \frac{1}{r}\frac{\partial u_\theta}{\partial \theta} + \frac{u_r}{r}, \quad e_{zz} = \frac{\partial u_z}{\partial z},$$

$$e_{r\theta} = \frac{1}{2}\left[\frac{1}{r}\frac{\partial u_r}{\partial \theta} + \frac{\partial u_\theta}{\partial r} - \frac{u_\theta}{r}\right], \quad e_{rz} = \frac{1}{2}\left[\frac{\partial u_z}{\partial r} + \frac{\partial u_r}{\partial z}\right], \quad (5.23)$$

$$e_{\theta z} = \frac{1}{2}\left[\frac{\partial u_\theta}{\partial z} + \frac{1}{r}\frac{\partial u_z}{\partial \theta}\right],$$

where (u_r, u_θ, u_z) are the cylindrical displacement components. On the other hand, the stress–strain laws (5.17) become

$$\tau_{rr} = \lambda \nabla \mathbf{u} + 2\mu e_{rr}, \quad \tau_{\theta\theta} = \lambda \nabla \mathbf{u} + 2\mu e_{\theta\theta}, \quad \tau_{zz} = \lambda \nabla \mathbf{u} + 2\mu e_{zz},$$
$$\tau_{r\theta} = 2\mu e_{r\theta}, \quad \tau_{rz} = 2\mu e_{rz}, \quad \tau_{\theta z} = 2\mu e_{\theta z}, \quad (5.24)$$

where

$$\nabla \mathbf{u} = \frac{1}{r}\frac{\partial}{\partial r}(r u_r) + \frac{\partial u_z}{\partial z}. \quad (5.25)$$

Substitution of the relations (5.24) into the elastostatic equations (5.22) and use of the strain–displacement conditions (5.23) now produce the following system:

$$\alpha \frac{\partial}{\partial r}\{\nabla \mathbf{u}\} + \Delta u_r - \frac{u_r}{r^2} = 0, \quad (5.26)$$

$$\alpha \frac{\partial}{\partial z}\{\nabla \mathbf{u}\} + \Delta u_z = 0, \quad (5.27)$$

$$\Delta u_\theta - \frac{u_\theta}{r^2} = 0, \quad (5.28)$$

where
$$\alpha = (\lambda + \mu)/\mu, \quad \Delta := \frac{\partial^2}{\partial r^2} + \frac{1}{r}\frac{\partial}{\partial r} + \frac{\partial^2}{\partial z^2} = L_1.$$

Note that the elastostatic system (5.26)–(5.28) is decoupled in so far as the torsional displacement term u_θ occurs only in the single equilibrium equation (5.28). However, Weinstein's correspondence principle may be adduced to link an important class of punch and crack problems governed by the pair of equations (5.26)–(5.27), subject to appropriate boundary conditions, to certain boundary value problems in which torsion only is involved. This association, which is due to Payne [99], is now described in detail.

In the first instance, if the plane $z = 0$ is assumed to be shear free, then the elastostatic equations reduce to consideration of (5.26) and (5.27) only. The solution of this system has the general representation (see Sneddon and Lowengrub [327])

$$u_r = -\frac{2}{r}\left[(1 - 2v)\psi + z\frac{\partial \psi}{\partial z}\right], \tag{5.29}$$

$$u_z = -4(1 - v)\phi + 2z\frac{\partial \phi}{\partial z}, \quad v = \lambda/[2(\lambda + \mu)] \tag{5.30}$$

where
$$\Delta \phi = 0, \tag{5.31}$$

and the auxiliary potential $\psi(r, z)$ is determined by the Stokes–Beltrami system

$$\begin{bmatrix} \phi \\ \psi \end{bmatrix}_r = \begin{bmatrix} 0 & -r^{-1} \\ r & 0 \end{bmatrix} \begin{bmatrix} \phi \\ \psi \end{bmatrix}_z. \tag{5.32}$$

Under this representation, the corresponding nonvanishing stresses are given by

$$\tau_{rr} + \tau_{\theta\theta} = -4\mu\left[(1 + 2v)\frac{\partial \phi}{\partial z} + z\frac{\partial^2 \phi}{\partial z^2}\right],$$

$$\tau_{rr} - \tau_{\theta\theta} = -4\mu\left[(1 - 2v)\frac{1}{r^2}\left(r\frac{\partial \psi}{\partial r} - 2\psi\right) - z\frac{\partial^2 \phi}{\partial r^2} + \frac{2z}{r}\frac{\partial \phi}{\partial r}\right], \tag{5.33}$$

$$\tau_{rz} = 4\mu z\frac{\partial^2 \phi}{\partial r \partial z}, \quad \tau_{zz} = -4\mu\left[\frac{\partial \phi}{\partial z} - z\frac{\partial^2 \phi}{\partial z^2}\right].$$

On the other hand, in the case of purely torsional deformation with circular symmetry,

$$u_r = u_z = 0, \tag{5.34}$$

5.2 PUNCH, CRACK, AND TORSION PROBLEMS

while the tangential displacement u_θ is independent of θ. Accordingly, the relations (5.23) reduce to

$$e_{rr} = e_{\theta\theta} = e_{zz} = 0,$$

$$e_{r\theta} = \frac{1}{2}\left[\frac{\partial u_\theta}{\partial r} - \frac{u_\theta}{r}\right], \qquad e_{rz} = 0, \qquad e_{\theta z} = \frac{1}{2}\frac{\partial u_\theta}{\partial z}, \tag{5.35}$$

whereas Hooke's law (5.17) shows that the corresponding stresses are given by

$$\tau_{rr} = \tau_{\theta\theta} = \tau_{zz} = 0,$$

$$\tau_{r\theta} = \mu\left[\frac{\partial u_\theta}{\partial r} - \frac{u_\theta}{r}\right], \qquad \tau_{rz} = 0, \qquad \tau_{\theta z} = \mu\frac{\partial u_\theta}{\partial z}. \tag{5.36}$$

Insertion of these relations into the elastostatic equations (5.22) leads to the single equilibrium condition

$$\frac{\partial}{\partial r}\left(r^3 \frac{\partial \Phi}{\partial r}\right) + \frac{\partial}{\partial z}\left(r^3 \frac{\partial \Phi}{\partial z}\right) = 0 \qquad (\Phi := r^{-1} u_\theta), \tag{5.37}$$

and introduction of the auxiliary potential $\Psi(r, z)$ such that

$$\frac{\partial \Psi}{\partial r} = r^3 \frac{\partial \Phi}{\partial z}, \qquad \frac{\partial \Psi}{\partial z} = -r^3 \frac{\partial \Phi}{\partial r}, \tag{5.38}$$

produces the generalized Stokes–Beltrami system

$$\begin{bmatrix}\Phi \\ \Psi\end{bmatrix}_r = \begin{bmatrix}0 & -r^{-3} \\ r^3 & 0\end{bmatrix}\begin{bmatrix}\Phi \\ \Psi\end{bmatrix}_z, \tag{5.39}$$

which is descriptive of torsional deformation. The nonvanishing stresses $\tau_{r\theta}$ and $\tau_{\theta z}$ are given in terms of Φ by

$$\tau_{r\theta} = \mu r \frac{\partial \Phi}{\partial r}, \qquad \tau_{\theta z} = \mu r \frac{\partial \Phi}{\partial z}. \tag{5.40}$$

In the above representation, Φ and ψ are axially symmetric potentials in a space of five dimensions, a fact first exploited by Arndt [316] in connection with the torsion of shafts of revolution. Subsequently, Weinstein [328] showed that an important boundary value problem involving a plane circular crack under torsion may be solved by use of the correspondence principle to reduce the problem to the determination of the electrostatic potential of a disk in seven dimensions. Here, Weinstein's correspondence principle is employed in a different manner to link wide classes of boundary value problems. The crack problem treated by Weinstein is readily solved as but one application of the method presented.

I An Axially Symmetric Punch Problem: The Associated Torsion Problem with Prescribed Shear Stress Interior to a Circular Region and Zero Tangential Displacement Outside

THE PUNCH PROBLEM The stress distribution is sought in the elastic half-space $z \geq 0$ due to indentation of the plane boundary $z = 0$ by a perfectly rigid axially symmetric punch. It is assumed both that the displacement of the surface in contact with the punch is known and that the indented surface varies only slightly from the original surface.

On the part of the boundary outside the region of contact, the condition of zero normal stress is imposed. The shear stress is assumed to vanish at all points of the plane $z = 0$. Accordingly, the relations (5.30) and (5.33) show that the appropriate boundary value problem requires the solution of (5.31) in the region $z > 0$, subject to the boundary conditions

$$\phi = f(r), \quad r \leq a, \qquad \frac{\partial \phi}{\partial z} = 0, \quad r > a \qquad \text{on} \quad z = 0. \tag{5.41}$$

In what follows, it proves convenient to introduce the oblate spheroidal coordinates ξ, η according to [336, 337]

$$z + ir = a \sinh(\xi + i\eta). \tag{5.42}$$

The disk $r \leq a$, $z = 0$ is then given by $\xi = 0$, while its exterior in the plane $z = 0$ is given by $\eta = \pi/2$.

A routine calculation shows that (5.31) admits a solution of the type

$$\phi = \sum_{n=0}^{\infty} A_n P_{2n}(\cos \eta) Q_{2n}(i \sinh \xi), \tag{5.43}$$

while from (5.32), the corresponding ψ is given by

$$\psi = \sum_{n=0}^{\infty} \frac{A_n}{2n(2n+1)} \cosh \xi \sin \eta \, P'_{2n}(\cos \eta) Q'_{2n}(i \sinh \xi). \tag{5.44}$$

Since

$$\frac{\partial}{\partial \xi} = a \cosh \xi \cos \eta \frac{\partial}{\partial z} + a \sinh \xi \sin \eta \frac{\partial}{\partial r}, \tag{5.45}$$

$$\frac{\partial}{\partial \eta} = -a \sinh \xi \sin \eta \frac{\partial}{\partial z} + a \cosh \xi \cos \eta \frac{\partial}{\partial r}, \tag{5.46}$$

it follows, in particular, that

$$\left. \frac{\partial}{\partial z} \right|_{z=0, r \leq a} = \frac{1}{a \cos \eta} \frac{\partial}{\partial \xi}, \tag{5.47}$$

$$\left. \frac{\partial}{\partial z} \right|_{z=0, r > a} = \frac{-1}{a \sinh \xi} \frac{\partial}{\partial \eta}. \tag{5.48}$$

5.2 PUNCH, CRACK, AND TORSION PROBLEMS

Use of the latter relation shows that

$$\left.\frac{\partial \phi}{\partial z}\right|_{z=0, r>a} = \frac{1}{a \sinh \xi} \sum_{n=0}^{\infty} A_n P'_{2n}(0) Q_{2n}(i \sinh \xi) = 0,$$

whence the second boundary condition of (5.41) holds automatically for the representation (5.43).

If it is now assumed that $f(r)$ may be expanded in the form

$$f(r) = \sum_{n=0}^{\infty} \bar{a}_n (a^2 - r^2)^n = \sum_{n=0}^{\infty} \bar{a}_n a^{2n} \cos^{2n} \eta, \quad (5.49)$$

identification of (5.43) at $\xi = 0$ with (5.49) determines the A_n and consequently the potential ϕ, as given by Payne [99], namely,

$$\phi = \frac{1}{2} \sum_{n=0}^{\infty} \bar{a}_n a^{2n} n! \Gamma\left(n + \frac{1}{2}\right) \sum_{m=0}^{n} \frac{4(n-m)+1}{m! \Gamma(2n-m+\frac{3}{2})}$$

$$\times \frac{P_{2(n-m)}(\cos \eta) Q_{2(n-m)}(i \sinh \xi)}{Q_{2(n-m)}(+0 \cdot i)}, \quad (5.50)$$

where

$$Q_{2(n-m)}(+0 \cdot i) = \frac{1}{2} i(-1)^{n-m+1} \frac{\Gamma(n-m+\frac{1}{2})}{\Gamma(n-m+1)}. \quad (5.51)$$

THE ASSOCIATED TORSION PROBLEM It is now shown that the above punch problem corresponds to a certain axially symmetric torsion boundary value problem. In this connection, the boundary condition

$$\phi\big|_{z=0, r \leq a} = f(r) \quad (5.52)$$

is seen by (5.32) to be equivalent to

$$\partial \psi / \partial z \big|_{z=0, r \leq a} = -rf'(r). \quad (5.53)$$

But, by Weinstein's correspondence principle,

$$\psi = Cr^2 \Phi, \quad (5.54)$$

where

$$L_3(\Phi) = 0, \quad (5.55)$$

and accordingly, the boundary condition (5.53) may be expressed as

$$\partial \Phi / \partial z \big|_{z=0, r \leq a} = -f'(r)/Cr \quad (5.56)$$

on Φ. Similarly, the boundary condition

$$\partial \phi/\partial z|_{z=0, r>a} = g(r)$$

leads to

$$\Phi|_{z=0, r>a} = \frac{1}{Cr^2} \int_0^r \rho g(\rho) \, d\rho. \tag{5.57}$$

Thus, associated with the solution ϕ of the mixed boundary value problem

$$\begin{aligned} L_1(\phi) &= 0, & z > 0, \\ \phi &= f(r), & z = 0, \quad r \leq a \\ \partial \phi/\partial z &= g(r), & z = 0, \quad r > a \end{aligned} \tag{5.58}$$

is the solution Φ of the torsion problem

$$L_3(\Phi) = 0, \quad z > 0,$$

$$\frac{\partial \Phi}{\partial z} = -\frac{f'(r)}{Cr}, \quad z = 0, \quad r \leq a, \tag{5.59}$$

$$\Phi = \frac{1}{Cr^2} \int_0^r \rho g(\rho) \, d\rho, \quad z = 0, \quad r > a,$$

where

$$\Phi = \psi/Cr^2 \tag{5.60}$$

and ψ is given by (5.44). The specialization $g(r) = 0$ gives the punch problem with prescribed displacement interior to the disk $r = a$, $z = 0$ and zero normal stress $r > a$, $z = 0$. Accordingly, the solution (5.50) of this problem generates the solution Φ of the pure torsion problem wherein the shear stress $\tau_{\theta z}$ is specified interior to the disk $r = a$, $z = 0$ while exterior to the disk, the displacement is zero. The particular case $f(r) = \frac{1}{2}(a^2 - r^2)$ corresponds to the half-space torsion problem discussed by Reissner and Sagoci [329]. The case $f = \text{const}$, $g = 0$ gives the problem of a penny-shaped crack under torsion, solved in [327] using a dual integral equation formulation.

II A Boussinesq Problem: The Associated Torsion Problem with Prescribed Tangential Displacement Interior to a Circular Region and Zero Displacement Outside

THE BOUSSINESQ PROBLEM   The stress distribution in the elastic half-space $z \geq 0$ is now sought when the normal pressure τ_{zz} is specified in terms of r over the circular disk $r \leq a$, $z = 0$ and is assumed to vanish for $r > a$,

5.2 PUNCH, CRACK, AND TORSION PROBLEMS

$z = 0$. Special cases of this normal loading problem were first considered by Boussinesq [330]. Subsequently, other *Boussinesq-type* problems were investigated by various authors (see Sneddon [331] and Gladwell [332]).

In the present case, the relevant boundary value problem is given by

$$L_1 \phi = 0, \qquad z > 0,$$

$$\frac{\partial \phi}{\partial z} = g(r), \qquad z = 0, \quad r \leq a, \tag{5.61}$$

$$\frac{\partial \phi}{\partial z} = 0, \qquad z = 0, \quad r > a.$$

Thus, ϕ corresponds to the potential of an electrified disk when the surface density of the electric charge is prescribed on the surface of the disk. Beltrami [333] showed that this potential is given by

$$\phi = -\int_0^\infty e^{-zt} J_0(rt) dt \int_0^a s J_0(st) g(s) ds, \tag{5.62}$$

and, in particular, in the case of a spherical distribution of pressure over the plane area $z = 0$, $r \leq a$ so that (Payne [99])

$$\frac{\partial \phi}{\partial z} = -\frac{k}{4\mu}[(R^2 - r^2)^{1/2} - (R^2 - a^2)^{1/2}], \qquad z = 0, \quad r \leq a,$$

$$\frac{\partial \phi}{\partial z} = 0, \qquad z = 0, \quad r > a, \tag{5.63}$$

the result (5.62) specializes to (Watson [334])

$$\phi = \frac{k}{8\mu} \sum_{n=1}^{\infty} \frac{(-1)^n \Gamma(n - \tfrac{1}{2}) a^{n+1} 2^n}{\Gamma(\tfrac{1}{2})(R^2 - a^2)^{n-(1/2)}}$$

$$\times \int_0^\infty e^{-zt} t^{-(n+1)} J_0(rt) J_{n+1}(at) dt. \tag{5.64}$$

THE ASSOCIATED TORSION PROBLEM Use of Weinstein's correspondence principle (5.54) indicates that the Boussinesq problem (5.61) is associated with the pure torsion boundary value problem determined by

$$L_3(\Phi) = 0, \qquad z > 0,$$

$$\Phi = \frac{1}{Cr^2} \int_0^r \rho g(\rho) d\rho, \qquad z = 0, \quad r \leq a, \tag{5.65}$$

$$\Phi = 0, \qquad z = 0, \quad r > a.$$

The solution Φ of (5.65), determined by the solution ϕ of the Boussinesq problem (5.61), corresponds to the torsion of the half-space $z \geq 0$ where the tangential displacement interior to the disk $r = a$, $z = 0$ is specified, while the displacement exterior to the disk is assumed to be zero. The case $g(r) = \frac{1}{2}(a^2 - r^2)$ was solved by Reissner [335].

III The Penny-Shaped Crack: The Associated Torsion Problem with Prescribed Tangential Displacement Interior to a Circular Region and Zero Shear Stress Outside

THE PENNY-SHAPED CRACK PROBLEM The boundary conditions for the determination of the stress in the half-space $z \geq 0$ due to the presence of a penny-shaped crack occupying the region $r \leq a$, $z = 0$ and opened by the application of an internal stress $\tau_{zz} = -p(r)$ applied at the crack are (Sneddon [336])

$$\begin{aligned}
\tau_{rz} &= 0, & z &= 0, \quad r \geq 0, \\
\tau_{zz} &= -p(r), & z &= 0, \quad r \leq a, \\
u_z &= 0, & z &= 0, \quad r > a.
\end{aligned} \quad (5.66)$$

In terms of the potential ϕ, the mixed boundary value problem is given by

$$\begin{aligned}
L_1(\phi) &= 0, & z &> 0, \\
\partial\phi/\partial z &= p(r)/4\mu, & z &= 0, \quad r \leq a, \\
\phi &= 0, & z &= 0, \quad r > a.
\end{aligned} \quad (5.67)$$

Payne [99] has shown that an appropriate representation for ϕ in this crack problem is

$$\phi = \sum_{n=0}^{\infty} A_{2n+1} P_{2n+1}(\cos\eta) Q_{2n+1}(i \sinh\xi), \quad (5.68)$$

where ξ, η are the oblate spheroidal coordinates introduced in (5.42). Thus, since $P_{2n+1}(0) = 0$, it follows that $\phi = 0$ when $\eta = \pi/2$, so that ϕ vanishes exterior to the disk $r = a$, $z = 0$ for the representation (5.68). Moreover, interior to the disk $r = a$, $z = 0$, on use of (5.47), it is seen that

$$\frac{\partial \phi}{\partial z} = \frac{1}{\cos\eta} \sum_{n=0}^{\infty} A_{2n+1} P_{2n+1}(\cos\eta) Q_{2n+1}(+i0), \quad (5.69)$$

and if $p^*(r) = p(r)/4\mu$ admits an expansion of the type

$$p^*(r) = \sum_{m=0}^{\infty} \bar{a}_m (a^2 - r^2)^m = \sum_{m=0}^{\infty} \bar{a}_m a^{2m} \cos^{2m}\eta, \quad (5.70)$$

5.2 PUNCH, CRACK, AND TORSION PROBLEMS

the expression (5.68) will be the solution of the mixed boundary value problem (5.67) if A_{2n+1} can be determined such that

$$\sum_{n=0}^{\infty} A_{2n+1} P_{2n+1}(\cos\eta) Q_{2n+1}(+i0) = \sum_{m=0}^{\infty} a_m \cos^{2m+1}\eta. \qquad (5.71)$$

Now, standard Legendre series theory produces the relations

$$A_{2n+1} Q_{2n+1}(+i0) = (2n + \tfrac{3}{2}) \sum_{m=0}^{\infty} a_m \int_{-1}^{1} x^{2m+1} P_{2n+1}(x)\, dx$$

$$= (2n + \tfrac{3}{2}) \sum_{m=n}^{\infty} \frac{(2m+1)!\,\Gamma(m-n+\tfrac{1}{2})\, a_m}{2^{2n+1}(2m-2n)!\,(m+n+\tfrac{5}{2})} \qquad (5.72)$$

whereby the A_{2n+1} may be calculated. In the special case of constant pressure p_0,

$$\sum_{n=0}^{\infty} A_{2n+1} P_{2n+1}(\cos\eta) Q_{2n+1}(+i0) = p_0^* P_1(\cos\eta), \qquad (p_0^* = p_0/4\mu),$$

so that

$$A_1 = p_0^*/Q_1(+i0),$$

while the remaining $A_3, A_5, A_7, \ldots$, are all zero. But, (Hobson [337])

$$Q_1(\mu \pm i0) = \mu \log[(1+\mu)/(1-\mu)] - 1 \pm i\pi\mu, \qquad |\mu| < 1$$

so that, proceeding to the limit $\mu \to 0$, we obtain $Q_1(+i0) = -1$, whence

$$\phi = -p_0^* P_1(\cos\eta) Q_1(i \sinh\xi) = -p_0(\cos\eta) Q_1(i \sinh\xi). \qquad (5.73)$$

In the case of a spherical distribution of pressure of the type $(5.63)_1$, the solution

$$\phi = \frac{k}{8\pi\mu} \sum_{n=0}^{\infty} \frac{a^{2n+1} \Gamma(n-\tfrac{1}{2}) \Gamma(n+\tfrac{3}{2})}{(R^2 - a^2)^{n-(1/2)}}$$

$$\times \sum_{m=0}^{n} \frac{(-1)^m [4(n-m)+3](n-m)!}{m!\,\Gamma(2n-m+\tfrac{5}{2})\Gamma(n-m+\tfrac{3}{2})} P_{2(n-m)+1}(\cos\eta) Q_{2(n-m)+1}(i\sinh\xi)$$

(5.74)

was obtained by Payne [99].

THE ASSOCIATED TORSION PROBLEM Weinstein's correspondence principle (5.54) shows that the solution ϕ of the penny-shaped crack problem (5.67) determines the solution Φ of the pure torsion mixed boundary value problem

$$L_3(\Phi) = 0, \qquad z > 0,$$

$$\Phi = \frac{1}{4\mu Cr^2} \int_0^r \rho p(\rho)\,d\rho, \qquad z = 0, \quad r \le a, \tag{5.75}$$

$$\frac{\partial \Phi}{\partial z} = 0, \qquad z = 0, \quad r > a.$$

This corresponds to a half-space torsion problem wherein the tangential displacement interior to the disk $r = a$ on the bounding surface is prescribed, while the shear stress exterior to the disk vanishes. The particular case $p(r) = 2\mu(a^2 - r^2)$ was solved by Weinstein [328], who used the correspondence principle to reduce the torsion problem to the determination of the electrostatic potential of a disk in a space of seven dimensions.

IV An Axially Symmetric Crack Problem with Prescribed Deformation: The Associated Half-Space Torsion Problem with Shear Stress Prescribed Interior to a Disk on the Boundary

THE CRACK PROBLEM WITH PRESCRIBED DISPLACEMENT In this case, our interest lies in ascertaining the distribution of pressure required to give each face of a disk-shaped crack a prescribed deformation. The appropriate boundary value problem is

$$\begin{aligned} L_1(\phi) &= 0, & z &> 0, \\ \phi &= g(r), & z &= 0, \quad r \le a, \\ \phi &= 0, & z &= 0, \quad r > a. \end{aligned} \tag{5.76}$$

Here, the value of ϕ is sought in the region $z > 0$. This function is then continued as an odd function across the boundary $z = 0$ to provide the value of ϕ in the region $z < 0$. Thus, ϕ corresponds to the potential of a magnetic disk with the distribution of magnetic moment prescribed on the disk.

If we set $\phi^* = \partial \phi/\partial z$, where ϕ is the solution of the Neumann problem (5.61), then since ϕ^* is harmonic, we see that it is the required solution of the associated Dirichlet problem (5.76), namely, on appeal to (5.62),

$$\phi^* = \int_0^\infty t e^{-zt} J_0(rt)\,dt \int_0^b s J_0(st) g(s)\,ds. \tag{5.77}$$

In particular, if $g(r) = k(a^2 - r^2)^q$, where q is real, then

$$\phi^* = 2^q a^{q+1} \Gamma(q+1) k \int_0^\infty e^{-zt} t^{-q} J_0(rt) J_{q+1}(at)\,dt. \tag{5.78}$$

5.3 ANTIPLANE CRACK AND CONTACT PROBLEMS

Payne [99] used this expression to derive the solution

$$\phi^* = -\frac{k}{8(1-v)} \sum_{n=1}^{\infty} \frac{(-1)^n 2^n \Gamma(n-\tfrac{1}{2}) a^{n+1}}{\Gamma(\tfrac{1}{2})(R^2-a^2)^{n-1/2}} \int_0^\infty e^{-zt} t^{-n} J_0(rt) J_{n+1}(at)\, dt, \quad (5.79)$$

corresponding to the case in which the cavity takes the shape of a symmetrical lens of spherical indentation with radius R.

THE ASSOCIATED TORSION PROBLEM Application of Weinstein's correspondence principle shows that associated with the crack problem determined by (5.76) is the pure torsion boundary value problem

$$L_3(\Phi) = 0, \qquad z > 0,$$

$$\frac{\partial \Phi}{\partial z} = -\frac{g'(r)}{Cr}, \qquad z = 0, \quad r \le a, \qquad (5.80)$$

$$\frac{\partial \Phi}{\partial z} = 0, \qquad z = 0, \quad r > a.$$

This corresponds to the torsion of a half-space $z \ge 0$ when the shear is specified interior to the disk $r = a$, $z = 0$ and is zero outside that region. The specialization $g(r) = \tfrac{1}{2}(a^2 - r^2)$ leads to a torsion problem solved by Weinstein [328].

5.3 ANTIPLANE CRACK AND CONTACT PROBLEMS IN LAYERED ELASTIC MEDIA. BÄCKLUND TRANSFORMATIONS AND THE BERGMAN SERIES METHOD

In this section, the stress distribution due to a mode III displacement of an *inhomogeneous* elastic solid is investigated. Such antiplane deformations have been discussed for *homogeneous* media by Sneddon and Lowengrub [327]. Here, the important Bergman series method is introduced and linked to results obtained earlier by Loewner's method. Two basic crack and contact problems for a layered elastic medium in mode III deformation are thereby solved.

In mode III displacement, the only nonzero component of displacement $u_z(x, y)$ is related to the nonzero components of stress τ_{xz} and τ_{yz} by [327]

$$\tau_{xz} = \mu\, \partial u_z / \partial x, \qquad (5.81)$$

$$\tau_{yz} = \mu\, \partial u_z / \partial y, \qquad (5.82)$$

where x, y, z are rectangular Cartesian coordinates. Accordingly, the single equilibrium equation

$$\frac{\partial}{\partial x}\left[\mu \frac{\partial u_z}{\partial x}\right] + \frac{\partial}{\partial y}\left[\mu \frac{\partial u_z}{\partial y}\right] = 0 \qquad (5.83)$$

results, that is,
$$\nabla^2 u + \Lambda(x, y)u = 0, \qquad (5.84)$$
where
$$u = \mu^{1/2} u_z, \qquad (5.85)$$
$$\Lambda = -\{\nabla^2 \mu^{1/2}\}/\mu^{1/2}, \qquad (5.86)$$
and the shear modulus μ is here assumed to be independent of z.

Bergman series type solutions of (5.84) are now sought in the form[†]
$$u = \sum_{n=0}^{\infty} \{\alpha_n(x, y)\phi_n(x, y) - \beta_n(x, y)\psi_n(x, y)\}, \qquad (5.87)$$
where α_n, β_n are determined by the inhomogeneity of the medium, while ϕ_n, ψ_n satisfy the Cauchy–Riemann equations
$$\begin{bmatrix} \phi_n \\ \psi_n \end{bmatrix}_y = \begin{bmatrix} 0 & -1 \\ 1 & 0 \end{bmatrix} \begin{bmatrix} \phi_n \\ \psi_n \end{bmatrix}_x, \qquad n = 0, 1, 2, \ldots, \qquad (5.88)$$
together with the recurrence relations
$$\partial \phi_n/\partial x = \phi_{n-1}, \qquad n = 1, 2, \ldots, \qquad (5.89)$$
$$\partial \psi_n/\partial x = \psi_{n-1}, \qquad n = 1, 2, \ldots. \qquad (5.90)$$
Substitution of the series (5.87) into the equilibrium condition (5.84) yields
$$\sum_{n=0}^{\infty} \left\{ (\nabla^2 \alpha_n)\phi_n + 2\frac{\partial \alpha_n}{\partial x}\frac{\partial \phi_n}{\partial x} + 2\frac{\partial \alpha_n}{\partial y}\frac{\partial \phi_n}{\partial y} + \Lambda \alpha_n \phi_n \right.$$
$$\left. - (\nabla^2 \beta_n)\psi_n - 2\frac{\partial \beta_n}{\partial x}\frac{\partial \psi_n}{\partial x} - 2\frac{\partial \beta_n}{\partial y}\frac{\partial \psi_n}{\partial y} - \Lambda \beta_n \psi_n \right\} = 0,$$
whence, on use of recurrence relations (5.89) and (5.90) are the following system of recurrence relations for α_n and β_n is obtained:
$$\frac{\partial \alpha_0}{\partial x} - \frac{\partial \beta_0}{\partial y} = 0, \qquad \frac{\partial \alpha_0}{\partial y} + \frac{\partial \beta_0}{\partial x} = 0,$$
$$2\left[\frac{\partial \alpha_n}{\partial x} - \frac{\partial \beta_n}{\partial y}\right] + \nabla^2 \alpha_{n-1} + \Lambda(x, y)\alpha_{n-1} = 0, \qquad (5.91)$$
$$(n = 1, 2, \ldots)$$
$$2\left[\frac{\partial \alpha_n}{\partial y} + \frac{\partial \beta_n}{\partial x}\right] + \nabla^2 \beta_{n-1} + \Lambda(x, y)\beta_{n-1} = 0.$$

[†] An introductory account of the theory of Bergman series and associated linear operators is to be found in the monograph by Bergman [338]. Modern developments have been reviewed recently by Kreyszig [339] and Bauer and Ruscheweyh [315].

5.3 ANTIPLANE CRACK AND CONTACT PROBLEMS

This system may be rewritten more compactly as

$$\frac{\partial \gamma_0}{\partial \bar{z}} = 0, \qquad \frac{\partial \gamma_n}{\partial \bar{z}} + \frac{\partial^2 \gamma_{n-1}}{\partial z \, \partial \bar{z}} + \Lambda^*(z, \bar{z}) \gamma_{n-1} = 0, \qquad n = 1, 2, \ldots, \quad (5.92)$$

where $z = x + iy$, $\bar{z} = x - iy$ and

$$\gamma_n(z, \bar{z}) = \alpha_n \left\{ \frac{z + \bar{z}}{2}, \frac{z - \bar{z}}{2i} \right\} + i\beta_n \left\{ \frac{z + \bar{z}}{2}, \frac{z - \bar{z}}{2i} \right\},$$

$$\Lambda^*(z, \bar{z}) = \frac{1}{4} \Lambda \left\{ \frac{z + \bar{z}}{2}, \frac{z - \bar{z}}{2i} \right\}, \qquad (5.93)$$

while, if we set

$$\Psi_n = \phi_n + i\psi_n, \qquad n = 0, 1, 2, \ldots, \qquad (5.94)$$

by virtue of (5.88)–(5.90), the Ψ_n are analytic and satisfy the recurrence relations

$$d\Psi_n/dz = \Psi_{n-1}, \qquad n = 1, 2, \ldots. \qquad (5.95)$$

Accordingly,

$$\Psi_n(z) = \int_{z_0}^{z} \frac{(z-t)^{n-1}}{(n-1)!} \Psi_0(t) \, dt, \qquad n = 1, 2, \ldots, \qquad (5.96)$$

so that on insertion in (5.87) and use of (5.85), it is seen that, if we set $g_0 = 1$, the displacement u_z is given by

$$u_z = \mu^{-1/2} \operatorname{Re} \left[\Psi_0(z) + \int_{z_0}^{z} \sum_{n=1}^{\infty} \gamma_n(z, \bar{z}) \frac{(z-t)^{n-1}}{(n-1)!} \Psi_0(t) \, dt \right], \qquad (5.97)$$

or

$$u_z = \mu^{-1/2} \operatorname{Re} \left[\Psi_0(z) + \int_{z_0}^{z} K(z, \bar{z}; t) \Psi_0(t) \, dt \right], \qquad (5.98)$$

where

$$K(z, \bar{z}; t) = \sum_{n=1}^{\infty} \gamma_n(z, \bar{z}) \frac{(z-t)^{n-1}}{(n-1)!} \qquad (5.99)$$

is the kernel of the *Bergman integral operator* (5.98). The latter may be interpreted as acting on the complex analytic function $\Psi_0(z)$, corresponding to the solution of a boundary value problem for deformation of a *homogeneous medium* to produce the displacement (5.98) associated with a linked deformation of an *inhomogeneous elastic material*.[†]

[†] In a gasdynamics context, such operators have been used extensively to link compressible and incompressible flows (Mises and Schiffer [340]). On the other hand, the operators also link important boundary value problems of filtration in homogeneous and inhomogeneous strata (Alferov and Ryashentsev [341]).

Substitution of (5.98) into the stress–displacement relations (5.81) and (5.82) shows that τ_{xz} and τ_{yz} are given respectively by

$$\tau_{xz} = \frac{1}{2}\mu^{1/2}[\Phi'(z) + \bar{\Phi}'(\bar{z})] - \frac{1}{4}\mu^{-1/2}\left[\frac{\partial\mu}{\partial z} + \frac{\partial\mu}{\partial \bar{z}}\right][\Phi(z) + \bar{\Phi}(\bar{z})], \qquad (5.100)$$

$$\tau_{yz} = \frac{1}{2}i\mu^{1/2}[\Phi'(z) - \bar{\Phi}'(\bar{z})] - \frac{1}{4}i\mu^{-1/2}\left[\frac{\partial\mu}{\partial z} - \frac{\partial\mu}{\partial \bar{z}}\right][\Phi(z) + \bar{\Phi}(\bar{z})], \qquad (5.101)$$

where

$$\Phi = \Psi_0 + \int_{z_0}^{z} K(z, \bar{z}; t)\Psi_0(t)\, dt. \qquad (5.102)$$

The two simplest cases for which the Bergman series terminate are now considered, and a link with Loewner's Bäcklund transformation method is established. Crack and contact boundary-value problems are then solved for a layered elastic medium.

Case I $(\gamma_0 \neq 0, \gamma_n = 0, n = 1, 2, \ldots)$ In this instance, relations (5.92) show that $\Lambda = 0$, whence

$$\nabla^2 \mu^{1/2} = 0, \qquad (5.103)$$

while the nonvanishing components of displacement and stress are given by

$$u_z = \frac{1}{2}\mu^{-1/2}[\Psi_0(z) + \bar{\Psi}_0(\bar{z})],$$

$$\tau_{xz} = \frac{1}{2}\mu^{1/2}[\Psi'_0(z) + \bar{\Psi}'_0(\bar{z})] - \frac{1}{4}\mu^{-1/2}\left[\frac{\partial\mu}{\partial z} + \frac{\partial\mu}{\partial \bar{z}}\right][\Psi_0(z) + \bar{\Psi}_0(\bar{z})], \qquad (5.104)$$

$$\tau_{yz} = \frac{1}{2}i\mu^{1/2}[\Psi'_0(z) - \bar{\Psi}'_0(\bar{z})] - \frac{1}{4}i\mu^{-1/2}[\Psi_0(z) + \bar{\Psi}_0(\bar{z})].$$

In the particular case of a layered elastic medium in which the shear modulus μ depends only on y, condition (5.103) shows that

$$\mu = (\alpha y + \beta)^2, \qquad (5.105)$$

while relations (5.104) become

$$u_z = (\alpha y + \beta)^{-1}\phi_1, \quad \tau_{xz} = (\alpha y + \beta)\frac{\partial \phi_1}{\partial x}, \quad \tau_{yz} = -\alpha\phi_1 + (\alpha y + \beta)\frac{\partial \phi_1}{\partial y}, \qquad (5.106)$$

where

$$\Psi_0(z) = \phi_1 + i\phi_2 \qquad (5.107)$$

and ϕ_1, ϕ_2 are real harmonic functions.

5.3 ANTIPLANE CRACK AND CONTACT PROBLEMS

It is noted that the above result may be obtained via a special case of Weinstein's correspondence principle, namely,

$$\psi\{0\} = \delta y \phi\{2\}.$$

Case II ($\gamma_0 \neq 0, \gamma_1 \neq 0, \gamma_n = 0, n = 2, 3, \ldots$) In this case, relations (5.92) yield

$$\frac{\partial}{\partial \bar{z}} \left\{ \frac{1}{\Lambda^*} \frac{\partial}{\partial z} [\Lambda^* \gamma_0] \right\} + \Lambda^* \gamma_0 = 0, \qquad \Lambda^* \neq 0, \tag{5.108}$$

while the nonvanishing components of displacement and stress are given by

$$u_z = \frac{1}{2} \mu^{-1/2} [\Psi_1(z) + \bar{\Psi}_1(\bar{z})],$$

$$\tau_{xz} = \frac{1}{2} \mu^{1/2} [\Psi'_1(z) + \bar{\Psi}'_1(\bar{z})] - \frac{1}{4} \mu^{-1/2} \left[\frac{\partial \mu}{\partial z} + \frac{\partial \mu}{\partial \bar{z}} \right] [\Psi_1(z) + \bar{\Psi}_1(\bar{z})], \tag{5.109}$$

$$\tau_{yz} = \frac{1}{2} i \mu^{1/2} [\Psi'_1(z) - \bar{\Psi}'_1(\bar{z})] - \frac{1}{4} i \mu^{-1/2} \left[\frac{\partial \mu}{\partial z} - \frac{\partial \mu}{\partial \bar{z}} \right] [\Psi_1(z) + \bar{\Psi}_1(\bar{z})],$$

where

$$\Psi_1(z) = \Psi_0 + \gamma_1 \int_{z_0}^{z} \Psi_0(t) \, dt \tag{5.110}$$

and

$$\frac{\partial \gamma_0}{\partial \bar{z}} = 0, \qquad \frac{\partial \gamma_1}{\partial \bar{z}} + \Lambda^*(z, \bar{z}) \gamma_0 = 0. \tag{5.111}$$

If the shear modulus varies only with y, condition (5.108), together with (5.86), yields

$$[\log \Lambda^*]_{yy} + 4\Lambda^* = 0, \tag{5.112}$$

$$[\mu^{1/2}]_{yy} + 4\mu^{1/2} \Lambda^* = 0, \tag{5.113}$$

and it may be readily shown that the representation (5.109)–(5.111) for the particular solution

$$\mu = (\alpha y + \beta)^{-2} \tag{5.114}$$

of this system can be derived alternatively via the Loewner-type Bäcklund transformations discussed in Chapter 3.

AN ANTIPLANE CRACK PROBLEM   The stress distribution is sought in an inhomogeneous half-space $x \geq 0$ under mode III displacement when there is a crack occupying the region $x = 0, |y| < 1$ (see Fig. 5.1). The shear modulus μ is assumed to be of type (5.105).

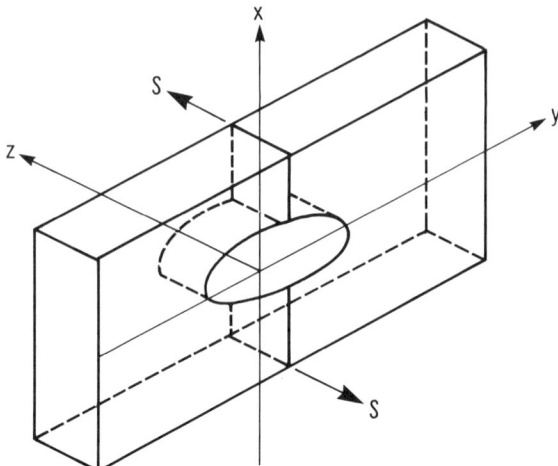

Fig. 5.1 Section of an elastic medium with a Griffith crack: mode III displacement.

The appropriate boundary value problem to be solved in this case is

$$\frac{\partial}{\partial x}\left[(\alpha y + \beta)^2 \frac{\partial u_z}{\partial x}\right] + \frac{\partial}{\partial y}\left[(\alpha y + \beta)^2 \frac{\partial u_z}{\partial y}\right] = 0, \quad x \geq 0,$$

$$\tau_{xz}(0, y) = 0, \quad |y| < 1, \quad (5.115)$$
$$\tau_{xz} \to S \quad \text{as} \quad x \to \infty,$$
$$u_z(0, y) = 0, \quad |y| > 1.$$

The representation (5.106) shows that

$$u_z = (\alpha y + \beta)^{-1} \phi_1 \quad (5.116)$$

is the solution of this mixed boundary value problem if the plane harmonic function ϕ_1 is such that

$$\frac{\partial \phi_1}{\partial x} = 0, \quad x = 0, \quad |y| < 1,$$

$$\frac{\partial \phi_1}{\partial x} \to \frac{S}{(\alpha y + \beta)} \quad \text{as} \quad x \to \infty, \quad (5.117)$$

$$\phi_1 = 0, \quad x = 0, \quad |y| > 1.$$

The appropriate ϕ_1 is readily seen to be given by

$$\phi_1 = S\left[\frac{x}{\alpha y + \beta} + \mathscr{F}_c[\psi(\xi)e^{-\xi x}; \xi \to y]\right], \quad (5.118)$$

5.3 ANTIPLANE CRACK AND CONTACT PROBLEMS

where $\psi(\xi)$ is required to satisfy the dual integral equations

$$\frac{d}{dy}\mathscr{F}_s[\psi(\xi); y] = \frac{1}{\alpha y + \beta}, \qquad 0 \le y < 1,$$
$$\mathscr{F}_c[\psi(\xi); y] = 0, \qquad y > 1, \tag{5.119}$$

where

$$\mathscr{F}_c[f(\xi,x); \xi \to y] := \sqrt{\frac{2}{\pi}} \int_0^\infty f(\xi,x) \cos(\xi y)\, d\xi \tag{5.120}$$

and

$$\mathscr{F}_s[f(\xi,x); \xi \to y] := \sqrt{\frac{2}{\pi}} \int_0^\infty f(\xi,x) \sin(\xi y)\, d\xi, \tag{5.121}$$

denote the Fourier cosine and sine transforms, respectively, of $f(\xi, x)$. Thus, (5.118) shows that

$$\phi_1 = 0, \qquad x = 0, \quad |y| > 1,$$

provided that

$$\mathscr{F}_c[\psi(\xi); y] = 0, \qquad y > 1,$$

while, since

$$\frac{\partial \phi_1}{\partial x} = S\left[\frac{1}{\alpha y + \beta} - \mathscr{F}_c[\xi\psi(\xi)e^{-\xi x}; \xi \to y]\right] \tag{5.122}$$

it follows that

$$\frac{\partial \phi_1}{\partial x} = S\left[\frac{1}{\alpha y + \beta} - \mathscr{F}_c[\xi\psi(\xi); y]\right]$$
$$= S\left[\frac{1}{\alpha y + \beta} - \frac{d}{dy}\mathscr{F}_s[\psi(\xi); y]\right] \qquad \text{at} \quad x = 0;$$

whence

$$\partial\phi_1/\partial x = 0, \qquad x = 0, \quad |y| < 1,$$

provided that

$$\frac{d}{dy}\mathscr{F}_s[\psi(\xi); y] = \frac{1}{\alpha y + \beta}, \qquad 0 \le y < 1.$$

Finally, (5.122) shows that

$$\frac{\partial \phi_1}{\partial x} \to \frac{S}{\alpha y + \beta} \qquad \text{as} \quad x \to \infty,$$

so that the required applied shear condition obtains.

5. BÄCKLUND TRANSFORMATIONS IN ELASTOSTATICS

The dual integral equations (5.119) are a particular case of the system

$$\frac{d}{dy}\mathscr{F}_S[\psi(\xi); y] = p(y), \qquad 0 \leq y < 1, \tag{5.123}$$

$$\mathscr{F}_C[\psi(\xi); y] = 0, \qquad y > 1, \tag{5.124}$$

which arises in the mode I displacement problem for a Griffith crack subjected to an internal pressure varying along its length. The complete solution of (5.123)–(5.124) was obtained by Sneddon and Elliot [342].[†] Thus, we see that (5.124) is identically satisfied if $\psi(\xi)$ has the integral representation

$$\psi(\xi) = \sqrt{\frac{\pi}{2}} \int_0^1 f(t) J_0(\xi t)\, dt. \tag{5.125}$$

Moreover, since

$$\int_0^\infty J_0(\xi t) \sin(\xi y)\, d\xi = \frac{H(y-t)}{\sqrt{y^2 - t^2}}, \tag{5.126}$$

we see that (5.123) is equivalent to the Abel integral equation

$$\frac{d}{dy} \int_0^y \frac{f(t)\, dt}{\sqrt{y^2 - t^2}} = p(y), \qquad 0 \leq y < 1, \tag{5.127}$$

with solution

$$f(t) = (2t/\pi) q(t), \tag{5.128}$$

where

$$q(t) = \int_0^t \frac{p(u)\, du}{\sqrt{t^2 - u^2}}. \tag{5.129}$$

Accordingly, the solution of the original antiplane crack problem is given by

$$\begin{aligned}
u_z &= (\alpha y + \beta)^{-1} S\left[\frac{x}{\alpha y + \beta} + \mathscr{F}_C[\psi(\xi)e^{-\xi x}; \xi \to y]\right], \\
\tau_{xz} &= S[1 - (\alpha y + \beta)\mathscr{F}_C[\xi\psi(\xi)e^{-\xi x}; \xi \to y]], \\
\tau_{yz} &= S\left[\frac{-2\alpha x}{\alpha y + \beta} - \alpha\mathscr{F}_C[\psi(\xi)e^{-\xi x}; \xi \to y]\right. \\
&\qquad \left. - (\alpha y + \beta)\mathscr{F}_S[\xi\psi(\xi)e^{-\xi x}; \xi \to y]\right]
\end{aligned} \tag{5.130}$$

[†] A more general pair of such dual integral equations was treated by Busbridge [343].

5.3 ANTIPLANE CRACK AND CONTACT PROBLEMS

where

$$\psi(\xi) = \sqrt{\frac{\pi}{2}} \int_0^1 \left\{ \frac{2t}{\pi} \int_0^t \frac{du}{(\alpha u + \beta)\sqrt{t^2 - u^2}} \right\} J_0(\xi t) \, dt. \quad (5.131)$$

AN ANTIPLANE CONTACT PROBLEM The stress distribution is sought in an inhomogeneous half-space $y \geq 0$ under the strip shear stress loading [101]

$$\tau_{yz} = \begin{cases} \tau_0, & |x| < a, \quad y = 0, \\ 0, & |x| > a, \quad y = 0, \end{cases} \quad (5.132)$$

where τ_0 is a constant. The variable shear modulus μ is again assumed to be of the type (5.105) so that the representations for the displacement u_z and shear stresses τ_{xz}, τ_{yz} become

$$u_z = \frac{1}{2}(\alpha y + \beta)^{-1}[\Psi_0(z) + \overline{\Psi}_0(\bar{z})], \quad (5.133)$$

$$\tau_{xz} = \frac{1}{2}(\alpha y + \beta)\left[\frac{\partial \Psi_0}{\partial z} + \frac{\partial \overline{\Psi}_0}{\partial \bar{z}}\right], \quad (5.134)$$

$$\tau_{yz} = \frac{1}{2}i(\alpha y + \beta)\left[\frac{\partial \Psi_0}{\partial z} - \frac{\partial \overline{\Psi}_0}{\partial \bar{z}}\right] - \frac{\alpha}{2}[\Psi_0 + \overline{\Psi}_0]. \quad (5.135)$$

The integral representation

$$\Psi_0(z) = \frac{1}{2\pi}\int_0^\infty A(p)\exp(ipz)\,dp \quad (5.136)$$

is now adopted for the analytic function $\Psi_0(z)$ so that the boundary condition (5.132) requires that

$$-\frac{1}{2\pi}\operatorname{Re}\int_0^\infty [\alpha + \beta p]A(p)\exp(ipx)\,dp = \begin{cases} \tau_0, & |x| < a, \\ 0, & |x| > a, \end{cases} \quad (5.137)$$

whence, from the inversion theorem for the Fourier cosine transform,

$$A(p) = -4\tau_0 p^{-1}(\alpha + \beta p)^{-1}\sin(pa). \quad (5.138)$$

Accordingly, the displacement and stress components have the integral representations

$$u_z = \frac{-2\tau_0}{\pi(\alpha y + \beta)}\int_0^\infty p^{-1}(\alpha + \beta p)^{-1}\sin(pa)\cos(px)\exp(-py)\,dp, \quad (5.139)$$

$$\tau_{xy} = \frac{2\tau_0(\alpha y + \beta)}{\pi}\int_0^\infty (\alpha + \beta p)^{-1}\sin(pa)\sin(px)\exp(-py)\,dp, \quad (5.140)$$

$$\tau_{yz} = \frac{2\tau_0}{\pi}\int_0^\infty p^{-1}(\alpha + \beta p)^{-1}(\alpha + \beta p + \alpha py)\sin(pa)\cos(px)\exp(-py)\,dp. \quad (5.141)$$

In particular, the displacement on the boundary $y = 0$ is given by

$$u_z|_{y=0} = -\frac{2\tau_0}{\pi\beta} \int_0^\infty p^{-1}(\alpha + \beta p)^{-1} \sin(pa)\cos(px)\,dp, \quad (5.142)$$

and if α is large enough that terms of order α^{-2} can be neglected, (5.142) reduces to

$$\begin{aligned} u_z|_{y=0} &= -\frac{2\tau_0}{\pi\beta\alpha} \int_0^\infty p^{-1} \sin(pa)\cos(px)\,dp \\ &= \begin{cases} -\dfrac{\tau_0}{\beta\alpha}, & |x| < a, \\ 0, & |x| > a. \end{cases} \end{aligned} \quad (5.143)$$

Thus, if terms of order α^{-2} are neglected, the displacement u_z on $y = 0$ is nonzero only in the strip where there is an applied shear load. Furthermore, the magnitude of this displacement on $y = 0$ tends to zero as $\alpha \to \infty$, in accordance with physical considerations, since the rigidity of the medium increases with α.

In conclusion, it is noted that other elastostatic crack problems for inhomogeneous media have been solved via the Bergman series method in [100, 102]. On the other hand, the method has been used extensively in elastodynamics and viscoelastodynamics where it was derived independently in the context of asymptotic wave front expansions [90-93, 344-350]. The application of simple Bäcklund transformations to the solution of initial-value problems involving wave propagation in inhomogeneous elastic media has been discussed by Clements and Rogers [89].

5.4 STRESS CONCENTRATION FOR SHEAR-STRAINED PRISMATIC BODIES WITH A NONLINEAR STRESS-STRAIN LAW

In his book "Kerbspannungslehre" [351] and in a subsequent paper [352], Neuber investigated the behavior of stress and strain concentrations for large antiplane deformation of notched prismatic bodies for which the classical Hooke's law is no longer applicable. Sokolovsky [353] subsequently introduced a nonlinear stress-deformation law for which a wide class of mode III deformation problems is readily solved once the solution of a corresponding harmonic boundary value problem for the linear Hooke material is known. Here, following the work of Clements and Rogers [96], new multiparameter nonlinear stress-deformation laws are obtained for which the Neuber pair of antiplane equations is reducible to an associated Cauchy-Riemann system by Bäcklund transformations. The stress-deformation laws

5.4 STRESS CONCENTRATION FOR SHEAR-STRAINED PRISMATIC BODIES

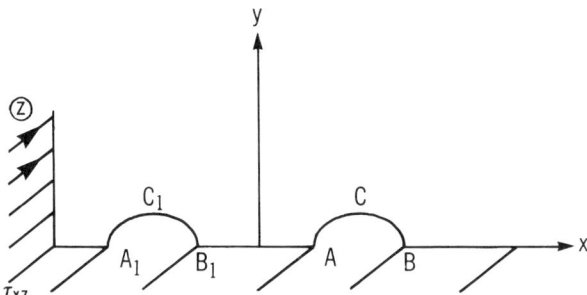

Fig. 5.2 The notched half-space $y \geq 0$ under antiplane deformation.

of Hooke and Sokolovsky emerge as special cases of the procedure that is here used to determine the stress distribution due to the antiplane deformation of a notched half-space of nonlinear elastic material.

Prismatic bodies under antiplane deformation are considered as shown in Fig. 5.2. If the coordinate system is chosen in such a way that the x and y axes lie in one of the cross sections of the prism, then, as in the preceding section, the only nonvanishing components of displacement and stress are u_z, τ_{xz}, and τ_{yz} which depend only on x and y with the shear stress components constrained by the single equilibrium equation

$$\frac{\partial}{\partial x}[\tau_{xz}] + \frac{\partial}{\partial y}[\tau_{yz}] = 0. \tag{5.144}$$

On the other hand, the nonvanishing components of the deformation tensor, namely, e_{xz} and e_{yz}, are subject to the strain–displacement relations

$$e_{xz} = \tfrac{1}{2} \partial u_z/\partial x, \tag{5.145}$$

$$e_{yz} = \tfrac{1}{2} \partial u_z/\partial y, \tag{5.146}$$

while, following Neuber [352], stress–strain relations

$$\tau_{xz} = (\tau/e)e_{xz}, \tag{5.147}$$

$$\tau_{yz} = (\tau/e)e_{yz} \tag{5.148}$$

are adopted, where

$$\tau^2 = \tau_{xz}^2 + \tau_{yz}^2, \tag{5.149}$$

$$e^2 = e_{xz}^2 + e_{yz}^2 \tag{5.150}$$

and τ, ε are subject to a *deformation law*

$$e = \Phi(\tau). \tag{5.151}$$

Introduction of the *warping* and *stress* functions $\phi(x, y)$ and $\psi(x, y)$, respectively, according to

$$\phi = ku_z, \qquad \tau_{xz} = \partial\psi/\partial y, \qquad \tau_{yz} = -\partial\psi/\partial x, \qquad (5.152)$$

shows that the antiplane deformations under consideration may be described by the matrix system

$$\begin{bmatrix} \phi \\ \psi \end{bmatrix}_x = \begin{bmatrix} 0 & 2k\Phi(\tau)/\tau \\ -\tau/(2k\Phi(\tau)) & 0 \end{bmatrix} \begin{bmatrix} \phi \\ \psi \end{bmatrix}_y, \qquad (5.153)$$

where it will be seen that the constant k represents a shear modulus in the Hookean range. If θ is the angle between the stress vector τ and the x axis, it follows that

$$\tau e^{i\theta} = \tau_{xz} + i\tau_{yz}, \qquad (5.154)$$

whence,

$$d\phi = 2k[e_{xz}\,dx + e_{yz}\,dy] = 2k\frac{e}{\tau}[\tau_{xz}\,dx + \tau_{yz}\,dy]$$
$$= 2k\Phi(\tau)[\cos\theta\,dx + \sin\theta\,dy], \qquad (5.155)$$
$$d\psi = -\tau_{yz}\,dx + \tau_{xz}\,dy = \tau[-\sin\theta\,dx + \cos\theta\,dy], \qquad (5.156)$$

or

$$(1/2k\Phi(\tau))\,d\phi = \cos\theta\,dx + \sin\theta\,dy, \qquad (5.157)$$

and

$$(1/\tau)\,d\psi = -\sin\theta\,dx + \cos\theta\,dy. \qquad (5.158)$$

Accordingly, if $z = x + iy$,

$$dz = dx + i\,dy$$
$$= [(1/2k\Phi(\tau))\cos\theta\,d\phi - (1/\tau)\sin\theta\,d\psi]$$
$$+ i[(1/2k\Phi(\tau))\sin\theta\,d\phi + (1/\tau)\cos\theta\,d\psi]$$
$$= e^{i\theta}[(1/2k\Phi(\tau))\,d\phi + i(1/\tau)\,d\psi], \qquad (5.159)$$

so that, on introduction of the Tschaplygin–Molenbroek transformation wherein τ and θ are taken as the new independent variables, it is seen that[†]

$$\partial z/\partial \tau = e^{i\theta}[(1/2k\Phi(\tau))\,\partial\phi/\partial\tau + i(1/\tau)\,\partial\psi/\partial\tau],$$
$$\partial z/\partial \theta = e^{i\theta}[(1/2k\Phi(\tau))\,\partial\phi/\partial\theta + i(1/\tau)\,\partial\psi/\partial\theta]. \qquad (5.160)$$

[†] This is analogous to the hodograph transformation of gasdynamics and elastodynamics, discussed in Chapters 3 and 4, respectively.

5.4 STRESS CONCENTRATION FOR SHEAR-STRAINED PRISMATIC BODIES

Application of the integrability condition $\partial^2 z/\partial \tau\, \partial\theta = \partial^2 z/\partial\theta\, \partial\tau$ now produces the Tschaplygin–Molenbroek system (Neuber [351]).

$$\begin{bmatrix} \phi \\ \psi \end{bmatrix}_\tau = \begin{bmatrix} 0 & -2k\Phi(\tau)/\tau^2 \\ \tau\Phi'(\tau)/(2k\Phi^2) & 0 \end{bmatrix} \begin{bmatrix} \phi \\ \psi \end{bmatrix}_\theta, \quad (5.161)$$

where it is assumed that

$$0 < |J(x,y;\tau,\theta)| < \infty. \quad (5.162)$$

Moreover, introduction of the new stress measure

$$s = \int_{\tau_0}^{\tau} \{\Phi'(\bar\sigma)/\bar\sigma\Phi\}^{1/2}\, d\bar\sigma \quad (5.163)$$

reduces (5.161) to the elliptic canonical form

$$\begin{bmatrix} \phi \\ \psi \end{bmatrix}_s = \begin{bmatrix} 0 & -K^{1/2}(s) \\ K^{-1/2}(s) & 0 \end{bmatrix} \begin{bmatrix} \phi \\ \psi \end{bmatrix}_\theta, \quad (5.164)$$

where

$$K = 4k^2\Phi^3/\tau^3\Phi'. \quad (5.165)$$

Loewner-type Bäcklund transformations may now be sought which reduce (5.164) to the Cauchy–Riemann system

$$\begin{bmatrix} \phi' \\ \psi' \end{bmatrix}_s = \begin{bmatrix} 0 & -1 \\ 1 & 0 \end{bmatrix} \begin{bmatrix} \phi' \\ \psi' \end{bmatrix}_\theta. \quad (5.166)$$

The results of Chapter 3 apply directly, whence we see that reduction to (5.166) may be achieved via the linear Bäcklund transformations

$$\begin{bmatrix} \phi' \\ \psi' \end{bmatrix}_s = \begin{bmatrix} a_1^1 & 0 \\ 0 & a_2^2 \end{bmatrix} \begin{bmatrix} \phi \\ \psi \end{bmatrix}_s + \begin{bmatrix} a_1^1 h_2^1 c_1^2/a_2^2 & 0 \\ 0 & a_2^2 h_1^2 c_2^1/a_1^1 \end{bmatrix} \begin{bmatrix} \phi \\ \psi \end{bmatrix},$$

$$\begin{bmatrix} \phi' \\ \psi' \end{bmatrix}_\theta = \begin{bmatrix} a_1^1 & 0 \\ 0 & a_2^2 \end{bmatrix} \begin{bmatrix} \phi \\ \psi \end{bmatrix}_\theta + \begin{bmatrix} 0 & c_2^1 \\ c_1^2 & 0 \end{bmatrix} \begin{bmatrix} \phi \\ \psi \end{bmatrix}, \quad (5.167)$$

where

$$h_2^1 = -K^{1/2}, \quad h_1^2 = K^{-1/2}, \quad a_1^1 a_2^2 = \text{const} = a, \quad (5.168)$$

and a_1^1 is such that

$$a_{1,s}^1 + \bar\alpha(a_1^1)^2 + \bar\beta = 0, \quad (5.169)$$

with $\bar\alpha, \bar\beta$ given in terms of the constants a, c_2^1, and c_1^2 by $\bar\alpha = c_2^1/a$, $\bar\beta = -c_1^2$.

Thus, the Tschaplygin–Molenbroek system (5.164) is reducible to the Cauchy–Riemann system (5.166) by the Bäcklund transformations (5.167) if the deformation function $\phi(\tau)$ is such that $K(\tau)$ as given by (5.165) adopts one of the following forms, corresponding in turn to the cases $\bar{\alpha} = 0$, $\bar{\beta} = 0$, $\bar{\beta}/\bar{\alpha} > 0$, and $\bar{\beta}/\bar{\alpha} < 0$:

(a) $\bar{\alpha} = 0$
$$K = a^2/[-\bar{\beta}s + \delta]^4,$$

(b) $\bar{\beta} = 0$
$$K = a^2[\bar{\alpha}s + \varepsilon]^4,$$

(c) $\bar{\beta}/\bar{\alpha} > 0$
$$K = (a\bar{\alpha}/\bar{\beta})^2 \cot^4(\bar{\beta}/\bar{\alpha})^{1/2}(-\bar{\alpha}s + \zeta),$$

(d) $\bar{\beta}/\bar{\alpha} < 0$
$$K = (a\bar{\alpha}/\bar{\beta})^2 \tanh^4(-\bar{\beta}/\bar{\alpha})^{1/2}(\bar{\alpha}s + \eta).$$

The approximations (a)–(d) to $K(\tau)$ each correspond to the following multiparameter deformation laws (which may be used to approximate empirical stress–strain relations):

(a) $\bar{\alpha} = 0$ In this case, if $F = 2k\Phi$, then
$$F^3/\tau^3 F' = a^2/[-\bar{\beta}s + \delta]^4 \tag{5.170}$$

and since
$$a\tau/ds = \{\tau F/F'\}^{1/2}, \tag{5.171}$$

we see that
$$d\tau/ds = a\tau^2/F(-\bar{\beta}s + \delta)^2, \tag{5.172}$$
$$dF/ds = F^2(-\bar{\beta}s + \delta)^2/a\tau, \tag{5.173}$$

which equations admit solutions corresponding to the parametric deformation laws
$$\begin{aligned} F &= 1/a[-\bar{\beta}(Ae^s + Be^{-s}) - (-\bar{\beta}s + \delta)(Ae^s - Be^{-s})], \\ \tau &= [-\bar{\beta}s + \delta]/[Ae^s + Be^{-s}], \end{aligned} \tag{5.174}$$

where A and B are arbitrary constants.

(b) $\bar{\beta} = 0$ In this case,
$$F^3/\tau^3 F' = a^2[\bar{\alpha}s + \varepsilon]^4,$$

whence, from (5.171),
$$d\tau/ds = a\tau^2(\bar{\alpha}s + \varepsilon)^2/F \tag{5.175}$$

and
$$dF/ds = F^2/a\tau(\bar{\alpha}s + \varepsilon)^2. \tag{5.176}$$

5.4 STRESS CONCENTRATION FOR SHEAR-STRAINED PRISMATIC BODIES

This pair of equations admits solutions corresponding to multiparameter deformation laws given by

$$F = [\bar{\alpha}s + \varepsilon]/[Ce^s + De^{-s}],$$
$$\tau = 1/a[\bar{\alpha}(Ce^s + De^{-s}) - (\bar{\alpha}s + \varepsilon)(Ce^s - De^{-s})],$$
(5.177)

where C, and D are arbitrary constants.

(c) $\bar{\beta}/\bar{\alpha} > 0$ Here, it is readily verified that

$$d\tau/ds = a\bar{\alpha}\tau^2 \cot^2(\bar{\beta}/\bar{\alpha})^{1/2}(-\bar{\alpha}s + \zeta)/\bar{\beta}F \qquad (5.178)$$

and

$$dF/ds = \bar{\beta}F^2 \tan^2(\bar{\beta}/\bar{\alpha})^{1/2}(-\bar{\alpha}s + \zeta)/a\bar{\alpha}\tau. \qquad (5.179)$$

These relations may be integrated numerically to approximate specific stress–strain laws over appropriate ranges of deformation.

(d) $\bar{\beta}/\bar{\alpha} < 0$ In this instance, the relations

$$d\tau/ds = -a\bar{\alpha}\tau^2 \tanh^2(-\bar{\beta}/\bar{\alpha})^{1/2}(\bar{\alpha}s + \eta)/\bar{\beta}F \qquad (5.180)$$

and

$$dF/ds = -\bar{\beta}F^2 \coth^2(-\bar{\beta}/\bar{\alpha})^{1/2}(\bar{\alpha}s + \eta)/a\bar{\alpha}\tau \qquad (5.181)$$

are obtained, and, as in case (c), recourse to numerical integration is required to determine the associated deformation laws.

Typical stress–deformation laws of types (a) and (b) are shown in Fig. 5.3. In particular, it is apparent that curves of type (a) could be used to approximate constitutive laws for materials which initially deform according to the

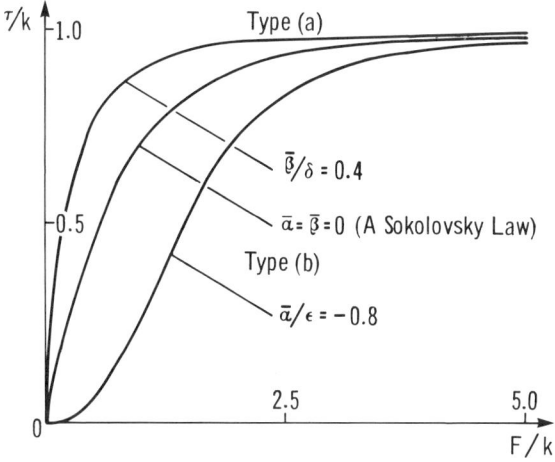

Fig. 5.3 Typical stress–deformation laws of types (a) and (b) (Clements and Rogers [96]).

law

$$\tau = e^n, \quad 0 < n < 1, \tag{5.182}$$

and then exhibit plastic behavior. Such constitutive laws have recently been adopted by Sih and MacDonald [354] in an investigation of the effect of material nonlinearity on crack propagation.

Included in Fig. 5.3 is a typical stress–deformation law with $\bar{\alpha} = \bar{\beta} = 0$. In this case the type (a) and type (b) laws coincide and the parameter s may be eliminated in (5.177) to give the *Sokolovsky Law* [353]

$$F = a\varepsilon^2\tau/[1 + 4a^2\varepsilon^2 CD\tau^2]^{1/2} \tag{5.183}$$

or, in the notation of Dobrovol'sky [355],

$$\tau = 2ke/[1 + (2me)^2]^{1/2}, \tag{5.184}$$

where $a^2\varepsilon^4 = +1$, $m^2\varepsilon^2 = -4CDK^2$, and k, m are parameters available for the approximation of the mechanical behavior of the material. Further Sokolovsky laws are shown in Fig. 5.4 for $k = 0.5$, 1.0, 5.0 and $k/m = 1.0$. Note that when $m = 0$ in (5.184), Hooke's law is retrieved.

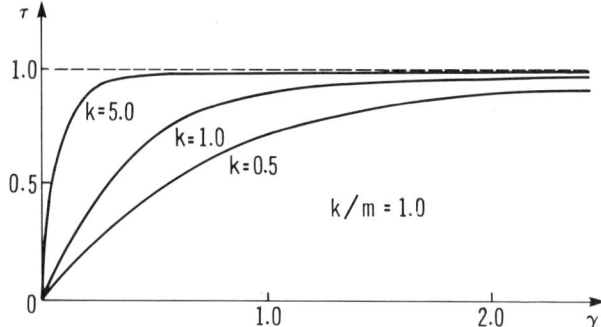

Fig. 5.4 Particular cases of Sokolovsky's law (Dobrovol'sky [355]).

Since Sokolovsky introduced his law in 1950, it has been used extensively to solve a wide range of antiplane notch problems for nonlinear elastic materials [352, 355]. In view of the importance of the Sokolovsky law, this section is concluded with an illustration of its use in the solution of a stress distribution problem involving a shear-strained, notched nonlinear elastic half-space. The basic procedure adopted is analogous to the Kármán–Tsien approximation of subsonic gasdynamics, in which, in the notation of Chapter 3, $q \leftrightarrow F(\tau)$, $\rho q/\rho_0 \leftrightarrow \tau$.

5.4 STRESS CONCENTRATION FOR SHEAR-STRAINED PRISMATIC BODIES

Thus, in the case of the Sokolovsky law (5.183), the Bäcklund transformation (5.167) reduces to

$$\begin{bmatrix} \phi' \\ \psi' \end{bmatrix}_s = \begin{bmatrix} a_1^1 & 0 \\ 0 & a_2^2 \end{bmatrix} \begin{bmatrix} \phi \\ \psi \end{bmatrix}_s,$$

$$\begin{bmatrix} \phi' \\ \psi' \end{bmatrix}_\theta = \begin{bmatrix} a_1^1 & 0 \\ 0 & a_2^2 \end{bmatrix} \begin{bmatrix} \phi \\ \psi \end{bmatrix}_\theta,$$
(5.185)

where $a_1^1 a_2^2 = a$ and $a_1^1 = \varepsilon^{-1}$. Accordingly, if we set $a = +1$, it follows that

$$\phi' = \varepsilon^{-1}\phi, \tag{5.186}$$

$$\psi' = \varepsilon\psi, \tag{5.187}$$

while the Cauchy–Riemann system (5.166) yields

$$\phi' + i\psi' = g(\bar{\sigma}), \tag{5.188}$$

where g is an analytic function of the complex variable $\bar{\sigma} = s - i\theta$. Thus,

$$\phi' = \tfrac{1}{2}[g + \bar{g}], \qquad \psi' = (1/2i)[g - \bar{g}],$$

so that

$$d\phi = \tfrac{1}{2}\varepsilon[dg + d\bar{g}], \qquad d\psi = (\varepsilon^{-1}/2i)[dg - d\bar{g}],$$

and substitution into the relation (5.159) yields

$$dz = \tfrac{1}{2}e^{i\theta}[(\varepsilon/F(\tau))(dg + d\bar{g}) + (1/\varepsilon\tau)(dg - d\bar{g})]. \tag{5.189}$$

At this stage, following a procedure analogous to the Kármán–Tsien method of subsonic gasdynamics, a new variable $F_H(F)$ is introduced according to[†]

$$dF_H/F_H = (\tau/F^2)\,dF, \tag{5.190}$$

where Sokolovsky's law (5.184) yields

$$\tau = \frac{E}{\sqrt{[1 + (mF/k)^2]}}, \tag{5.191}$$

and we have set $\varepsilon = +1$. Substitution of (5.191) into (5.190) and integration gives

$$F_H = \frac{cF}{(k/m)[1 + \sqrt{1 + (mF/k)^2}]}, \tag{5.192}$$

[†] An account of the important Kármán–Tsien procedure in a gasdynamics context is given by Shapiro [356].

where c is an arbitrary constant of integration. If we require that $F = F_H$ correspond to Hooke's law ($m = 0$), we set $c = 2k/m$, and the $F_H(F)$ relation (5.192) becomes

$$F_H = \frac{2F}{[1 + \sqrt{1 + (mF/k)^2}]}, \tag{5.193}$$

whence, on inversion,

$$F = \frac{4(k/m)^2 F_H}{[4(k/m)^2 - F_H^2]}. \tag{5.194}$$

Combination of (5.191) and (5.194) now yields

$$\tau = \frac{4(k/m)^2 F_H}{[4(k/m)^2 + F_H^2]}, \tag{5.195}$$

so that, on substitution of (5.194) and (5.195) into (5.189), we see that

$$dz = F_H^{-1} e^{i\theta} dg - \tfrac{1}{4}(m/k)^2 F_H e^{i\theta} d\bar{g}. \tag{5.196}$$

The analytic function $g(\bar{\sigma})$ corresponds to an antiplane deformation of a Hooke material in a certain ζ plane where

$$d\zeta = F_H^{-1} e^{i\theta} dg, \tag{5.197}$$

and it follows that (5.196) may be written as

$$\begin{aligned} dz &= d\zeta - \tfrac{1}{4}(m/k)^2 F_H e^{i\theta} d\bar{g} = d\zeta - \tfrac{1}{4}(m/k)^2 (d\bar{g}/d\bar{\zeta}) d\bar{g} \\ &= d\zeta - \tfrac{1}{4}(m/k)^2 (d\bar{g}/d\bar{\zeta})^2 d\bar{\zeta}, \end{aligned} \tag{5.198}$$

whence,

$$z = \zeta - \frac{1}{4}\left(\frac{m}{k}\right)^2 \int \left(\frac{d\bar{g}}{d\bar{\zeta}}\right)^2 d\bar{\zeta}, \tag{5.199}$$

up to an arbitrary constant of integration.

Expression (5.199) relates the antiplane deformation of a Hooke material to an associated deformation of a nonlinear elastic medium that can be modeled by the Sokolovsky law (5.184). An illustration of such an association is presented below.

Antiplane Deformation of a Nonlinear Elastic Notched Half-space

The problem considered here is that of the antiplane deformation of a nonlinear elastic prism, the cross section of which is a semiplane with two oval notches (Fig. 5.2). The complex potential associated with the correspond-

5.4 STRESS CONCENTRATION FOR SHEAR-STRAINED PRISMATIC BODIES

ing Hookean problem is (Dobrovol'sky [355])

$$g(\zeta) = T\left[\zeta + \beta\left(\frac{1}{\zeta - \alpha} + \frac{1}{\zeta + \alpha}\right)\right], \quad (5.200)$$

where T, α, and β are real parameters. Insertion of (5.200) into (5.199) and integration yield

$$z = \zeta - \frac{1}{4}\left(\frac{m}{k}\right)^2 \Big[\bar{\zeta} + 2\beta(\bar{v}_1 + \bar{v}_2)$$

$$-\beta^2\left(\frac{\bar{\zeta}}{\alpha^2}\bar{v}_1\bar{v}_2 + \frac{1}{3}\bar{v}_1^3 + \frac{1}{3}\bar{v}_2^3 + \frac{1}{2\alpha^3}\ln\left(\frac{\bar{v}_2}{\bar{v}_1}\right)\right)\Big], \quad (5.201)$$

where, if $\zeta = \xi + i\eta$,

$$v_1 = \lambda_1 + i\mu_1, \qquad v_2 = \lambda_2 + i\mu_2,$$

$$\lambda_1 = \frac{\xi - \alpha}{[(\xi - \alpha)^2 + \eta^2]}, \qquad \lambda_2 = \frac{\xi + \alpha}{[(\xi + \alpha)^2 + \eta^2]}, \quad (5.202)$$

$$\mu_1 = \frac{-\eta}{[(\xi - \alpha)^2 + \eta^2]}, \qquad \mu_2 = \frac{-\eta}{[(\xi + \alpha)^2 + \eta^2]}.$$

The warping and stress functions ϕ, ψ are given by

$$\phi = T[\xi + \beta(\lambda_1 + \lambda_2)], \quad (5.203)$$

$$\psi = T[\eta + \beta(\mu_1 + \mu_2)], \quad (5.204)$$

while the longitudinal displacement is

$$u_z = (T/k)[\xi + \beta(\lambda_1 + \lambda_2)]. \quad (5.205)$$

The parametric equation of the boundary in the physical plane is obtained by substitution of the values of ξ, η such that $\psi(\xi,\eta) = 0$ into (5.201). Use of (5.204) shows that the rectilinear sections of the boundary in Fig. 5.2 correspond to $\eta = 0$, while the curvilinear arcs correspond to the Persey curves

$$[(\xi - \alpha)^2 + \eta^2][(\xi + \alpha)^2 + \eta^2] = 2\beta[\xi^2 + \eta^2 + \alpha^2], \quad (5.206)$$

where the constraint

$$\alpha > \sqrt{\tfrac{1}{2}\beta(1 + \sqrt{2})} \quad (5.207)$$

on the parameters α, β ensures that the curves do not intersect.

In Fig. 5.5, the stress distribution along the boundary contour is shown for $T = 1.0$, $m/k = 0.4$, $\alpha = 2.1224$, and $\beta = 1.1155$, where the values of α and β are chosen so that the abscissas of A and B are 1.0 and 3.0, respectively, [355].

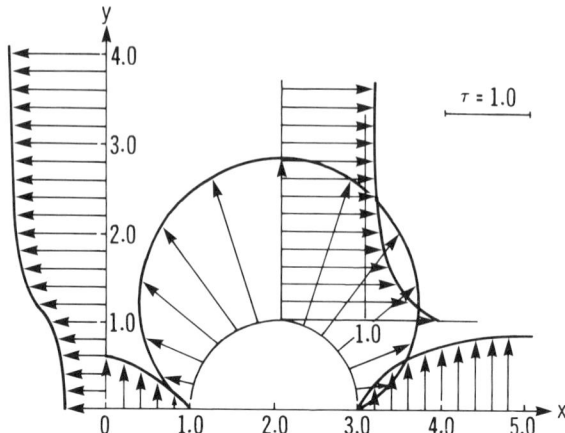

Fig. 5.5 The stress distribution along the boundary contour in the nonlinear elastic antiplane notch problem (Dobrovol'sky [355]).

Dobrovol'sky generalized the above discussion to the case in which the complex potential of the notch problem for the Hookean material adopts the form

$$g(\zeta) = T\left[\zeta + \sum_{s=1}^{N} \frac{\beta_s}{\zeta - \alpha_s}\right], \qquad (5.208)$$

where $T, \alpha_s, \beta_s, s = 1, \ldots, n$, are arbitrary real parameters. In this case, substitution of (5.208) into (5.199) and integration yield

$$z = \zeta - \frac{1}{4}\left(\frac{m}{k}\right)^2 \left[\bar{\zeta} + 2\sum_{s=1}^{N}\beta_s \bar{v}_s - \frac{1}{3}\sum_{s=1}^{N}\beta_s^2 \bar{v}_s^2 \right. \\
\left. - \sum_{s,r=1, s\neq r} \left\{\frac{2\bar{\zeta} - (\alpha_s + \bar{\alpha}_r)}{(\bar{\alpha}_s - \bar{\alpha}_r)^2}\beta_s\beta_r\bar{v}_s\bar{v}_r + \frac{2\beta_s\beta_r}{(\bar{\alpha}_r - \bar{\alpha}_s)^3}\ln\left(\frac{\bar{v}_s}{\bar{v}_r}\right)\right\}\right], \quad (5.209)$$

where

$$v_s = \lambda_s + i\mu_s, \qquad \alpha_2 = \alpha'_s + i\alpha''_s, \qquad s = 1, 2, \ldots, N,$$

$$\lambda_s = \frac{\xi - \alpha'_s}{[(\xi - \alpha'_s)^2 + (\eta - \alpha''_s)^2]}, \qquad \mu_s = \frac{\alpha''_s - \eta}{[(\xi - \alpha'_s)^2 + (\eta - \alpha''_s)^2]}. \qquad (5.210)$$

The warping and stress functions ϕ, ψ are given by

$$\phi = T\left[\xi + \sum_{s=1}^{N}\beta_s\lambda_s\right] \qquad (5.211)$$

5.4 STRESS CONCENTRATION FOR SHEAR-STRAINED PRISMATIC BODIES

and

$$\psi = T\left[\eta + \sum_{s=1}^{N} \beta_s \mu_s\right], \tag{5.212}$$

while the longitudinal displacement becomes

$$u_z = (T/k)\left[\xi + \sum_{s=1}^{N} \beta_s \lambda_s\right]. \tag{5.213}$$

APPENDIX **I**

Properties of the Hirota Bilinear Operators

$$D_t^n D_x^m a \circ b := \left(\frac{\partial}{\partial t} - \frac{\partial}{\partial t'}\right)^n \left(\frac{\partial}{\partial x} - \frac{\partial}{\partial x'}\right)^m a(t,x)b(t',x')\bigg|_{x'=x,\,t'=t} \quad (\text{I.1})$$

$$D_x^m a \circ b = (-1)^m D_x^m (b \circ a),$$

$$\frac{\partial^2}{\partial x\,\partial t}(\log f) = \frac{1}{2f^2} D_x D_t f \circ f, \quad (\text{I.2})$$

$$\frac{\partial^4}{\partial x^4}(\log f) = \frac{1}{2f^2} D_x^4 f \circ f - 6\left[\frac{1}{2f^2} D_x^2 f \circ f\right]^2, \quad (\text{I.3})$$

$$(D_x^2 a \circ b)cd - ab(D_x^2 c \circ d) = D_x[(D_x a \circ d) \circ cb + ad \circ (D_x c \circ b)] \quad (\text{I.4})$$

$$(D_x^4 a \circ a)cc - aa(D_x^4 c \circ c) = 2D_x(D_x^3 a \circ c) \circ ca$$
$$\qquad + 6D_x(D_x^2 a \circ c) \circ (D_x c \circ a), \quad (\text{I.5})$$

$$D_t(D_x^2 a \circ b) \circ ab = D_x[(D_x D_t a \circ b) \circ ab + (D_t a \circ b) \circ (D_x a \circ b)], \quad (\text{I.6})$$

$$D_t[(D_x a \circ b) \circ cd + ab \circ (D_x c \circ d)]$$
$$= (D_t D_x a \circ d)cb - ad(D_t D_x c \circ b)$$
$$+ (D_t a \circ d)(D_x c \circ b) - (D_x a \circ d)(D_t c \circ b), \quad (\text{I.7})$$

$$bbD_xD_ta \circ a - aaD_xD_tb \circ b = 2D_x(ba) \circ (D_tb \circ a), \tag{I.8}$$

$$\begin{aligned}bbD_x^6 a \circ a - aaD_x^6 b \circ b = {}& D_x[2(ba) \circ (D_x^5 b \circ a) - 10(D_x b \circ a) \circ (D_x^4 b \circ a) \\ & + 20(D_x^2 b \circ a) \circ (D_x^3 b \circ a)] \\ & - 5[(D_x^2 b \circ b)(D_x^4 a \circ a) - (D_x^4 b \circ b)(D_x^2 a \circ a)],\end{aligned} \tag{I.9}$$

$$\begin{aligned}bbD_x^6 a \circ a - aaD_x^6 b \circ b = {}& D_x^3[2(ba) \circ (D_x^3 b \circ a) - 6(D_x b \circ a) \circ (D_x^2 b \circ a)] \\ & + 3[(D_x^2 b \circ b)(D_x^4 a \circ a) - (D_x^4 b \circ b)(D_x^2 a \circ a)],\end{aligned} \tag{I.10}$$

$$\begin{aligned}D_x^3(ba) & \circ (D_x^3 b \circ a) + D_x^3(D_x b \circ a) \circ (D_x^2 b \circ a) \\ & = D_x[(ba) \circ (D_x^5 b \circ a) - (D_x b \circ a) \circ (D_x^4 b \circ a) \\ & \quad - 2(D_x^2 b \circ a) \circ (D_x^3 b \circ a)],\end{aligned} \tag{I.11}$$

$$cd(D_x a \circ b) - ab(D_x c \circ d) = bd(D_x a \circ c) - ac(D_x b \circ d), \tag{I.12}$$

$$cd(D_x a \circ b) - ab(D_x c \circ d) = D_x ad \circ bc, \tag{I.13}$$

$$(D_x a \circ b)c - (D_x a \circ c)b = -a(D_x b \circ c). \tag{I.14}$$

APPENDIX **II**

Differential Forms

II.1 TANGENT SPACES AND VECTOR FIELDS ON $\mathbb{R}^k$

Suppose $x^i, i = 1, \ldots, k$, are coordinates on $\mathbb{R}^k$ and that f is a differentiable function on $\mathbb{R}^k$. If **V** is a vector at $\mathbf{x} \in \mathbb{R}^k$ with components $V^1, \ldots, V^k$, then the *directional derivative* of f at **x** in the direction of **V** is given by

$$\mathbf{V}(f)(\mathbf{x}) = V^i \frac{\partial f}{\partial x^i}(\mathbf{x}). \tag{II.1}$$

Thus, we may identify the real k-dimensional vector space with basis $\{\partial/\partial x^i|_\mathbf{x}; i = 1, \ldots, k\}$, with the collection of directional derivatives at **x**. This vector space is called the *tangent space* to $\mathbb{R}^k$ at **x** and will be denoted by $T_\mathbf{x}\mathbb{R}^k$. In what follows, it will be convenient to write the basis for $T_\mathbf{x}\mathbb{R}^k$ more concisely as $\{\partial/\partial x^i\}$.

If $X^1, \ldots, X^k$ are smooth functions on $\mathbb{R}^k$, then to each point $\mathbf{x} \in \mathbb{R}^k$ one may assign an operator $\mathbf{X}(\mathbf{x})$ given by

$$\mathbf{X}(\mathbf{x}) := X^i(\mathbf{x}) \, \partial/\partial x^i. \tag{II.2}$$

An assignment of such an operator to each point in $\mathbb{R}^k$ is called a *vector field on* $\mathbb{R}^k$. This provides a map from $C^\infty(\mathbb{R}^k)$ to $C^\infty(\mathbb{R}^k)$ by sending $f \in C^\infty(\mathbb{R}^k)$ to

DIFFERENTIAL FORMS

$X(f) \in C^\infty(\mathbb{R}^k)$, where

$$X(f)(\mathbf{x}) = X^i(\mathbf{x}) \frac{\partial f}{\partial x^i}(\mathbf{x}). \tag{II.3}$$

II.2 DIFFERENTIAL p-FORMS ON $\mathbb{R}^k$

We now introduce 1-*forms* at a point $\mathbf{x}$ in $\mathbb{R}^k$ as vectors in the dual space to $T_\mathbf{x}\mathbb{R}^k$, denoted by $T_\mathbf{x}^*\mathbb{R}^k$. The space $T_\mathbf{x}^*\mathbb{R}^k$ is thus a real vector space of dimension k and has a basis dual to the basis $\{\partial/\partial x^i; i = 1, \ldots, k\}$, which for reasons that will become apparent is denoted by $\{dx^1, dx^2, \ldots, dx^k\}$. The canonical pairing of vectors in $T_\mathbf{x}\mathbb{R}^k$ and $T_\mathbf{x}^*\mathbb{R}^k$ is denoted by $\langle\,,\,\rangle$, so that

$$\langle \partial/\partial x^i, dx^j \rangle = \delta_i^j. \tag{II.4}$$

Following Flanders [122], for each $p = 0, 1, 2, \ldots, k$, a new vector space $\wedge^p T_\mathbf{x}^*\mathbb{R}^k$ over $\mathbb{R}$ may be constructed. Thus, if we set

$$\wedge^0 T_\mathbf{x}^*\mathbb{R}^k = \mathbb{R}^k, \tag{II.5}$$

$$\wedge^1 T_\mathbf{x}^*\mathbb{R}^k = T_\mathbf{x}^*\mathbb{R}^k, \tag{II.6}$$

then $\wedge^p T_\mathbf{x}^*\mathbb{R}^k$ is the space consisting of all finite formal sums (*p-forms*)

$$\sum c(\eta_1 \wedge \eta_2 \wedge \cdots \wedge \eta_p), \tag{II.7}$$

where $c \in \mathbb{R}$, $\eta_i \in T_\mathbf{x}^*\mathbb{R}^k$, and the exterior (*wedge*) product $\wedge$ is subject to the rules

$$(a\xi + b\zeta) \wedge \eta_2 \wedge \cdots \wedge \eta_p = a(\xi \wedge \eta_2 \wedge \cdots \wedge \eta_p) + b(\zeta \wedge \eta_2 \wedge \cdots \wedge \eta_p), \tag{II.8}$$

$$\eta_1 \wedge \eta_2 \wedge \cdots \wedge \eta_p = 0 \quad \text{if for some} \quad i \neq j, \quad \eta_i = \eta_j, \tag{II.9}$$

$\eta_1 \wedge \eta_2 \wedge \cdots \wedge \eta_p$ changes sign if any two η_i are interchanged. (II.10)

It follows from (II.8) and (II.10) that the product $\wedge$ is linear in each factor. Thus, for example,

$$\eta_1 \wedge (a\xi + b\zeta) \wedge \eta_2 \wedge \cdots \wedge \eta_p$$
$$= a(\eta_1 \wedge \xi \wedge \eta_2 \wedge \cdots \wedge \eta_p) + b(\eta_1 \wedge \zeta \wedge \cdots \wedge \eta_p).$$

Since $\{dx^1, \ldots, dx^k\}$ is a basis for $\wedge^1 T_\mathbf{x}^*\mathbb{R}^k$, it follows by a short computation that each p-form $\eta \in \wedge^p T_\mathbf{x}^*\mathbb{R}^k$ is a linear combination of terms of the form

$$\xi_{i_1} dx^{i_1} \wedge \eta_{i_2} dx^{i_2} \wedge \cdots \wedge \zeta_{i_p} dx^{i_p}. \tag{II.11}$$

But from (II.8), this is the same as

$$\xi_{i_1} \eta_{i_2} \cdots \zeta_{i_p} dx^{i_1} \wedge dx^{i_2} \wedge \cdots \wedge dx^{i_p}, \tag{II.12}$$

and from (II.9)–(II.10) we conclude that any element of $\wedge^p T^*_x \mathbb{R}^k$ may be written in the form

$$\sum \xi_{i_1 \ldots i_p} dx^{i_1} \wedge \cdots \wedge dx^{i_p}, \quad 1 \leq i_1 < i_2 \cdots < i_p \leq k, \quad \text{(II.13)}$$

and that the p-forms $\{dx^{i_1} \wedge \cdots \wedge dx^{i_p}\}$, $1 \leq i_1 < i_2 \cdots < i_p \leq k$, comprise a basis for the space $\wedge^p T^*_x \mathbb{R}^k$ of dimension $\binom{k}{p}$. Note that when $p = k$, this basis consists of only one element, namely,

$$dx^1 \wedge dx^2 \wedge \cdots \wedge dx^k,$$

so that $\wedge^k T^*_x \mathbb{R}^k$ is one dimensional. If $p > k$, then there must be at least one term dx^{a_i} that appears twice in the sum (II.13), whence from (II.9), it follows that any p-form with $p > k$ is zero.

Use may be made of the fact that $\wedge^k T^*_x \mathbb{R}^k$ is one dimensional to give a basis-independent definition of the determinant of a linear transformation on $\mathbb{R}^k$ [122]. Thus if $A: \mathbb{R}^k \to \mathbb{R}^k$ is such a transformation, then there is an alternating linear map from $\wedge^k T^*_x \mathbb{R}^k$ to itself defined by

$$\xi_1 \wedge \xi_2 \wedge \cdots \wedge \xi_k \to A\xi_1 \wedge A\xi_2 \wedge \cdots \wedge A\xi_k.$$

Since $\wedge^k T^*_x \mathbb{R}^k$ is one dimensional, it follows that there is a real number $|A|$ such that

$$A\xi_1 \wedge A\xi_2 \wedge \cdots \wedge A\xi_k = |A| \, \xi_1 \wedge \xi_2 \wedge \cdots \wedge \xi_k.$$

The determinant of A is defined to be the number $|A|$. The reader may verify that if A is given in a basis by the matrix $[a_{ij}]$, then $|A|$ is the determinant of this matrix.

A p-form of the type $\xi = \xi_1 \wedge \xi_2 \wedge \cdots \wedge \xi_p$ is called a *monomial*. Each monomial p-form ξ determines a functional on

$$\underbrace{T_x \mathbb{R}^k \times T_x \mathbb{R}^k \times \cdots \times T_x \mathbb{R}^k}_{p\text{-copies}}$$

by

$$\xi(X_1, \ldots, X_p) = \begin{vmatrix} \langle \xi_1, X_1 \rangle & \langle \xi_2, X_1 \rangle & \cdots & \langle \xi_p, X_1 \rangle \\ \langle \xi_1, X_2 \rangle & \langle \xi_2, X_2 \rangle & \cdots & \langle \xi_p, X_2 \rangle \\ \vdots & \vdots & & \vdots \\ \langle \xi_1, X_p \rangle & \langle \xi_2, X_p \rangle & \cdots & \langle \xi_p, X_p \rangle \end{vmatrix}$$

where $X_i \in T_x \mathbb{R}^k$, $i = 1, \ldots, p$. It follows from the properties of determinants that ξ is a p-linear functional and is totally antisymmetric. Thus

$$\xi(X_1, \ldots, \lambda X_i + \mu Y, \ldots, X_p)$$
$$= \lambda \xi(X_1, \ldots, X_i, \ldots, X_p) + \mu \xi(X_1, \ldots, Y, \ldots, X_p) \quad \text{for each} \quad i,$$

and $\xi(X_1, \ldots, X_p)$ changes sign if any two X_i are interchanged.

If η is another monomial p-form $\eta \in \wedge^p T_{\mathbf{x}}^* \mathbb{R}^k$ and a, b are constants, then we may define $(a\xi + b\eta)(X_1, \ldots, X_p)$ by

$$(a\xi + b\eta)(X_1, \ldots, X_p) = a\xi(X_1, \ldots, X_p) + b\eta(X_1, \ldots, X_p).$$

Thus every p-form determines a p-linear totally antisymmetric functional on $T_{\mathbf{x}}\mathbb{R}^k \times \cdots \times T_{\mathbf{x}}\mathbb{R}^k$.

A *differential p-form* may be defined as a smooth assignment of a p-form at each $\mathbf{x} \in \mathbb{R}^k$. Thus, for example, if f, η_i, ξ_{ij}, $i,j = 1, \ldots, k$, are smooth functions on $\mathbb{R}^k$, then f, $\eta := \eta_i dx^i$ and $\xi := \sum_{i<j} \xi_{ij} dx^i \wedge dx^j$ represent, in turn, a differential 0-form, a differential 1-form, and a differential 2-form. We denote by $F^p(\mathbb{R}^k)$ the space of differential p-forms on $\mathbb{R}^k$. If $\xi \in F^p(\mathbb{R}^k)$, then the *degree* of ξ is the number deg $\xi = p$.

The functionals determined by p-forms may now be used to combine differential p-forms ξ and p vector fields $X_1, \ldots, X_p$ to form functions $\xi(X_1, \ldots, X_p)$ on $\mathbb{R}^k$ according to

$$\xi(X_1, \ldots, X_p): \mathbf{x} \to \xi(X_1(\mathbf{x}), \ldots, X_p(\mathbf{x})).$$

In particular, for $p = 1$, if X is any vector field and ξ is any differential 1-form, $\xi(X)$ is the real-valued function on $\mathbb{R}^k$ given by

$$\xi(X): \mathbf{x} \to \langle X(\mathbf{x}), \xi(\mathbf{x}) \rangle.$$

If $X = X^i \partial/\partial x^i$ and $\xi = \xi_j dx^j$, then the component functions of X and ξ are conveniently picked out by

$$\langle X, dx^i \rangle = X^i, \quad \text{and} \quad \langle \frac{\partial}{\partial x^j}, \xi \rangle = \xi_j.$$

II.3 THE EXTERIOR DERIVATIVE

An operator $d: F^p(\mathbb{R}^k) \to F^{p+1}(\mathbb{R}^k)$, called the *exterior derivative*, is now introduced. If $f \in F^0(\mathbb{R}^k)$, then df is the 1-form defined by the requirement that for any vector field X,

$$\langle X, df \rangle = X(f). \tag{II.14}$$

Thus, if df is given in coordinates by $df = f_j dx^j$, it follows from (II.4) and (II.14) that

$$\langle \frac{\partial}{\partial x^i}, f_j dx^j \rangle = f_i = \frac{\partial f}{\partial x^i}.$$

Accordingly,

$$df = \frac{\partial f}{\partial x^i} dx^i, \tag{II.15}$$

and hence our choice of the notation dx^i for the basis 1-forms. Note the consistency of this notation with the usual convention in that x^i may be regarded as a function on $\mathbb{R}^k$ and

$$d(x^i) = \frac{\partial x^i}{\partial x^j} dx^j = dx^i.$$

We now extend d to an operation on p-forms for $p > 0$ by the requirements

$$d(\xi + \eta) = d\xi + d\eta, \tag{II.16}$$

$$d(f\xi) = df \wedge \xi + f d\xi, \tag{II.17}$$

$$d(\xi \wedge \eta) = d\xi \wedge \eta + (-1)^{\deg \xi} \xi \wedge d\eta, \tag{II.18}$$

$$d(d\eta) = 0 \tag{II.19}$$

for all forms ξ, η and functions f. Existence and uniqueness properties of d based on conditions (II.16)–(II.19) are established in Flanders [122].

In view of properties (II.16) and (II.17), it is sufficient to exhibit the effect of d on a p-form

$$\xi = \sum \xi_{i_1 \ldots i_p} dx^{i_1} \wedge \cdots \wedge dx^{i_p}$$

since an arbitrary p-form will consist of linear combinations of such terms over $F^0(\mathbb{R}^k)$.

Now (II.18) implies that

$$d\xi = \sum d(\xi_{i_1 \ldots i_p} dx^{i_1} \wedge \cdots \wedge dx^{i_p})$$
$$= \sum d\xi_{i_1 \ldots i_p} \wedge (dx^{i_1} \wedge \cdots \wedge dx^{i_p})$$
$$+ (-1)^0 \xi_{i_1 \ldots i_p} d(dx^{i_1} \wedge \cdots \wedge dx^{i_p}),$$

and since

$$d(dx^{i_1} \wedge \cdots \wedge dx^{i_p}) = 0,$$

it follows that

$$d\xi = \sum \frac{\partial \xi_{i_1 \ldots i_p}}{\partial x^j} dx^j \wedge (dx^{i_1} \wedge \cdots \wedge dx^{i_p}). \tag{11.20}$$

Example II.1 The operators gradient, curl, and divergence of vector calculus are readily recovered in terms of the exterior derivative d when $k = 3$.

Since the space of 1-forms on $\mathbb{R}^3$ is three dimensional, we may identify the basis 1-forms dx^1, dx^2, and dx^3 with the usual basis vectors $\mathbf{e}^1, \mathbf{e}^2, \mathbf{e}^3$ in $\mathbb{R}^3$.[†] The space of 2-forms on $\mathbb{R}^3$ is again three dimensional, and if the basis forms $dx^2 \wedge dx^3, dx^3 \wedge dx^1$, and $dx^1 \wedge dx^2$ are associated, in turn, with the

---

[†] There is a good reason for this apparently arbitrary choice. The Hodge star operator (Flanders [122]) gives an isomorphism between $F^1(\mathbb{R}^3)$ and $F^2(\mathbb{R}^3)$ which maps dx^1, dx^2, and dx^3 to $dx^2 \wedge dx^3$, $dx^3 \wedge dx^1$, and $dx^1 \wedge dx^2$, respectively.

DIFFERENTIAL FORMS

basis vectors $\mathbf{e}^1$, $\mathbf{e}^2$, $\mathbf{e}^3$, then $\wedge$ corresponds to the vector product $\times$ in $\mathbb{R}^3$ in that

$$dx^2 \wedge dx^3 \leftrightarrow \mathbf{e}^2 \times \mathbf{e}^3 = \mathbf{e}^1,$$
$$dx^3 \wedge dx^1 \leftrightarrow \mathbf{e}^3 \times \mathbf{e}^1 = \mathbf{e}^2,$$
$$dx^1 \wedge dx^2 \leftrightarrow \mathbf{e}^1 \times \mathbf{e}^2 = \mathbf{e}^3.$$

Suppose that f is a 0-form on $\mathbb{R}^3$. Then from (II.15),

$$df = \frac{\partial f}{\partial x^1} dx^1 + \frac{\partial f}{\partial x^2} dx^2 + \frac{\partial f}{\partial x^3} dx^3,$$

so that the components of df are the components of

$$\operatorname{grad} f = \frac{\partial f}{\partial x^1} \mathbf{e}^1 + \frac{\partial f}{\partial x^2} \mathbf{e}^2 + \frac{\partial f}{\partial x^3} \mathbf{e}^3.$$

Similarly, if η is a 1-form on $\mathbb{R}^3$,

$$\eta = P \, dx^1 + Q \, dx^2 + R \, dx^3,$$

then (II.17) shows that the 2-form $d\eta$ is given by

$$d\eta = dP \wedge dx^1 + dQ \wedge dx^2 + dR \wedge dx^3$$
$$= \frac{\partial P}{\partial x^i} dx^i \wedge dx^1 + \frac{\partial Q}{\partial x^i} dx^i \wedge dx^2 + \frac{\partial R}{\partial x^i} dx^i \wedge dx^3$$
$$= \left(\frac{\partial R}{\partial x^2} - \frac{\partial Q}{\partial x^3} \right) dx^2 \wedge dx^3$$
$$+ \left(\frac{\partial P}{\partial x^3} - \frac{\partial R}{\partial x^1} \right) dx^3 \wedge dx^1$$
$$+ \left(\frac{\partial Q}{\partial x^1} - \frac{\partial P}{\partial x^2} \right) dx^1 \wedge dx^2.$$

Hence, the components of $d\eta$ in this basis are the same as the components of

$$\operatorname{curl} \mathbf{F} = \begin{vmatrix} \mathbf{e}_1 & \mathbf{e}_2 & \mathbf{e}_3 \\ \dfrac{\partial}{\partial x^1} & \dfrac{\partial}{\partial x^2} & \dfrac{\partial}{\partial x^3} \\ P & Q & R \end{vmatrix},$$

where $\mathbf{F} = P\mathbf{e}_1 + Q\mathbf{e}_2 + R\mathbf{e}_3$.

Finally, if ζ is a 2-form given by

$$\zeta = A \, dx^2 \wedge dx^3 + B \, dx^3 \wedge dx^1 + C \, dx^1 \wedge dx^2,$$

then

$$d\zeta = d(A\,dx^2) \wedge dx^3 - A\,dx^2 \wedge d(dx^3)$$
$$+ d(B\,dx^3) \wedge dx^1 - B\,dx^3 \wedge d(dx^1)$$
$$+ d(C\,dx^1) \wedge dx^2 - C\,dx^1 \wedge d(dx^2)$$
$$= \left(\frac{\partial A}{\partial x^1} + \frac{\partial B}{\partial x^2} + \frac{\partial C}{\partial x^3}\right) dx^1 \wedge dx^2 \wedge dx^3.$$

Since $F^3(\mathbb{R}^3)$ is one dimensional, $dx^1 \wedge dx^2 \wedge dx^3$ is a basis, and the component of $d\zeta$ is

$$\text{div } \mathbf{G} = \frac{\partial A}{\partial x^1} + \frac{\partial B}{\partial x^2} + \frac{\partial C}{\partial x^3},$$

where $\mathbf{G} = A\mathbf{e}_1 + B\mathbf{e}_2 + C\mathbf{e}_3$. ∎

II.4 PULL-BACK MAPS

One of the most useful features of differential forms is their behavior under mappings of the space in which they live. Thus, if ϕ is a smooth map from $\mathbb{R}^k$ to $\mathbb{R}^l$, then there is a naturally defined map of differential forms

$$\phi^* : F^p(\mathbb{R}^l) \to F^p(\mathbb{R}^k)$$

called the *pull-back* map. If $f \in F^0(\mathbb{R}^l)$, then $\phi^* f$ is the 0-form on $\mathbb{R}^k$ defined by

$$(\phi^* f)(\mathbf{x}) = f(\phi(\mathbf{x})),$$

so that

$$\phi^* f = f \circ \phi. \tag{II.21}$$

The following properties of the pull-back map apply for arbitrary smooth maps $\phi : \mathbb{R}^k \to \mathbb{R}^l$, $\psi : \mathbb{R}^m \to \mathbb{R}^k$, functions $f, g \in F^0(\mathbb{R}^l)$, and p-forms $\xi, \eta \in F^p(\mathbb{R}^l)$:

$$\phi^*(f\xi + g\eta) = (\phi^* f)(\phi^* \xi) + (\phi^* g)(\phi^* \eta), \tag{II.22}$$

$$\phi^* d\xi = d(\phi^* \xi), \tag{II.23}$$

$$\phi^*(\xi \wedge \eta) = \phi^* \xi \wedge \phi^* \eta, \tag{II.24}$$

$$(\phi \circ \psi)^* \xi = (\psi^* \circ \phi^*) \xi. \tag{II.25}$$

If y^i are the coordinate functions on $\mathbb{R}^l$ and $\phi : \mathbb{R}^k \to \mathbb{R}^l$ is given by $y^i = y^i(x^1, \ldots, x^k)$, then

$$\phi^* dy^i = d(\phi^* y^i) = \frac{\partial y^i}{\partial x^j} dx^j. \tag{II.26}$$

DIFFERENTIAL FORMS

Example II.2 If $\phi:\mathbb{R}^2 \to \mathbb{R}^2$ is given by
$$y^1 = y^1(x^1, x^2) = x^1 - ax^2, \qquad y^2 = y^2(x^1, x^2) = x^2,$$
then, from (II.26),
$$\phi^* dy^1 = dx^1 - a\,dx^2, \qquad \phi^* dy^2 = dx^2,$$
while, on use of (II.22),
$$\begin{aligned}\phi^*(f(y^1, y^2)\,dy^1) &= (\phi^* f)(\phi^* dy^1) \\ &= f(y^1(x^1, x^2), y^2(x^1, x^2))(dx^1 - a\,dx^2).\end{aligned} \qquad \blacksquare$$

II.5 THE INTERIOR PRODUCT OF VECTOR FIELDS AND DIFFERENTIAL p-FORMS

An *interior product*, denoted by $\lrcorner$, of vector fields and differential p-forms may now be introduced. If X is any vector field and f is any 0-form, then we set
$$X \lrcorner f := 0. \tag{II.27}$$
More generally, if ξ is any p-form, then $X \lrcorner \xi$ is defined as the $(p-1)$-form such that
$$(X \lrcorner \xi)(Y_1, \ldots, Y_{p-1}) = \xi(X, Y_1, \ldots, Y_{p-1}) \tag{II.28}$$
for all vector fields Y_i. It follows from the linearity properties of p-forms that
$$(fX + gY) \lrcorner \xi = fX \lrcorner \xi + gY \lrcorner \xi, \tag{II.29}$$
and
$$X \lrcorner (f\xi + g\eta) = fX \lrcorner \xi + gX \lrcorner \eta, \tag{II.30}$$
where f, g are arbitrary 0-forms. In addition, it is readily shown that
$$X \lrcorner (\xi \wedge \eta) = (X \lrcorner \xi) \wedge \eta + (-1)^{\deg \xi} \xi \wedge (X \lrcorner \eta), \tag{II.31}$$
$$X \lrcorner Y \lrcorner \xi = -Y \lrcorner X \lrcorner \xi, \tag{II.32}$$
and
$$\xi(X_1, \ldots, X_p) = X_p \lrcorner X_{p-1} \lrcorner \cdots \lrcorner X_1 \lrcorner \xi. \tag{II.33}$$
In the case $p = 1$, $X \lrcorner \xi$ is a 0-form
$$X \lrcorner \xi = \xi(X) = \langle X, \xi \rangle,$$
and if $X = X^i \partial/\partial x^i$, on use of (II.4), one sees that
$$X \lrcorner dx^i = X^i. \tag{II.34}$$

The calculation of $X \lrcorner \xi$ may be extended to p-forms ξ, $p > 1$, by use of the result (II.31) to show that

$$X \lrcorner dx^{i_1} \wedge \cdots \wedge dx^{i_p}$$
$$= X^{i_1} dx^{i_2} \wedge \cdots \wedge dx^{i_p} - X^{i_2} dx^{i_1} \wedge dx^{i_3} \wedge \cdots \wedge dx^{i_p}$$
$$+ \cdots + (-1)^{l-1} X^{i_l} dx^{i_1} \wedge \cdots \wedge dx^{i_{l-1}} \wedge dx^{i_l} \wedge \cdots \wedge dx^{i_p}$$
$$+ \cdots + (-1)^{p-1} X^{i_p} dx^{i_1} \wedge \cdots \wedge dx^{i_{p-1}}.$$

Example II.3 Let

$$\xi = \xi_1 dx^1 + \xi_2 dx^2 + \xi_3 dx^3$$

and

$$\eta = A_1 dx^2 \wedge dx^3 + A_2 dx^3 \wedge dx^1 + A_3 dx^1 \wedge dx^2.$$

If $X = X^i \partial/\partial x^i$, then

$$X \lrcorner \xi = X \lrcorner (\xi_1 dx^1 + \xi_2 dx^2 + \xi_3 dx^3)$$
$$= \xi_1 (X \lrcorner dx^1) + \xi_2 (X \lrcorner dx^2) + \xi_3 (X \lrcorner dx^3)$$
$$= \xi_1 X^1 + \xi_2 X^2 + \xi_3 X^3$$

and

$$X \lrcorner \eta = (A_2 X^3 - A_3 X^2) dx^1 + (A_3 X^1 - A_1 X^3) dx^2$$
$$+ (A_1 X^2 - A_2 X^1) dx^3.$$

Combination of these results shows that

$$Y \lrcorner (X \lrcorner \eta) = (A_2 X^3 - A_3 X^2) Y^1 + (A_3 X^1 - A_1 X^3) Y^2$$
$$+ (A_1 X^2 - A_2 X^1) Y^3,$$

and hence,

$$Y \lrcorner (X \lrcorner \eta) = \eta(X, Y). \blacksquare$$

II.6 THE LIE DERIVATIVE OF DIFFERENTIAL FORMS

The operators d and $\lrcorner$ may be used to define an operator called the Lie derivative of a differential form with respect to a vector field. Thus, if X is a vector field on $\mathbb{R}^n$ and ξ is a p-form, the *Lie derivative* $L_X \xi$ of ξ with respect to X is defined as the p-form given by

$$X(\xi) = L_X \xi := X \lrcorner d\xi + d(X \lrcorner \xi).^{\dagger} \qquad (\text{II}.35)$$

---

† In general, the notation $X(\xi)$ for the Lie derivative proves more convenient in the sequel than the more usual $L_X \xi$.

DIFFERENTIAL FORMS

It is readily shown that

$$L_X f = \langle X, df \rangle, \quad f \in F^0(\mathbb{R}^l), \tag{II.36}$$

$$L_X(\xi \wedge \eta) = L_X(\xi) \wedge \eta + \xi \wedge L_X(\eta), \tag{II.37}$$

$$L_X(d\xi) = d(L_X\xi). \tag{II.38}$$

The coordinate expression for $L_X \xi$ may be deduced from (II.36) and (II.37) in terms of the 1-forms $L_X(dx^i)$ given by, if $X = X^j \partial/\partial x^j$,

$$L_X(dx^i) = (X^j \frac{\partial}{\partial x^j}) \lrcorner d(dx^i) + d(X^j \frac{\partial}{\partial x^j} \lrcorner dx^i)$$

$$= d(X^i) = \frac{\partial X^i}{\partial x^j} dx^j. \tag{II.39}$$

Thus if $\xi = \xi_i dx^i$,

$$L_X \xi = (L_X \xi_i) dx^i + \xi_i L_X(dx^i)$$

$$= (X^i \frac{\partial \xi_j}{\partial x^i} + \xi_i \frac{\partial X^i}{\partial x^j}) dx^j, \tag{II.40}$$

and the coordinate expression for $L_X \xi$ with ξ an arbitrary p-form may now be deduced from this result and (II.37).

Example II.4 If

$$\xi = f\, dx^1 \wedge dx^2,$$

then

$$L_X \xi = X^i \frac{\partial f}{\partial x^i} dx^1 \wedge dx^2 + f \left[\frac{\partial X^1}{\partial x^i} dx^i \wedge dx^2 + \frac{\partial X^2}{\partial x^i} dx^1 \wedge dx^i \right]. \blacksquare$$

APPENDIX **III**

Differential Forms on Jet Bundles

III.1 PRELIMINARIES

At this juncture, we list certain facts about differential forms on jet bundles which prove useful for later calculations. In particular, we introduce a basis better adapted to calculations involving pull-backs of forms on $J^k(M, N)$ to forms on M by k-jet extensions than is the basis $dx^a, dz^A, \ldots, dz^A_{a_1 \cdots a_k}$.

If ξ is any 1-form on $J^k(M, N)$, then it follows from (II.25), together with the definitions of π_k^{k+1} and $j^k f$, that

$$j^k f^* \xi = (\pi_k^{k+1} \circ j^{k+1} f)^* \xi = j^{k+1} f^* (\pi_k^{k+1 *} \xi), \qquad \text{(III.1)}$$

so that the pull-back of ξ by $j^k f$ is the same as the pull-back of $\pi_k^{k+1 *} \xi$ by $j^{k+1} f$.

In view of the fact that the 1-forms $\pi_k^{k+1 *} dz^A_{a_1 \cdots a_k}$ are given by

$$\pi_k^{k+1 *} dz^A_{a_1 \cdots a_k} = \theta^A_{a_1 \cdots a_k} + z^A_{a_1 \cdots a_k b} dx^b,$$

where $\theta^A_{a_1 \cdots a_k}$ are the contact forms introduced in Section 2.2, it follows that the 1-form $\pi_k^{k+1 *} \xi$ may be expressed in the form

$$\pi_k^{k+1 *} \xi = \xi_a dx^a - \bar{\xi}, \qquad \text{(III.2)}$$

where $\bar{\xi} \in \Omega^{k+1}(M, N)$ and ξ_a are functions on $J^{k+1}(M, N)$. Thus, once ξ has been pulled up to $J^{k+1}(M, N)$ it may be expressed in terms of the 1-forms dx^a and 1-forms in the contact module. Moreover,

$$j^{k+1}f^*\pi_k^{k+1*}\xi = j^{k+1}f^*(\xi_a dx^a) + j^{k+1}f^*\bar{\xi} = (\xi_a \circ j^{k+1}f)dx^a,$$

so that from (III.1),

$$j^k f^*\xi = (\xi_a \circ j^{k+1}f)dx^a. \tag{III.3}$$

Hence, in view of (III.2), it follows that for any 1-form on $J^k(M, N)$ there are functions ξ_a on $J^{k+1}(M, N)$ such that

$$\pi_k^{k+1*}\xi \equiv \xi_a dx^a \quad \mod \Omega^{k+1}(M, N).$$

Use of (III.2) and the fact that

$$D_b^{(k+1)} \lrcorner \, \theta = 0, \quad \forall \theta \in \Omega^{k+1}(M, N),$$

shows that

$$(\pi_k^{k+1*}\xi)(D_b^{(k+1)}) = D_b^{(k+1)} \lrcorner (\pi_k^{k+1*}\xi) = \xi_a D_b^{(k+1)} \lrcorner \, dx^a,$$

whence, the functions ξ_a may be expressed in terms of the total derivative operator $D_b^{(k+1)}$ according to

$$\xi_b = (\pi_k^{k+1*}\xi)(D_b^{(k+1)}). \tag{III.4}$$

Example III.1 If $M = \mathbb{R}^2$, $N = \mathbb{R}$, and ξ is the 1-form on $J^1(M, N)$ given by

$$\xi = f_a dx^a + f\theta + f^a dz_a,$$

where f_a, f, and f^a are smooth functions, then

$$\pi_1^{2*}\xi \equiv (f_a + f^b z_{ba})dx^a \quad \mod \Omega^2(M, N).$$

On the other hand,

$$(\pi_1^{2*}\xi)(D_b^{(2)}) = D_b^{(2)} \lrcorner \, \pi_1^{2*}\xi$$

$$= \left(\frac{\partial}{\partial x^b} + z_b \frac{\partial}{\partial z} + z_{bc} \frac{\partial}{\partial z_c}\right) \lrcorner (f_a dx^a + f\theta + f^a dz_a)$$

$$= f_b + f^c z_{bc}. \quad \blacksquare$$

Let $\mathscr{I}(\Omega^l(M, N))$ denote the exterior *ideal* generated by $\Omega^l(M, N)$. Thus $\mathscr{I}(\Omega^l(M, N)) := \{\Sigma \, \xi \wedge \phi \, | \, \xi$ is any p-form $p = 0, 1, \ldots, \phi \in \Omega^l(M, N)\}$. It is a simple matter now to show that if ξ is any p-form on $J^k(M, N)$, then there are functions $\xi_{a_1 \cdots a_p}$ on $J^{k+1}(M, N)$ such that

$$\pi_k^{k+1*}\xi = \sum_{1 \leq a_1 < \cdots < a_p \leq m} \xi_{a_1 \cdots a_p} dx^{a_1} \wedge \cdots \wedge dx^{a_p} + \bar{\xi}, \tag{III.5}$$

where $\bar{\xi}$ is in $\mathscr{I}(\Omega^{k+1}(M, N))$. Again, the functions $\xi_{a_1 \cdots a_p}$ may be expressed in terms of the total derivative operators, for if $\bar{\xi}$ is a p-form in $\mathscr{I}(\Omega^{k+1}(M, N))$, it follows that

$$\bar{\xi}(D_{a_1}^{(k+1)}, \ldots, D_{a_p}^{(k+1)}) = D_{a_p}^{(k+1)} \, \lrcorner \, D_{a_{p-1}}^{(k+1)} \cdots D_{a_1}^{(k+1)} \, \lrcorner \, \bar{\xi} = 0,$$

whence

$$\xi_{a_1 \cdots a_p} = (\pi_k^{k+1} * \xi)(D_{a_1}^{(k+1)}, \ldots, D_{a_p}^{(k+1)}). \tag{III.6}$$

Example III.2 If $M = \mathbb{R}^2$, $N = \mathbb{R}$, and ξ is the 2-form on $J^1(M, N)$ given by

$$\xi = dz_1 \wedge dx^2 + dz_2 \wedge dx^1 - f(z) \, dx^1 \wedge dx^2,$$

then

$$\begin{aligned}\pi_1^{2*}\xi(D_1^{(2)}, D_2^{(2)}) &= D_2^{(2)} \, \lrcorner \, D_1^{(2)} \, \lrcorner \, (dz_1 \wedge dx^2 + dz_2 \wedge dx^1 - f(z) \, dx^1 \wedge dx^2) \\ &= D_2^{(2)} \, \lrcorner \, (z_{11} \, dx^2 + z_{12} \, dx^1 - dz_2 - f(z) \, dx^2) \\ &= z_{11} - z_{22} - f(z),\end{aligned}$$

so that

$$\pi_1^{2*}\xi \equiv (z_{11} - z_{22} - f(z)) \, dx^1 \wedge dx^2 \qquad \mod \mathscr{I}(\Omega^2(M, N)). \quad \blacksquare$$

Finally, there are two cases of particular interest for the discussion of conservation laws, namely, the cases $p = m$ and $p = m - 1$. In this connection, if ξ is an m-form on $J^k(M, N)$, then we may write $\pi_k^{k+1}*\xi$ in terms of the volume m-form w on M and the ideal generated by $\Omega^{k+1}(M, N)$.[†] Thus, in this case, there is a single function f_ξ on $J^{k+1}(M, N)$ such that

$$\pi_k^{k+1}*\xi \equiv f_\xi w \qquad \mod \mathscr{I}(\Omega^{k+1}(M, N)), \tag{III.7}$$

and

$$f_\xi = \pi^{k+1}*\xi(D_1^{(k+1)}, D_2^{(k+1)}, \ldots, D_m^{(k+1)}).$$

If ξ is an $(m-1)$-form, then

$$\pi_k^{k+1}*\xi \equiv \xi^a w_a \qquad \mod \mathscr{I}(\Omega^{k+1}(M, N)).[‡] \tag{III.8}$$

III.2 EXTERIOR DIFFERENTIAL SYSTEMS ON JET BUNDLES

Let $\Sigma = \{\sigma_\lambda\}$ be a collection of differential forms on $J^k(M, N)$, $k \geq 0$. Such a collection is referred to as an *exterior differential system*. A *solution* of this system is a map $f \in C^\infty(M, N)$ which satisfies

$$j^k f * \sigma_\lambda = 0 \qquad \forall \sigma_\lambda \in \Sigma. \tag{III.9}$$

[†] The volume m-form w on M is given in local coordinates by $w = dx^1 \wedge dx^2 \wedge \cdots \wedge dx^m$.
[‡] The $(m-1)$-forms w_a are defined by $w_a := D_a^{(1)} \, \lrcorner \, \alpha^* w$.

DIFFERENTIAL FORMS ON JET BUNDLES 289

If f is a solution of Σ, then the image of $j^k f$ is a submanifold of $J^k(M, N)$ on which the differential forms σ_λ all vanish. For this reason, $\operatorname{im} j^k f$ is called a *solution submanifold* for Σ; it provides a solution to the system of exterior differential equations

$$\sigma_\lambda = 0.$$

It follows from the properties of the pull-back map that if f is a solution of Σ, then

$$j^k f^*(\xi \wedge \sigma) = 0$$

for any differential form ξ and any $\sigma \in \Sigma$. Thus if we define $\mathscr{I}(\Sigma)$ as the exterior ideal generated by Σ, that is, $\mathscr{I}(\Sigma) := \{\sum_i \xi_i \wedge \sigma | \xi_i$ is any p-form, $\sigma \in \Sigma; p = 0, 1, \ldots, \dim J^k(M, N)\}$, then f is also a solution of $\mathscr{I}(\Sigma)$. Conversely, since $\Sigma \subset \mathscr{I}(\Sigma)$ it follows that every solution of $\mathscr{I}(\Sigma)$ is a solution of Σ. Thus, we may regard the exterior systems Σ and $\mathscr{I}(\Sigma)$ as equivalent, since f is a solution of Σ iff f is a solution of $\mathscr{I}(\Sigma)$. The exterior system Σ is said to be *closed* if, for every $\sigma \in \Sigma$, $d\sigma \in \mathscr{I}(\Sigma)$. In this case, $\mathscr{I}(\Sigma)$ is called a *differential ideal*.

If Σ is an exterior system and X is a vector field which satisfies

$$X \lrcorner \sigma = 0, \qquad X \lrcorner d\sigma \in \Sigma$$

$\forall \sigma \in \Sigma$, then X is called a *characteristic vector field for* Σ. The collection of all characteristic vector fields for Σ is denoted by $\operatorname{Char}(\Sigma)$. We note that if Σ is closed, since $X \lrcorner \sigma = 0$ $\forall \sigma \in \Sigma$ it follows that $X \lrcorner d\sigma \in \Sigma$.

III.3 THE SYSTEM OF DIFFERENTIAL EQUATIONS ASSOCIATED WITH AN EXTERIOR DIFFERENTIAL SYSTEM

We conclude this appendix with a simple construction that associates with any exterior differential system on $J^k(M, N)$ a system of differential equations with the same solutions.

Let $\Sigma = \{\sigma_\lambda\}$ and define the set of functions $F_{a_1 \cdots a_p \lambda}$ by

$$F_{a_1 \cdots a_p \lambda} := \{(\pi_k^{k+1*}\sigma_\lambda)(D_{a_1}^{(k+1)}, \ldots, D_{a_p}^{(k+1)}) | p = \deg \sigma_\lambda\}. \quad (\text{III.10})$$

The submanifold R^{k+1} of $J^{k+1}(M, N)$, determined by the constraint equations

$$F_{a_1 \cdots a_p \lambda} = 0, \quad (\text{III.11})$$

is defined as the *system of differential equations associated with* Σ.

That (III.11) has the same solutions as Σ may readily be seen. Thus, if σ is any p-form in Σ, it follows from (III.5) and (III.6) that

$$\pi_k^{k+1*}\sigma \equiv \sum_{a_1 < \cdots < a_p} (\pi_k^{k+1*}\sigma)(D_{a_1}^{(k+1)}, \ldots, D_{a_p}^{(k+1)}) \, dx^{a_1} \wedge \cdots \wedge dx^{a_p}$$
$$\operatorname{mod} \mathscr{I}(\Omega^{k+1}(M, N)), \quad (\text{III.12})$$

and if f is a solution of the associated system equations, then

$$j^{k+1}f^*\{(\pi_k^{k+1}{}^*\sigma)(D_{a_1}^{(k+1)},\ldots,D_{a_p}^{(k+1)})\} = 0. \tag{III.13}$$

However, by (III.12), it is seen that (III.13) holds iff

$$j^{k+1}{}^*\pi_k^{k+1}{}^*\sigma = 0,$$

and thus, iff

$$j^k f^*\sigma = 0.$$

Consequently, f is a solution of Σ iff it is a solution of R^{k+1}.

Example III.3 Let $M = \mathbb{R}^2$ and $N = \mathbb{R}$. If Σ is the exterior system on $J^1(M, N)$ generated by

$$\sigma_1 = dz \wedge dx^1 + dz_1 \wedge dx^2 - 2zz_1\, dx^1 \wedge dx^2,$$
$$\sigma_2 = dz \wedge dx^2 - z_1\, dx^1 \wedge dx^2,$$

then Σ is associated with the Burgers' equation

$$z_2 + 2zz_1 - z_{11} = 0.$$

Thus, since $D_a^{(2)} = \partial/\partial x^a + z_a \partial/\partial z + z_{ab} \partial/\partial z_b$, we see that

$$(\pi_1^2{}^*\sigma_1)(D_1^{(2)}, D_2^{(2)}) = D_2^{(2)} \,\lrcorner\, (z_1\, dx^1 - dz + z_{11}\, dx^2 - 2zz_1\, dx^2)$$
$$= -z_2 + z_{11} - 2zz_1.$$

On the other hand, since $\sigma_2 = \theta \wedge dx^2$, this 2-form has the vacuous associated equation $0 = 0$. ∎

III.4 AN EXTERIOR DIFFERENTIAL SYSTEM OF m-FORMS ON $J^k(M, N)$ ASSOCIATED WITH A QUASI-LINEAR EQUATION ON $J^{k+1}(M, N)$.

If R^{k+1} is any quasi-linear system of equations on $J^{k+1}(M, N)$, then one may construct an exterior differential system of m-forms on $J^k(M, N)$ which has R^{k+1} as its associated equation (Pirani et al. [48]). We describe the construction for the case in which R^{k+1} is given by a single constraint equation $F = 0$; the extension to systems is straightforward.

Since R^{k+1} is quasi-linear,

$$F = F_A^{a_1 \cdots a_{k+1}} z_{a_1 \cdots a_{k+1}}^A + G, \tag{III.14}$$

where $F_A^{a_1 \cdots a_{k+1}}$ and G are functions independent of $z_{a_1 \cdots a_{k+1}}^A$. A map $f \in C^\infty(M, N)$ is a solution of R^{k+1} iff $j^{k+1}f^*F = 0$ and thus iff

$$j^{k+1}f^*(Fw) = 0, \tag{III.15}$$

since
$$j^{k+1}f^*(Fw) = (j^{k+1}f^*F)j^{k+1}f^*(w) = (j^{k+1}f^*F)w.$$

It follows from (III.14) that
$$j^{k+1}f^*(Fw) = j^{k+1}f^*(F_A^{a_1\cdots a_{k+1}} dz^A_{a_1\cdots a_{k+1}} + Gw), \quad \text{(III.16)}$$

and from the definition of w_a in Section III.2 that
$$F_A^{a_1\cdots a_{k+1}} z^A_{a_1\cdots a_{k+1}} w = F_A^{a_1\cdots a_{k+1}}(dz^A_{a_1\cdots a_k} \wedge w_{a_{k+1}} - \theta^A_{a_1\cdots a_k} \wedge w_{a_{k+1}}). \quad \text{(III.17)}$$

Combination of (III.16) and (III.17) shows that
$$j^{k+1}f^*(Fw) = j^{k+1}f^*(F_A^{a_1\cdots a_k b} dz^A_{a_1\cdots a_k} \wedge w_b + Gw),$$

and since $F_A^{a_1\cdots a_k b}$ and G are functions on $J^k(M, N)$, it follows that
$$j^{k+1}f^*(Fw) = j^k f^*(F_A^{a_1\cdots a_k b} dz^A_{a_1\cdots a_k} \wedge w_b + Gw)$$

where w_b are the $(m-1)$-forms defined in Section III.1.

If we now define the m-form σ by
$$\sigma := F_A^{a_1\cdots a_k b} dz^A_{a_1\cdots a_k} \wedge w_b + Gw, \quad \text{(III.18)}$$

then f is a solution of the equation R^{k+1} iff
$$j^k f^* \sigma = 0. \quad \text{(III.19)}$$

Let Σ^k be the exterior differential system generated by the contact m-forms $\{\theta^A \wedge w_a, \ldots, \theta^A_{a_1\cdots a_{(k-1)}} \wedge w_a\}$ together with the m-form σ. It is readily verified that if $s \in C^\infty(M, J^k(M, N))$ is such that $\alpha \circ s = \mathrm{id}_M$, then
$$s^*\phi = 0 \quad \forall \phi \in \Sigma^k \quad \text{iff} \quad s = j^k f,$$

where $f = \beta \circ s$ is a solution of R^{k+1}. Thus, *the solutions of the exterior differential system Σ^k are in one-to-one correspondence with the solutions of the equation R^{k+1}*. We may therefore consider Σ^k and R^{k+1} as entirely equivalent. There are many instances for which it is natural and very convenient to make this identification. Thus, for example, there is a simple geometric interpretation of conservation laws and symmetries in this context and because of the reduction in the number of variables in passing from $J^{k+1}(M, N)$ to $J^k(M, N)$, it is frequently easier to compute the symmetries of Σ^k than it is to compute those of R^{k+1} directly.

Note that we may pull back the exterior system Σ^k to an exterior system $\pi_k^{k+1*}\Sigma^k$ on $J^{k+1}(M, N)$ and that $\pi_k^{k+1*}\Sigma^k$ is generated by $\Omega_{(m)}^{k+1}$ and Fw. Thus $\pi_k^{k+1*}\Sigma^k$ is contained in the ideal generated by the 1-forms in Ω^{k+1} and by the m-form Fw. Finally, it is immediate from this observation and the fact that $D_a^{k+1} \lrcorner \theta = 0$ for $\theta \in \Omega^{k+1}(M, N)$ that the equation associated with Σ^k is just $F = 0$.

APPENDIX **IV**

The Derivation of the Equations That Define $B^{k+1,1}(\psi)$

Equations (2.45) and (2.46), which define $B^{k+1,1}(\psi)$, may be obtained by imitating the procedure described in Appendix III.3 to associate with an exterior system Σ on $J^k(M,N) \underset{M}{\times} J^0(M,N')$ a submanifold of $J^{k+1}(M,N) \underset{M}{\times} J^1(M,N')$. All that is required is that the total derivatives D_a^{k+1} be replaced by the vector fields $D_a^{k+1,1}$. Thus, if Σ is an exterior differential system on $J^k(M,N) \underset{M}{\times} J^0(M,N')$ and $\{\sigma_\lambda\}$ are its generators, we associate with Σ the submanifold S of $J^{k+1}(M,N) \underset{M}{\times} J^0(M,N')$, defined as the zero set of the functions

$$\{(\pi^{k+1,1*}\sigma_\lambda)(D_{a_1}^{k+1,1},\ldots,D_{a_p}^{k+1,1}); p = \text{degree of } \sigma_\lambda\}. \tag{IV.1}$$

It is readily verified that if f and g are two maps $f \in C^\infty(M,N)$, $g \in C^\infty(M,N')$, then the map $j^k f \underset{M}{\times} j^0 g$ is a solution of Σ iff the image of the map $j^{k+1}f \underset{M}{\times} j^1 g$ lies in S.

The foregoing procedure is now applied to the exterior system Σ_ψ associated with a Bäcklund map $\psi: J^k(M,N) \underset{M}{\times} J^0(M,N') \to J^1(M,N')$. The generators of Σ_ψ are the 1-forms $\{\psi^*\theta'^\mu\}$ and the 2-forms $\{\psi^* d\theta'^\mu\}$, where

THE DERIVATION OF EQUATIONS THAT DEFINE $B^{k+1,1}(\psi)$

$\{\theta'^\mu\}$ is a basis for the contact module $\Omega^1(M, N')$. Since $\psi^*\theta'^\mu = dy^\mu - \psi_b^\mu dx^b$ and $\psi^* d\theta'^\mu = d(\psi^*\theta'^\mu)$, it follows that the submanifold $B^{k+1,1}(\psi)$ associated with Σ_ψ is the zero set of the functions

$$(\pi^{k+1,1*}(dy^\mu - \psi_b^\mu dx^b))(D_a^{k+1,1}) \tag{IV.2}$$

and

$$(\pi^{k+1,1*}(dx^c \wedge d\psi_c^\mu))(D_a^{k+1,1}, D_b^{k+1,1}). \tag{IV.3}$$

From the definition of $D_a^{k+1,1}$, we see that these functions are just

$$y_a^\mu - \psi_a^\mu \tag{IV.4}$$

and

$$D_b^{k+1,1}\psi_a^\mu - D_a^{k+1,1}\psi_b^\mu, \tag{IV.5}$$

respectively. Thus, the submanifold $B^{k+1,1}(\psi)$ is determined by the equations

$$y_a^\mu = \psi_a^\mu, \qquad D_a^{k+1,1}\psi_b^\mu = D_b^{k+1,1}\psi_a^\mu. \tag{IV.6}$$

APPENDIX V

Symmetries of Differential Equations and Exterior Systems

The topic of symmetries of differential equations and exterior differential systems is introduced in a jet-bundle context. We do not attempt to describe the most general setting for the discussion of symmetries. Thus, in the case of differential equations we confine ourselves to consideration of extended point transformations since these are sufficient for our purposes. On the other hand, it was noted in Chapter 3 that the reciprocal relations in gas-dynamics and magnetogasdynamics may be treated as generalized symmetries of a natural exterior differential system. Correspondingly, we consider symmetries of exterior differential systems on $J^0(M, N)$, which are more general than point transformations, namely, those which arise as diffeomorphisms of $J^0(M, N)$.

V.1 POINT TRANSFORMATIONS

A map $\phi: \mathbb{R}^k \to \mathbb{R}^k$ that is C^∞ and that has a C^∞ inverse is called a *diffeomorphism* of $\mathbb{R}^k$. If ϕ and Φ are diffeomorphisms of M and $J^0(M, N) = M \times N$, respectively, and we have a commutative diagram (Fig. V.1), then Φ is said

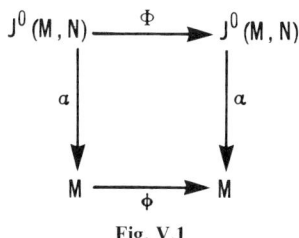

Fig. V.1

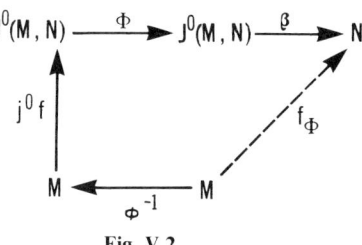

Fig. V.2

to be compatible with ϕ and (Φ, ϕ) is said to be a *point transformation* of $J^0(M, N)$. If ϕ is given in coordinates by

$$x'^a = \phi^a(x^b), \tag{V.1}$$

then Φ is given by

$$x'^a = \phi^a(x^b), \quad z'^A = \Phi^A(x^b, z^B), \tag{V.2}$$

for some functions Φ^A.

If f is any map in $C^\infty(M, N)$ and (Φ, ϕ) is a point transformation, we may combine f and (Φ, ϕ) to construct a new map $f_\Phi \in C^\infty(M, N)$ defined by

$$f_\Phi := \beta \circ \Phi \circ j^0 f \circ \phi^{-1}, \tag{V.3}$$

as shown in Fig. V.2. If (Ψ, ψ) is another point transformation, then use of (V.3) shows that

$$(f_\Phi)_\Psi = f_{\Psi \circ \Phi} \tag{V.4}$$

and, in particular,

$$(f_\Phi)_{\Phi^{-1}} = f_{\Phi^{-1} \circ \Phi} = f \tag{V.5}$$

for any $f \in C^\infty(M, N)$.

Any point transformation (Φ, ϕ) may be extended to a diffeomorphism $p^l \Phi$ of $J^l(M, N)$ that is uniquely determined by the requirement that

$$p^l \Phi \circ j^l f \circ \phi^{-1} = j^l f_\Phi \tag{V.6}$$

for every $f \in C^\infty(M, N)$. It follows from (V.5) and (V.6) that

$$p^l(\Phi^{-1}) = (p^l \Phi)^{-1}, \tag{V.7}$$

and it is readily deduced that $p^l \Phi$ preserves the contact module $\Omega^l(M, N)$, that is,

$$(p^l \Phi)^* \Omega^l(M, N) = \Omega^l(M, N). \tag{V.8}$$

In terms of coordinates, the diffeomorphism $p^l \Phi$ may be given by

$$x'^a = \phi^a(x), \quad z'^A = \Phi^A(x^b, z^B),$$
$$z'^A_{a_1 \cdots a_j} = \Phi^A_{a_1 \cdots a_j}(x^b, \ldots, z^B_{b_1 \cdots b_j}), \quad j = 1, \ldots, l-1, \tag{V.9}$$

and

$$\partial_a \phi^b z'^A_{a_1 \cdots a_{t-1} b} = D_a^{(l)} \Phi^A_{a_1 \cdots a_{t-1}}. \tag{V.10}$$

Suppose now that $\{(\Phi_t, \phi_t)\}$ is a 1-parameter group of point transformations, that is, for each $t \in \mathbb{R}$, (Φ_t, ϕ_t) is a point transformation such that

$$\Phi_{t_1} \circ \Phi_{t_2} = \Phi_{t_1 + t_2}, \qquad \phi_{t_1} \circ \phi_{t_2} = \phi_{t_1 + t_2} \qquad \forall t_1, t_2 \in \mathbb{R} \tag{V.11}$$

and $\Phi_0 = \mathrm{id}_{M \times N}$, $\phi_0 = \mathrm{id}_M$. These conditions imply that

$$\Phi_t \circ \Phi_{-t} = \mathrm{id}_{M \times N}, \qquad \phi_t \circ \phi_{-t} = \mathrm{id}_M, \tag{V.12}$$

so that (Φ_{-t}, ϕ_{-t}) is the point transformation inverse to (Φ_t, ϕ_t).

Example V.1 Let $M = \mathbb{R}^2$, $N = \mathbb{R}$, and let (Φ_t, ϕ_t) be the 1-parameter group of point transformations given by

$$\Phi_t : (x^1, x^2, z) \to (e^t x^1, e^{-t} x^2, z),$$
$$\phi_t : (x^1, x^2) \to (e^t x^1, e^{-t} x^2),$$

that is, by

$$z' = z, \qquad x'^a = \phi^a = \delta^a_1 e^t x^1 + \delta^a_2 e^{-t} x^2, \qquad a = 1, 2.$$

The coordinate version of the first prolongation of (Φ_t, ϕ_t) is determined by the single set of equations

$$(\partial_a \phi^b) z'_b = D_a^{(1)} z',$$

or

$$(\partial_a \phi^1) z'_1 + (\partial_a \phi^2) z'_2 = z_a, \qquad a = 1, 2.$$

Hence

$$z'_1 = e^{-t} z_1, \qquad z'_2 = e^t z_2,$$

so that $p^1 \Phi_t$ is given by

$$p^1 \Phi_t : (x^1, x^2, z, z_1, z_2) \to (e^t x^1, e^{-t} x^2, z, e^{-t} z_1, e^t z_2). \qquad \blacksquare$$

If (Φ, ϕ) and (Φ', ϕ) are point transformations, we may construct a diffeomorphism $p^{k,l}(\Phi \times \Phi')$ of $J^k(M, N) \underset{M}{\times} J^l(M, N')$ given by

$$p^{k,l}(\Phi \times \Phi') := p^k \Phi \times p^l \Phi', \tag{V.13}$$

and it is readily verified that $p^{k,l}(\Phi \times \Phi')$ has the property

$$p^{k,l}(\Phi \times \Phi')^* \Omega^{k,l} = \Omega^{k,l}. \tag{V.14}$$

Note that the definition (V.13) applies more generally to 1-parameter groups of point transformations (Φ_t, ϕ_t), (Φ'_t, ϕ_t) with

$$p^{k,l}(\Phi_t \times \Phi'_t) := p^k \Phi_t \times p^l \Phi'_t. \tag{V.15}$$

V.2 SYMMETRIES

If R^k is a system of differential equations, a point transformation (Φ, ϕ) is said to be a *symmetry* of R^k whenever the condition

$$p^k \Phi(R^k) = R^k \tag{V.16}$$

is satisfied. This formulation of symmetry in purely geometric terms is thus made without any specific reference to the space of solutions of R^k. However, that a symmetry (Φ, ϕ) does indeed send solutions to solutions may be shown as follows:

Let $f \in C^\infty(M, N)$ be a solution of R^k and suppose that the condition (V.16) holds. Then since $\operatorname{im}(j^k f) \subset R^k$, it follows that

$$\operatorname{im}(p^k \Phi \circ j^k f \circ \phi^{-1}) \subset p^k \Phi(R^k) \subset R^k, \tag{V.17}$$

whence

$$\operatorname{im}(j^k f_\Phi) \subset R^k, \tag{V.18}$$

so that f_Φ is another solution of R^k.

If $\{(\Phi_t, \phi_t)\}$ is a 1-parameter group of point transformations, it is said to be a 1-parameter group of symmetries of R^k whenever

$$p^k \Phi_t R^k = R^k. \tag{V.19}$$

The fact that for each t, (Φ_t, ϕ_t) is a point transformation implies that if f is a solution of R^k, then f_{Φ_t} is a 1-parameter family of solutions.

Example V.2 A straightforward calculation shows that the second prolongation of the group of point transformations of Example V.1 is given by

$$p^2 \Phi_t : (x^1, x^2, z, z_1, z_2, z_{11}, z_{12}, z_{22}) \to$$
$$(e^t x^1, e^{-t} x^2, z, e^{-t} z_1, e^t z_2, e^{-2t} z_{11}, z_{12}, e^{2t} z_{22}).$$

It follows, in particular, that every point $\mathbf{q} \in J^2(M, N)$ for which $z_{12} = \sin z$ is mapped under $p^2 \Phi_t$ to a point $\mathbf{q}'$ with $z'_{12} = \sin z'$. Thus $\{(\Phi_t, \phi_t)\}$ is a 1-parameter group of symmetries of the sine-Gordon equation. ■

Symmetries of exterior differential systems may be defined in a manner similar to symmetries of R^k (see, for example, Shadwick [407]). For our purposes it is sufficient to deal with exterior systems Σ defined on $J^0(M, N)$ while broadening somewhat the allowable types of transformations. Thus, it is sometimes convenient to consider arbitrary diffeomorphisms of $J^0(M, N)$ rather than restrict attention to point transformations. In the general case, a diffeomorphism $\Phi : J^0(M, N) \to J^0(M, N)$ is given in coordinates by

$$x'^a = \phi^a(x^b, z^B), \quad z'^A = \Phi^A(x^b, z^B) \tag{V.20}$$

[cf. (V.2) for the corresponding equations for a point transformation]. One consequence of this generality is that there is no natural way to construct out of Φ and a map $f \in C^{\infty}(M, N)$ another map in $C^{\infty}(M, N)$ without the imposition of additional conditions. Thus if $f \in C^{\infty}(M, N)$, then Φ may be composed with f to obtain a map $\Phi \circ f : M \to J^0(M, N)$, but it will not necessarily happen that $\Phi \circ f = j^0 g$ for any $g \in C^{\infty}(M, N)$.

If Φ is a diffeomorphism of $J^0(M, N)$ and $\phi^* \Sigma = \Sigma$, then Φ will be referred to as a *generalized symmetry* of Σ. In order that Φ take solutions of Σ to solutions of Σ, it is required that whenever f is a solution of Σ, the map $\Phi \circ j^0 f$ have the form $j^0 g \circ \phi$ for some $g \in C^{\infty}(M, N)$ and some diffeomorphism $\phi : M \to M$. Thus, if this is the case,

$$(\Phi \circ j^0 f \circ \phi^{-1})^* \Sigma = 0, \qquad (V.21)$$

and

$$\Phi \circ j^0 f \circ \phi^{-1} = j^0 g, \qquad (V.22)$$

so that g is another solution of Σ. This situation will occur if it is required that the map $\phi := \alpha \circ \Phi \circ j^0 f$ be a diffeomorphism of M or, equivalently, that the 1-forms $\phi^* dx^a$ be linearly independent. In this case, the map $\Phi \circ j^0 f \circ \phi^{-1}$ is the graph of a map from M to N, that is, (V.22) holds for some $g \in C^{\infty}(M, N)$.

As an illustration of various aspects of the above discussion, the interested reader is referred to Section 3.4 on reciprocal relations in magnetogasdynamics.

APPENDIX **VI**

Composition of Symmetries and Bäcklund Maps

Here we show that if $\tilde{\psi}$ is the map constructed in Section 2.7 from a symmetry and a Bäcklund map ψ for R^{k+1}, then $\tilde{\psi}$ is also a Bäcklund map for R^{k+1}. To do this, it suffices to show that the constraint equations describing $B^{k+1,1}(\tilde{\psi})$ are satisfied on R^{k+1}.

Let $\Sigma_{\tilde{\psi}}$ be the exterior system associated with $\tilde{\psi}$. Then $\Sigma_{\tilde{\psi}}$ is generated by $\tilde{\psi}^*\theta'^\mu$ and $\tilde{\psi}^*d\theta'^\mu$. Moreover, from the definition of $\tilde{\psi}$ and (II.25), we have

$$\tilde{\psi}^* = p^{k,0}(\Phi \times \Phi')^* \circ \psi^* \circ p^1\Phi'^{*-1}, \qquad (VI.1)$$

and since $p^1\Phi'^*\{\theta'^\mu\} = \{\theta'^\mu\}$, it follows that $\Sigma_{\tilde{\psi}}$ is generated by

$$\{p^{k,0}(\Phi \times \Phi')^* \circ \psi^*\theta'^\mu\} \quad \text{and} \quad \{p^{k,0}(\Phi \times \Phi')^* \circ \psi^* d\theta'^\mu\}.$$

Thus the equations associated with $\Sigma_{\tilde{\psi}}$ are determined by the functions

$$(\pi^{k+1,1*}p^{k,0}(\Phi \times \Phi')^* \circ \psi^*\theta'^\mu)(D_a^{k+1,1}) \qquad (VI.2)$$

and

$$(\pi^{k+1,1*} \circ p^{k,0}(\Phi \times \Phi')^* \circ \psi^* d\theta'^\mu)(D_a^{k+1,1}, D_b^{k+1,1}). \qquad (VI.3)$$

The zero set of (VI.2) is just im($\tilde{\psi}$) and thus $B^{k+1,1}(\tilde{\psi}) \supset R^{k+1} \times$ im $\tilde{\psi}$ provided R^{k+1} is contained in the zero set of (VI.3).

Since, on $\psi^* d\theta'^\mu$,

$$\pi^{k+1,1} * p^{k,0}(\Phi \times \Phi')^* = p^{k+1,1}(\Phi \times \Phi')^*, \qquad (\text{VI.4})$$

(VI.3) may be written as

$$(p^{k+1,1}(\Phi \times \Phi')^* \circ \psi^*(d\theta'^\mu))(D_a^{k+1,1}, D_b^{k+1,1}). \qquad (\text{VI.5})$$

Moreover, it follows from the definition of $\Omega^{k+1,1}$ that

$$\psi^* d\theta'^\mu \equiv [D_a^{k+1,1} \psi_b^\mu - D_b^{k+1,1} \psi_a^\mu] dx^a \wedge dx^b \quad \mod \mathscr{I}(\Omega^{k+1,1}), \quad (\text{VI.6})$$

so that since $p^{k+1,1} \Phi \times \Phi'$ preserves $\mathscr{I}(\Omega^{k+1,1})$,

$$\begin{aligned}
(p^{k+1,1}&(\Phi \times \Phi')^* \circ \psi^*(d\theta'^\mu)) \\
&\equiv p^{k+1,1}(\Phi \times \Phi')^* [D_a^{k+1,1} \psi_b^\mu - D_b^{k+1,1} \psi_a^\mu] \\
&\quad \times p^{k+1,1}(\Phi \times \Phi')^*(dx^a \wedge dx^b) \quad \mod \mathscr{I}(\Omega^{k+1,1}).
\end{aligned} \qquad (\text{VI.7})$$

But

$$\begin{aligned}
p^{k+1,1}(\Phi \times \Phi')^* dx^a \wedge dx^b &= \frac{\partial \phi^a}{\partial x^c} \frac{\partial \phi^b}{\partial x^e} dx^c \wedge dx^e \\
&= \frac{1}{2} \left[\frac{\partial \phi^a}{\partial x^c} \frac{\partial \phi^b}{\partial x^e} - \frac{\partial \phi^a}{\partial x^e} \frac{\partial \phi^b}{\partial x^c} \right] dx^c \wedge dx^e, \qquad (\text{VI.8})
\end{aligned}$$

and thus,

$$\begin{aligned}
p^{k+1,1}&(\Phi \times \Phi')^*(\psi^* d\theta'^\mu)(D_c^{k+1,1}, D_e^{k+1,1}) \\
&\equiv p^{k+1,1}(\Phi \times \Phi') [D_a^{k+1,1} \psi_b^\mu - D_b^{k+1,1} \psi_a^\mu] \left[\frac{\partial \phi^a}{\partial x^c} \frac{\partial \phi^b}{\partial x^e} - \frac{\partial \phi^a}{\partial x^e} \frac{\partial \phi^b}{\partial x^c} \right] \\
&\qquad \mod \mathscr{I}(\Omega^{k+1,1}). \qquad (\text{VI.9})
\end{aligned}$$

Since ϕ is invertible,

$$\frac{\partial \phi^a}{\partial x^c} \frac{\partial \phi^b}{\partial x^e} - \frac{\partial \phi^a}{\partial x^e} \frac{\partial \phi^b}{\partial x^c} \neq 0 \quad \text{for} \quad a \neq b,$$

so that the zero set of the functions (VI.5) is just the zero set of the functions

$$\begin{aligned}
p^{k+1,1}&(\Phi \times \Phi')^*(D_a^{k+1,1} \psi_b^\mu - D_b^{k+1,1} \psi_a^\mu) \\
&= [D_a^{k+1,1} \psi_b^\mu - D_b^{k+1,1} \psi_a^\mu] \circ p^{k+1,1}(\Phi \times \Phi'). \qquad (\text{VI.10})
\end{aligned}$$

COMPOSITION OF SYMMETRIES AND BÄCKLUND MAPS 301

But Φ is a symmetry of R^{k+1}; whence if $\mathbf{q} \in R^{k+1}$, it follows that

$$\tilde{\mathbf{q}} = p^{k+1,1}(\Phi \times \Phi')(\mathbf{q})$$

is such that

$$[D_a^{k+1,1}\psi_b^\mu - D_b^{k+1,1}\psi_a^\mu](\tilde{\mathbf{q}}) = 0 \quad \text{for} \quad \tilde{\mathbf{q}} \in R^{k+1}.$$

Thus, R^{k+1} is contained in the zero set of the functions (VI.10), and we conclude that $\tilde{\psi}$ is indeed a Bäcklund map for R^{k+1}.

References

1. Bäcklund, A. V., Einiges über Curven-und Flächentransformationen, *Lunds Univ. Årsskr. Avd.* **X**, För Ar (1873); Afdelningen för Mathematik och Naturetenskap, *ibid.* 1–12 (1873–74).
2. Bäcklund, A. V., Ueber Flächentransformationen, *Math. Ann.* **IX**, 297–320 (1876).
3. Bäcklund, A. V., Zür Theorie der Partiellen Differentialgleichungen erster Ordnung, *Math. Ann.* **XVII**, 285–328 (1880).
4. Bäcklund, A. V., Zür Theorie der Flächentransformationen, *Math. Ann.* **XIX**, 387–422 (1882).
5. Bäcklund, A. V., Om Ytor Med Konstant Negativ Krökning, *Lunds Univ. Arsskr. Avd.* **19** (1883).
6. Josephson, B. D., Supercurrents Through Barriers, *Adv. Phys.* **14**, 419–451 (1965).
7. Lebwohl, P. and Stephen, M. J., Properties of Vortex Lines in Superconducting Barriers, *Phys. Rev.* **163**, 367–379 (1967).
8. Scott, A. C., "Active and Non-linear Wave Propagation in Electronics." Wiley, New York, 1970.
9. Scott, A. C., Propagation of Magnetic Flux on a Long Josephson Junction, *Nuovo Cimento B* **69**, 241–261 (1970).
10. Seeger, A., Donth, H., and Kochendorfer, A., Theorie der Versetzungen in Eindimensionalen Atomreihen: II Beliebig Angeordnete und Beschleunigte Versotzungen, *Z. Phys.* **127**, 533–550 (1951).
11. Skyrme, T. H. R., A Nonlinear Theory of Strong Interactions, *Proc. Roy. Soc. London Ser. A* **247**, 260–278 (1958).

12. Skyrme, T. H. R., Particle States of a Quantized Meson Field, *Proc. Roy. Soc. London Ser. A* **262**, 237–245 (1961).
13. Enz, U., Discrete Mass, Elementary Length, and a Topological Invariant as a Consequence of a Relativistic Invariant Variational Principle, *Phys. Rev.* **131**, 1392–1394 (1963).
14. Rubinstein, J., Sine-Gordon Equation, *J. Math. Phys.* **11**, 258–266 (1970).
15. Scott, A. C., A Nonlinear Klein–Gordon Equation, *Amer. J. Phys.* **37**, 52–61 (1969).
16. Bean, C. P. and deBlois, R. W., Ferromagnetic Domain Wall as a Pseudorelativistic Entity *Bull. Am. Phys. Soc.* **4**, 53 (1959).
17. Lamb, G. L., Jr., Propagation of Ultrashort Optical Pulses, *Phys. Lett. A* **25**, 181–182 (1967).
18. Lamb, G. L., Jr., Pulse Propagation in a Lossless Amplifier, *Phys. Lett. A* **29**, 507–508 (1969).
19. Lamb, G. L., Jr., Propagation of Ultrashort Optical Pulses, *in* "In Honor of Philip M. Morse", (H. Feshbach and K. U. Ingard, eds.), pp. 88–105. MIT Press, Cambridge, Massachusetts, 1969.
20. Lamb, G. L., Jr., Scully, M. O., and Hopf, F. A., Higher Conservation Laws for Coherent Optical Pulse Propagation in an Inhomogeneously Broadened Medium, *Appl. Opt.* **11**, 2572–2575 (1972).
21. Lamb, G. L., Jr., Analytical Descriptions of Ultrashort Optical Pulse Propagation in a Resonant Medium, *Rev. Modern Phys.* **43**, 99–124 (1971).
22. Barnard, T. W., $2N\pi$ Ultrashort Light Pulses, *Phys. Rev. A* **7**, 373–376 (1973).
23. Coleman, S., Quantum Sine-Gordon Equation as the Massive Thirring Model, *Phys. Rev. D* **11**, 2088–2097 (1975).
24. Morris, H. C., A Generalized Prolongation Structure and the Bäcklund Transformation of the Anticommuting Massive Thirring Model, *J. Math. Phys.* **19**, 85–87 (1978).
25. Leibbrandt, G., Exact Solutions of the Elliptic Sine Equation in Two Space Dimensions with Application to the Josephson Effect, *Phys. Rev. B* **15**, 3353–3361 (1977).
26. Leibbrandt, G., Soliton-like Solutions of the Elliptic Sine-Cosine Equation by Means of Harmonic Functions, *J. Math. Phys.* **19**, 960–966 (1978).
27. Leibbrandt, G., "New Exact Solutions of the Classical Sine-Gordon Equation in 2 + 1 and 3 + 1 Dimensions." Harvard Preprint HUTP 78/A016, 1978.
28. Miura, R. M., Korteweg–deVries Equation and Generalizations: I. A Remarkable Explicit Nonlinear Transformation, *J. Math. Phys.* **9**, 1202–1204 (1968).
29. Whitham, G. B., "Linear and Nonlinear Waves". Wiley, New York, 1974.
30. Korteweg, D. J. and deVries, G., On the Change of Form of Long Waves Advancing in a Rectangular Channel, and on a New Type of Long Stationary Wave, *Philos. Mag.* **39**, 422–443 (1895).
31. Gardner, C. S. and Morikawa, G. K., Courant Inst. Math. Sci. Rep. NYU-9082, New York Univ. New York, 1960.
32. Kruskal, M. D., Asymptotology in Numerical Computation: Progress and Plans on the Fermi–Pasta–Ulam Problem, *Proc. IBM Sci. Comput. Symp. Large-Scale Prob. Phys.*, IBM Data Processing Division, White Plains, New York, 1965.
33. Zabusky, N. J., "A Synergetic Approach to Problems of Nonlinear Dispersive Wave Propagation and Interaction, *in* "Nonlinear Partial Differential Equations" (W. F. Ames, ed.). Academic Press, New York, 1967.
34. Washimi, H. and Taniuti, T., Propagation of Ion-acoustic Solitary Waves of Small Amplitude, *Phys. Rev. Lett.* **17**, 996–998 (1966).
35. Lamb, G. L., Jr., Bäcklund Transformations for Certain Nonlinear Evolution Equations, *J. Math. Phys.* **15**, 2157–2165 (1974).

36. Wahlquist, H. D. and Estabrook, F. B., Bäcklund Transformations for Solutions of the Korteweg–deVries Equation, *Phys. Rev. Lett.* **31**, 1386–1390 (1973).
37. Miura, R. M., Gardner, C. S., and Kruskal, M. D., Korteweg–deVries Equation and Generalizations: II. Existence of Conservation Laws and Constants of Motion, *J. Math. Phys.* **9**, 1204–1209 (1968).
38. Su, C. H. and Gardner, C. S., Korteweg–deVries Equation and Generalizations: III. Derivation of the Korteweg–deVries Equation and Burgers' Equation, *J. Math. Phys.* **10**, 536–539 (1969).
39. Hopf, E., The Partial Differential Equation $u_t + uu_x = \mu u_{xx}$, *Comm. Pure Appl. Math.* **3**, 201–230 (1950).
40. Cole, J. D., On a Quasilinear Parabolic Equation Occurring in Aerodynamics, *Quart. Appl. Math.* **9**, 225–236 (1951).
41. Chu, C-w., A Class of Reducible Systems of Quasilinear Partial Differential Equations, *Quart. Appl. Math.* **23**, 275–278 (1965).
42. Calogero, F., A Method to Generate Solvable Nonlinear Evolution Equations, *Lett. Nuovo Cimento* **14**, 443–447 (1975).
43. Chiu, S. C. and Ladik, J. F., Generating Exactly Soluble Nonlinear Discrete Evolution Equations by a Generalized Wronskian Technique, *J. Math. Phys.* **18**, 690–700 (1977).
44. Case, K. M. and Chiu, S. C., Some Remarks on the Wronskian Technique and the Inverse Scattering Transform, *J. Math. Phys.* **18**, 2044–2052 (1977).
45. Bruschi, M., Levi, D., and Ragnisco, O., Toda Lattice and Generalized Wronskian Technique, *J. Phys. A* **13**, 2531–2533 (1980).
46. Bruschi, M. and Ragnisco, O., Nonlinear Differential-Difference Equations, Associated Bäcklund Transformations and Lax Technique. *J. Phys. A* **14**, 1075–1081 (1981).
47. Pirani, F. A. E. and Robinson, D. C., Definition of Bäcklund Transformations, *CR Acad. Sci. Ser. A* **285**, 581–583 (1977).
48. Pirani, F. A. E., Robinson, D. C., and Shadwick, W. F., "Local Jet-bundle Formulation of Bäcklund Transformations." Reidel, Dordrecht, (1979).
49. Cartan, E., "Les Systèmes Différentials Extérieures et Leurs Applications Géométriques." Hermann, Paris, 1946.
50. Rogers, C., On Applications of Generalized Bäcklund Transformations in Continuum Mechanics, *in* "Lecture Notes in Mathematics 515 Bäcklund Transformations, Proceedings" (R. M. Miura, ed.), pp 106–135. Springer-Verlag, Berlin, 1976.
51. Haar, A., Über Adjungierte Variationsprobleme und Adjungierte Extremalflächen, *Math. Ann.* **100**, 481–502 (1928).
52. Chaplygin, S. A., On Gas Jets, *Sci. Mem. Moscow Univ. Math. Phys.* **21**, 1–121 (1904).
53. Bateman, H., The Lift and Drag Functions for an Elastic Fluid in Two-dimensional Irrotational Flow, *Proc. Nat. Acad. Sci. U.S.A.* **24**, 246–251 (1938).
54. Tsien, H. S., Two-dimensional Subsonic Flow of Compressible Fluids, *J. Aeronaut. Sci.* **6**, 399–407 (1939).
55. von Mises, R., "Mathematical Theory of Compressible Fluid Flow." Academic Press, New York, 1958.
56. Bateman, H., The Transformation of Partial Differential Equations, *Quart. Appl. Math.* **1**, 281–295 (1943–44).
57. Prim, R. C., A Note on the Substitution Principle for Steady Gas Flow, *J. Appl. Phys.* **20**, 448–450 (1949).
58. Prim, R. C., Steady Rotational Flow of Ideal Gases, *Arch. Rational Mech. Anal.* **1**, 425–497 (1952).
59. Power, G. and Smith, P., Application of a Reciprocal Property to Subsonic Flow, *Appl. Sci. Res. A* **8**, 386–392 (1959).

REFERENCES

60. Power, G. and Smith, P., Reciprocal Properties of Plane Gas Flows, *J. Math. Mech.* **10**, 349–361 (1961).
61. Rogers, C., The Construction of Invariant Transformations in Plane Rotational Gasdynamics, *Arch. Rational Mech. Anal.* **47**, 36–46 (1972).
62. Rogers, C., Castell, S. P., and Kingston, J. G., On Invariance Properties of Conservation Laws in Nondissipative Planar Magnetogasdynamics, *J. Mécanique* **13**, 343–354 (1974).
63. Rogers, C., Kingston, J. G., and Shadwick, W. F., On Reciprocal-type Transformations in Magnetogasdynamics, *J. Math. Phys.* **21**, 395–397 (1980).
64. Loewner, C., A Transformation Theory of Partial Differential Equations of Gasdynamics, *Nat. Advis. Comm. Aeronat. Tech. Notes* **2065**, 1–56 (1950).
65. Loewner, C., Generation of Solutions of Systems of Partial Differential Equations by Composition of Infinitesimal Bäcklund Transformations, *J. Anal. Math.* **2**, 219–242 (1952).
66. Power, G., Rogers, C., and Osborn, R. A., Bäcklund and Generalized Legendre Transformations in Gasdynamics, *Z. Angew. Math. Mech.* **49**, 333–340 (1969).
67. von Kármán, T., Compressibility Effects in Aerodynamics, *J. Aeronaut. Sci.* **8**, 337–356 (1941).
68. Coburn, N., The Kármán–Tsien Pressure–Volume Relation in the Two-dimensional Supersonic Flow of Compressible Fluids, *Quart. Appl. Math.* **3**, 106–116 (1945).
69. Pérès, J., Quelques Transformations des Équations du Mouvement d'un Fluide Compressible, *CR Acad. Sci. Ser. A.* **219**, 501–504 (1944).
70. Sauer, R., Unterschallströmungen um Profile bei Quadratisch Approximierter Adiabate, *Bayer. Akad. Wiss. Math.-Natur. Kl. Sitzungsber.* **9**, 65–71 (1951).
71. Dombrovskii, G. A., Approximation Methods in the Theory of Plane Adiabatic Gas Flow, Moscow, 1964.
72. Grad, H., Reducible Problems in Magneto-fluid Dynamic Steady Flows, *Rev. Modern Phys.* **32**, 830–847 (1960).
73. Jeffrey, A. and Taniuti, T., "Nonlinear Wave Propagation with Applications to Physics and Magnetohydrodynamics." Academic Press, New York and London, 1964.
74. Dragos, L., "Magnetofluid Dynamics", Abacus Press, Tunbridge Wells, 1975.
75. Nikol'skii, A. A., Invariant Transformation of the Equations of Motion of an Ideal Monatomic Gas and New Classes of Their Exact Solutions, *Prikl. Math. Meh.* **27**, 740–756 (1963).
76. Tomilov, E. D., On the Method of Invariant Transformations of the Gasdynamics Equations, *Prikl. Math. Meh.* **29**, 959–960 (1965).
77. Movsesian, L. A., On an Invariant Transformation of Equations of One-dimensional Unsteady Motion of an Ideal Compressible Fluid, *Prikl. Math. Meh.* **31**, 137–141 (1967).
78. Ustinov, M. D., Transformation and Some Solutions of the Equation of Motion of an Ideal Gas, *Izv. Akad. Nauk SSSR Meh. Zidk. Gaza* **3**, 68–74 (1966).
79. Ustinov, M. D., Ideal Gas Flow Behind an Infinite Amplitude Shock Wave, *Izv. Akad. Nauk SSSR Meh. Zidk. Gaza* **4**, 88–90 (1967).
80. Rykov, V. A., On an Exact Solution of the Equations of Magnetogasdynamics of Finite Conductivity, *Prikl. Math. Meh.* **29**, 178–181 (1965).
81. Rogers, C., Reciprocal Relations in Non-steady One-dimensional Gasdynamics, *Z. Angew. Math. Phys.* **19**, 58–63 (1968).
82. Rogers, C., Invariant Transformations in Non-steady Gasdynamics and Magnetogasdynamics, *Z. Angew. Math. Phys.* **20**, 370–382 (1969).
83. Castell, S. P. and Rogers, C., Application of Invariant Transformations in One-dimensional Nonsteady Gasdynamics, *Quart. Appl. Math.* **32**, 241–251 (1974).

84. Sedov, L. I., On the Integration of the Equations of a One-dimensional Motion of a Gas, *Dokl. Akad. Nauk SSSR* **90**, 5 (1953).
85. Cekirge, H. M. and Varley, E., Large Amplitude Waves in Bounded Media: I. Reflexion and Transmission of Large Amplitude Shockless Pulses at an Interface, *Philos. Trans. Roy. Soc. London Ser. A* **273**, 261–313 (1973).
86. Rogers, C., Iterated Bäcklund Transformations and the Propagation of Disturbances in Nonlinear Elastic Materials, *J. Math. Anal. Appl.* **49**, 638–648 (1975).
87. Rogers, C. and Clements, D. L., On the Reduction of the Hodograph Equations for One-dimensional Elastic-plastic Wave Propagation, *Quart. Appl. Math.* **37**, 469–474 (1975).
88. Cekirge, H. M. and Rogers, C., On Elastic-plastic Wave Propagation; Transmission of Elastic-plastic Boundaries, *Arch. Mech.* **29**, 125–141 (1977).
89. Clements, D. L. and Rogers, C., On Wave Propagation in Inhomogeneous Elastic Media, *Internat. J. Solids and Structures* **10**, 661–669 (1974).
90. Karal, F. C., Jr. and Keller, J. B., Elastic Wave Propagation in Homogeneous and Inhomogeneous Media, *J. Acoust. Soc. Amer.* **31**, 694–705 (1959).
91. Lewis, R. M. and Keller, J. B., "Asymptotic Methods for Partial Differential Equations; the Reduced Wave Equation and Maxwell's Equations." Courant Inst. Math. Sci. Rep. EM-194, New York Univ., New York, 1964.
92. Luneburg, R. K., "Mathematical Theory of Optics," Univ. of California Press, Berkeley, 1964.
93. Clements, D. L., Moodie, T. B., and Rogers, C., On the Propagation of Waves from Spherical and Cylindrical Cavities in Anisotropic Materials, *Internat. J. Engrg. Sci.* **15**, 429–445 (1977).
94. Kazakia, J. Y. and Venkataraman, R., Propagation of Electromagnetic Waves in a Nonlinear Dielectric Slab, *Z. Angew. Math. Phys.* **26**, 61–76. (1975).
95. Rogers, C., Cekirge, H. M., and Askar, A., Electromagnetic Wave Propagation in Nonlinear Dielectric Media, *Acta Mech.* **26**, 59–73 (1977).
96. Clements, D. L. and Rogers, C., On the Theory of Stress Concentration for Shear-strained Prismatical Bodies with a Nonlinear Stress–Strain Law, *Mathematika* **22**, 34–42 (1975).
97. Clements, D. L. and Rogers, C., On the Application of a Bäcklund Transformation to Linear Isotropic Elasticity, *J. Inst. Math. Appl.* **14**, 23–30 (1974).
98. Weinstein, A., Generalized Axially Symmetric Potential Theory, *Bull. Amer. Math. Soc.* **59**, 20–28 (1953).
99. Payne, L. E., On Axially Symmetric Punch, Crack, and Torsion Problems, *Soc. Ind. Appl. Math. J. Appl. Math.* **1**, 53–71 (1953).
100. Clements, D. L., Atkinson, C., and Rogers, C., Antiplane Crack Problems for an Inhomogeneous Elastic Material, *Acta Mech.* **29**, 199–211 (1978).
101. Clements, D. L. and Rogers, C., On the Bergman Operator Method and Anti-plane Contact Problems Involving an Inhomogeneous Halfspace, *Soc. Ind. Appl. Math. J. Appl. Math.* **34**, 764–773 (1978).
102. C. Rogers and D. L. Clements, Bergman's Integral Operator Method in Inhomogeneous Elasticity, *Quart. Appl. Math.* **40**, 315–321 (1978).
103. Lie, S., Zür Theorie der Flächen Konstanter Krümmung III, *Arch. Math. Naturvidensk.* V Heft 3, 282–306 (1880).
104. Lie, S., Zür Theorie der Flächen Konstanter Krümmung IV, *Arch. Math. Naturvidensk.* V Heft 3, 328–358 (1880).
105. Goursat, E., "Lecons sur L'intégration des Equations aux Dérivées Partielles du Second Ordre," Vol. II. 1902.

106. Clairin, J., Sur les Transformations de Baecklund, *Ann. Sci. Ećole Norm. Sup. 3ᵉ*, **27**, 451–489 (1910).
107. Forsyth, A. R., "Theory of Differential Equations," Vol. VI. Dover New York, 1959.
108. Lie, S., "Theorie der Transformationsgruppen." Chelsea, New York, 1970.
109. Goursat, E., Le Problème de Bäcklund, *in* "Mémorial des Sciences Mathématiques," Fasc. VI. Gauthier-Villars, Paris, (1925).
110. Anderson, R. L. and Ibragimov, N. H., "Lie–Backlund Transformations in Applications" (Studies in Applied Mathematics). Soc. Ind. Appl. Math., Philadelphia, 1979.
111. Clairin, J., Sur les Transformations de Bäcklund, *Ann. Sci. Ećole Norm. Sup. 3ᵉ Suppl.* **19**, S1–63 (1902).
112. Dodd, R. K. and Bullough, R. K., Bäcklund Transformations for the Sine-Gordon Equations, *Proc. Roy. Soc. London Ser. A* **351**, 499–523 (1976).
113. Bianchi, L., Ricerche Sulle Superficie a Curvatura Constante e Sulle Elicoidi, *Ann. Scuola Norm. Sup. Pisa Sci. Fis. Mat.* **II**, 285 (1879).
114. Bianchi, L., "Lezioni di Geometria Differenziale," Vol. I. Enrico Spoerri, Pisa, 1922.
115. Darboux, G., "Lecons sur la Theorie Générale des Surfaces, Vol. III, pp. 438–444. Gauthier–Villars Paris, 1894.
116. Gibbs, H. M. and Slusher, R. E., Peak Amplification and Pulse Breakup of a Coherent Optical Pulse in a Simple Atomic Absorber, *Phys. Rev. Lett.* **24**, 638–641 (1970).
117. Burgers, J. M., "The Nonlinear Diffusion Equation: Asymptotic Solutions and Statistical Problems," Reidel, Dordrecht, 1974.
118. Hirota, R., Exact Solution of the Korteweg–deVries Equation for Multiple Collisions of Solitons, *Phys. Rev. Lett.* **27**, 1192–1194 (1971).
119. Wadati, M., The Modified Korteweg–deVries Equation, *J. Phys. Soc. Japan* **34**, 1289–1296 (1973).
120. Zakharov, V. E. and Shabat, A. B., Exact Theory of Two-dimensional Self-focusing and One-dimensional Self-modulation of Waves in Nonlinear Media, *Soviet Phys. JETP* **34**, 62–69 (1972).
121. Ablowitz, M. J., Kaup, D. J., Newell, A. C., and Segur, H., Nonlinear Evolution Equations of Physical Significance, *Phys. Rev. Lett.* **31**, 125–127 (1973).
122. Flanders, H., "Differential Forms: with Applications to the Physical Sciences." Academic Press, New York, 1963.
123. Adamson, I. T., "Elementary Rings and Modules." Oliver and Boyd, Edinburgh, 1972.
124. Johnson, H. H., Classical Differential Invariants and Applications to Partial Differential Equations, *Math. Ann.* **148**, 308–329 (1962).
125. Power, G., and Smith, P., A Modified Tangent-gas Approximation for Two-dimensional Steady Flow, *J. Fluid Mech.* **4**, 600–606 (1958).
126. Glauert, H., The Effect of Compressibility on the Lift of an Airfoil, *Proc. Roy. Soc. London, Ser. A* **118**, 113–119 (1928).
127. Müller, W., "Gasströmungen bei Quadratisch Angeñaherter Adiabate." Diss. Th., München, 1953.
128. Sauer, R., Unterschallströmungen um Profile bei Quadratisch Approximierter Adiabate, *Bayer. Akad. Wiss. Math.-Natur. Kl. Sitzungsber.* **9**, 65–71 (1951).
129. Frankl', F. L., On Chaplygin's Problem for Mixed Sub- and Supersonic Flows, *Izv. Akad. Nauk SSSR* **9**, 121–143 (1945).
130. Tomotika, S. and Tamada, K., Studies on Two-dimensional Transonic Flows of Compressible Fluid, I. *Quart. Appl. Math.* **7** 381–397 (1950); II. *ibid.* **8**, 127–136 (1951); III, *ibid.* **9**, 129–147 (1952).
131. Khristianovich, S. A., On Supersonic Gas Flows, *Tr. TSAGI* 543, 1941.

132. Rogers, C., Transformation Theory in Gas and Magnetogasdynamics, Ph.D. Thesis, Univ. of Nottingham, 1969.
133. Iur'ev, I. M., On a Solution to the Equations of Magnetogasdynamics, J. Appl. Math. Mech. **24**, 233–237 (1960).
134. Seebass, R., On Transcritical and Hypercritical Flows in Magnetogasdynamics, Quart. Appl. Math. **19**, 231–237 (1960).
135. Tamada, K., Transonic Flow of a Perfectly Conducting Gas with an Aligned Magnetic Field, Phys. Fluids **5**, 871–876 (1962).
136. Smith, P., An Extension of the Substitution Principle to Certain Unsteady Gas Flows, Arch. Rational Mech. Anal. **15**, 147–153 (1964).
137. Rogers, C. and Kingston, J. G., Reciprocal Properties in Quasi One-dimensional Nonsteady Oblique Magnetogasdynamics, J. Mécanique **15**, 185–192 (1976).
138. Martin, M. H., The Propagation of a Plane Shock Into a Quiet Atmosphere, Canad. J. Math. **5**, 37–39 (1953).
139. Hayes, W. D., On Hypersonic Similitude, Quart. Appl. Math. **5**, 105–106 (1947).
140. Goldsworthy, F. A., Two-dimensional Rotational Flow at High Mach Number Past Thin Aerofoils, Quart. J. Mech. Appl. Math. **5**, 54–63 (1952).
141. Kazakia, J. Y. and Varley, E., Large Amplitude Waves in Bounded Media: II. The Deformation of an Impulsively Loaded Slab: the First Reflexion, Philos. Trans. Roy. Soc. London Ser A **277**, 191–237 (1974).
142. Kazakia, J. Y. and Varley, E., Large Amplitude Waves in Bounded Media: III. The Deformation of an Impulsively Loaded Slab: the Second Reflexion, Philos. Trans. Roy. Soc. London Ser. A **277**, 239–250 (1974).
143. Cekirge, H. M., Propagation of a Shock Wave in a Locking (Strain Hardening) Material, Lett. Appl. Eng. Sci. **3**, 203–212 (1975).
144. Mortell, M. P. and Seymour, B. R., Nonlinear Forced Oscillations in a Closed Tube: Continuous Solutions of a Functional Equation, Proc. Roy. Soc. London Ser. A **367**, 253–270 (1979).
145. Eringen, A. C. and Suhubi, E. S., "Elastodynamics, Finite Motions," (Vol. I). Academic Press, New York and London, 1974.
146. Critescu, N., "Dynamic Plasticity." Wiley, New York, 1964.
147. Bell, J. F., "The Physics of Large Deformations of Crystalline Solids," Springer-Verlag, Berlin and New York, 1968.
148. Courant, R. and Friedrichs, K. O., "Supersonic Flow and Shock Waves," Wiley (Interscience), New York, 1948.
149. Jeffrey, A., Non-dispersive Wave Propagation in Nonlinear Dielectrics, Z. Angew. Math. Phys. **19**, 741–745 (1968).
150. Broer, L. J. F., Wave Propagation in Nonlinear Media, Z. Angew. Math. Phys. **16**, 18–26 (1965).
151. Jeffrey, A. and Korobeinikov, V. P., Formation and Decay of Electromagnetic Shock Waves, Z. Angew. Math. Phys. **20**, 440–447 (1969).
152. Venkataraman, R. and Rivlin, R. S., Propagation of First Order Electromagnetic Discontinuities in an Isotropic Medium. Tech. Rep. CAM-110-9 (BRLCR No 7), Lehigh U., 1970.
153. De Martini, F., Townes, C. H., Gustafson, T. K., and Kelley, P. L., Self-steepening of Light Pulses, Phys. Rev. **164**, 312–323 (1967).
154. Donato, A. and Fusco, D., Some Applications of the Riemann Method to Electromagnetic Wave Propagation in Nonlinear Media, Z. Angew. Math. Mech. **60**, 539–542 (1980).
155. Born, M. and Wolf, E.: "Principles of Optics." Pergamon, Oxford, 1970.

156. Owczarek, J. A., "Fundamentals of Gasdynamics." Int. Textb., Scranton, 1964.
157. Kulik, I. O., Wave Propagation in a Josephson Tunnel Junction in the Presence of Vortices and the Electrodynamics of Weak Superconductivity, *Soviet Phys. JETP* **24**, 1307–1317 (1967).
158. Scott, A. C., Steady Propagation on Long Josephson Junctions, *Bull. Am. Phys. Soc.* **12**, 308–309 (1967).
159. Scott, A. C. and Johnson, W. J., Internal Flux Motion in Large Josephson Junctions, *Appl. Phys. Lett.* **14**, 316–318 (1969).
160. Frenkel, J. and Kontoteva, T., On the Theory of Plastic Deformation and Twinning, *Soviet Phys. J.* (USSR) **1**, 137–139 (1939).
161. Köchendörfer, A. and Seeger, A., Theorie der Versetzungen in Eindimensionalen Atomreihen: I. Periodisch Angeordnete Versetzungen, *Z. Phys.* **127**, 533–550 (1950).
162. Seeger, A. Donth, H., and Köchendörfer, A., Theorie der Versetzungen in Eindimensionalen Atomreihen: III. Versetzungen, Eigenbewegungen und Ihre Wechselwirkung *Z. Phys.* **134**, 173–193 (1953).
163. Rosen, N. and Rosenstock, H. B., The Force Between Particles in a Nonlinear Field Theory, *Phys. Rev.* **85**, 257–259 (1952).
164. Fergason, J. L. and Brown, G. H., Liquid Crystals and Living Systems, *J. Am. Oil Chem. Soc.* **45**, 120–127 (1968).
165. Döring, W., Über die Trägheit der Wande Zwischen Weisschen Bezirken, *Z. Naturforsch.* **31**, 373–379 (1948).
166. Enz, U., Die Dynamic der Blochshen Wand, *Helv. Phys. Acta* **37**, 245–251 (1964).
167. Maki, K. and Tsuneto, T., Longitudinal Nuclear Magnetic Resonance in Superfluid He as Internal Josephson Effect, *Progr. Theoret. Phys.* **52**, 773–776 (1974).
168. Eisenhart, L. P., "A Treatise on the Differential Geometry of Curves and Surfaces." Dover, New York, 1960.
169. Ames, W. F., "Nonlinear Partial Differential Equations in Engineering," Vol. II. Academic Press, New York, 1972.
170. McLaughlin, D. W. and Scott, A. C., A Restricted Bäcklund Transformation, *J. Math. Phys.* **14**, 1817–1828 (1973).
171. Shadwick, W. F., The Bäcklund Problem for the Equation $\partial^2 z/\partial x^1 \, \partial x^2 = f(z)$, *J. Math. Phys.* **19**, 2312–2317 (1978).
172. Martin, M. H., A New Approach to Problems in Two Dimensional Flow, *Quart. Appl. Math.* **8**, 137–150 (1951).
173. Gibbon, J. D., James, I. N., and Moroz, I. M., An Example of Soliton Behaviour in a Rotating Baroclinic Fluid, *Proc. Roy. Soc. London Ser. A* **367**, 219–237 (1979).
174. Bateman, H., Some Recent Researches on the Motion of Fluids, *Mon. Weather Rev.* **43**, 163–170 (1915).
175. Burgers, J. M., "A Mathematical Model Illustrating the Theory of Turbulence." Academic Press, New York, 1948.
176. Lighthill, M. J., Viscosity Effects in Sound Waves of Finite Amplitude *in* "Surveys in Mechanics," (G. K. Batchelor and R. M. Davies, eds.) Cambridge Univ. Press, London and New York, 1956.
177. Blackstock, D. T., Thermoviscous Attenuation of Plane, Periodic, Finite Amplitude Sound Waves, *J. Acoust. Soc. Amer.* **36**, 534–542 (1964).
178. Karpman, V. I., "Nonlinear Waves in Dispersive Media," Pergamon, Oxford, 1975.
179. Sachdev, P. L., A Generalised Cole-Hopf Transformation for Nonlinear Parabolic and Hyperbolic Equations, *Z. Angew. Math. Phys.* **29**, 963–970 (1978).
180. Tasso, H. and Teichmann, J., On the Generalization of the Cole-Hopf Transformation, *Z. Angew. Math. Phys.* **30**, 1023–1024 (1979).

181. Parker, D. F., The Decay of Sawtooth Solutions of Burgers' Equation, *Proc. Roy. Soc. London Ser. A* **369**, 409–424 (1980).
182. Scott-Russell, J., Report on Waves, *Proc. Roy. Soc. Edinburgh*, 319–320 (1844).
183. Jeffrey, A. and Kakutani, T., Weak Nonlinear Dispersive Waves: a Discussion Centered Around the Korteweg-de Vries Equation, *Soc. Ind. Appl. Math. Rev.* **14**, 582–643 (1972).
184. Miura, R. M., The Korteweg–deVries Equation: a Survey of Results, *Soc. Ind. Appl. Math. Rev.* **18**, 412–459 (1976).
185. Scott, A. C., Chu, F. Y. F., and McLaughlin, D. W., The Soliton: a New Concept in Applied Science, *Proc. IEEE* **61**, 1444–1483 (1973).
186. Leibovich, S. and Seebass, A., eds., "Nonlinear Waves." Cornell Univ. Press, Ithaca, New York, 1974.
187. Lonngren, K. and Scott, A. C., eds., "Solitons in Action." Academic Press, New York, 1978.
188. Zabusky, N. J. and Kruskal, M. D., Interaction of Solitons in a Collisionless Plasma and the Recurrence of Initial States, *Phys. Rev. Lett.* **15**, 240–243 (1965).
189. Fermi, E., Pasta, J., and Ulam, S., Studies of Nonlinear Problems: I. Los Alamos Report. Los Alamos, New Mexico, 1940.
190. Berezin, Y. A. and Karpman, V. I., Theory of Non-stationary Finite Amplitude Waves in a Low Density Plasma, *Soviet Phys. JETP* **19**, 1265–1271 (1964).
191. Berezin, Y. A. and Karpman, V. I., Nonlinear Evolution of Disturbances in Plasmas and other Dispersive Media, *Soviet Phys. JETP* **24**, 1049–1056 (1967).
192. Shen, M. C., Asymptotic Theory of Unsteady Three-dimensional Waves in a Channel of Arbitrary Cross Section, *Soc. Ind. Appl. Math. J. Appl. Math.* **17**, 260–271 (1969).
193. Leibovich, S., Weakly Nonlinear Waves in Rotating Fluids, *J. Fluid Mech.* **42**, 803–822 (1970).
194. van Wijngaarden, L., On the Equations of Motion for Mixtures of Liquid and Gas Bubbles, *J. Fluid Mech.* **33**, 465–474 (1968).
195. Nariboli, G. A., Nonlinear Longitudinal Dispersive Waves in Elastic Rods, Iowa State Univ. Engrg. Rev. Inst., Preprint 442, 1969.
196. Tappert, F. and Varma, C. M., Asymptotic Theory of Self Trapping of Heat Pulses in Solids, *Phys. Rev. Lett.* **25**, 1108–1111 (1970).
197. Taniuti, T. and Wei, C. C., Reductive Perturbation Method in Nonlinear Wave Propagation: I, *J. Phys. Soc. Japan* **24**, 941–946 (1968).
198. Philips, O. M., "Dynamics of the Upper Ocean." Cambridge Univ. Press, London and New York, 1967.
199. Peregrine, D. H., Equations for Water Waves and the Approximation Behind Them, *in* "Waves on Beachs" (R. E. Meyer, ed.). Academic Press, New York, 1972.
200. Daikoku, K., Mizushima, Y., and Tamama, T., Computer Experiments on New Lattice Solitons Propagating in Volterra's System, *Jpn. J. Appl. Phys.* **14**, 367–376 (1975).
201. Hirota, R. and Satsuma, J., A Simple Structure of Superposition Formula of the Bäcklund Transformation, *J. Phys. Soc. Japan* **45**, 1741–1750 (1978).
202. Herrmann, R., "The Geometry of Non-Linear Differential Equations, Bäcklund Transformations and Solitons," Part A. Math. Sci. Press, Brookline, Maine, 1976.
203. Kakutani, T. and Ono, H., Weak Nonlinear Hydromagnetic Waves in a Cold Collisionless Plasma, *J. Phys. Soc. Japan* **26**, 1305–1318 (1969).
204. Hirota, R., Exact Solution of the Modified Korteweg–deVries Equation for Multiple Collisions of Solitons, *J. Phys. Soc. Japan* **33**, 1456–1458 (1972).
205. Bespalov, V. I. and Talanov, V. I., Filamentary Structure of Light Beams in Nonlinear Liquids, *JETP Lett. Engl. Transl.* **3**, 307–310 (1966).

206. Kelley, P. L., Self Focusing of Optic Beams, *Phys. Rev. Lett.* **15**, 1005–1008 (1965).
207. Talanov, V. I., Self Focusing of Wave Beams in Nonlinear Media, *JETP Lett. Engl. Transl.* **2**, 138–141 (1965).
208. Taniuti, T. and Washimi, H., Self Trapping and Instability of Hydromagnetic Waves along the Magnetic Field in a Cold Plasma, *Phys. Rev. Lett.* **21**, 209–212 (1968).
209. Asano, N., and Taniuti, T., and Yajima, N., Perturbation Method for Nonlinear Wave Modulation: II, *J. Math. Phys.* **10**, 2020–2024 (1969).
210. Karpman, V. I. and Kruskal, E. M., Modulated Waves in Nonlinear Dispersive Media *Soviet Phys. JETP* **28**, 277–281 (1969).
211. Hasegawa, A. and Tappert, F., Transmission of Stationary Nonlinear Optical Pulses in Dispersive Dielectric Fibers: I. Anomalous Dispersion *Appl. Phys. Lett.* **23**, 142–144 (1973).
212. Fried, B. D. and Ichikawa, Y. H., On the Nonlinear Schrödinger Equation for Langmuir Waves, *J. Phys. Soc. Japan* **34**, 1073–1082 (1973).
213. Ichikawa, Y. H., Imamura, T., and Taniuti, T., Nonlinear Wave Modulation in Collisionless Plasma, *J. Phys. Soc. Japan* **33**, 189–197 (1972).
214. Shimizu, K. and Ichikawa, Y. H., Automodulation of Ion Oscillation Modes in Plasma, *J. Phys. Soc. Japan* **33**, 789–792 (1972).
215. deGennes, P. G., "Superconductivity of Metals and Alloys." Benjamin, New York, 1966.
216. Tsuzuki, T., Nonlinear Waves in the Pitaevsky-Gross Equation, *J. Low Temp. Phys.* **4**, 441–457 (1971).
217. Hasimoto, H., A Soliton on a Vortex Filament, *J. Fluid Mech.* **51**, 477–485 (1972).
218. Yuen, H. C., "Waves on Vortex Filaments." Ph. D. Thesis, California Inst. of Technology, 1973.
219. Chen, H. H. and Liu, C. S., Nonlinear Wave and Soliton Propagation in Media with Arbitrary Inhomogeneities, *Phys. Fluids* **21**, 377–380 (1978).
220. Motz, H., Pavlenko, V. P., and Weiland, J., Acceleration and Slowing Down of Nonlinear Wave Packets in a Weakly Nonuniform Plasma, *Phys. Lett. A* **76**, 131–133 (1980).
221. Zakharov, V. E., Stability of Periodic Waves of Finite Amplitude on the Surface of a Deep Fluid, *J. Appl. Mech. Tech. Phys.* **9**, 86–94 (1968).
222. Hasimoto, H. and Ono, H., Nonlinear Modulation of Gravity Waves, *J. Math. Soc. Japan* **33**, 805–811 (1972).
223. Davey, A., The Propagation of a Weak Nonlinear Wave, *J. Fluid Mech.* **53**, 769–781 (1972).
224. Benney, D. J. and Roskes, G., Wave Instabilities, *Stud. Appl. Math.* **48**, 377–385 (1969).
225. Davey, A. and Stewartson, K., On Three-dimensional Packets of Surface Waves, *Proc. Roy. Soc. London* **338**, 101–110 (1974).
226. Yuen, H. C. and Lake, B. M., Nonlinear Wave Concepts Applied to Deep Water Waves *in* "Solitons in Action" (K. Lonngren and A. Scott, eds.) Academic Press, New York 1978.
227. Grimshaw, R., Slowly Varying Solitary Waves: II. Nonlinear Schrödinger Equation, *Proc. Roy. Soc. London Ser. A* **368**, 377–388 (1979).
228. Zakharov, V. E., Collapse of Langmuir Waves, *Soviet Phys. JETP* **35**, 908–914 (1972).
229. Konno, K. and Wadati, M., Simple Derivation of Bäcklund Transformation from Riccati Form of Inverse Method, *Progr. Theoret. Phys.* **53**, 1652–1656 (1975).
230. Gerdzhikov, V. S. and Kulish, P. P., Derivation of the Bäcklund Transformation in the Formalism of the Inverse Scattering Problem, *Teoret. Mat. Fiz.* **39**, 69–74 (1979).
231. Konopelchenko, B. G., The Linear Spectral Problem of Arbitrary Order: The General Form of the Integrable Equations and Their Bäcklund Transformations, *Phys. Lett. A* **75**, 447–450 (1980).

232. Boussinesq, J., Théorie de L'intumescence Liquide Appelée onde Solitaire ou de Translation se Propageant dans un Canal Rectangulaire, *Comptes Rendus* **72**, 755–759 (1871).
233. Hirota, R., A New Form of Bäcklund Transformation and its Relation to the Inverse Scattering Problem, *Progr. Theoret. Phys.* **52**, 1498–1512 (1974).
234. Hirota, R., Direct Method of Finding Exact Solutions of Nonlinear Evolution Equations, *in* "Lecture Notes in Mathematics 515, Bäcklund Transformations" (R. M. Miura ed.), pp. 40–68. Springer-Verlag, Berlin, 1976.
235. Hirota, R. and Satsuma, J., Nonlinear Evolution Equations Generated from the Bäcklund Transformation for the Boussinesq Equation, *Progr. Theoret. Phys.* **57**, 797–807 (1977).
236. Kaup, D. J., A Higher-order Water-wave Equation and the Method for Solving It, *Progr. Theoret. Phys.* **54**, 396–408 (1975).
237. Ursell, F., The Long-wave Paradox in the Theory of Gravity Waves, *Proc. Cambridge Philos. Soc.* **49**, 685–694 (1953).
238. Grimshaw, R., Slowly Varying Solitary Waves: I. Korteweg–deVries Equation, *Proc. Roy. Soc. London Ser. A* **368**, 359–375 (1979).
239. Hirota, R., Exact Solutions to the Equation Describing Cylindrical Solitons, *Phys. Lett. A* **71**, 393–394 (1979).
240. Hirota, R., The Bäcklund and Inverse Scattering Transform of the K-dV Equation with Nonuniformities, *J. Phys. Soc. Japan* **46**, 1681–1682 (1979).
241. Lax, P. D., Integrals of Nonlinear Equations of Evolution and Solitary Waves, *Comm. Pure Appl. Math.* **21**, 467–490 (1968).
242. Caudrey, P. J., Dodd, R. K., and Gibbon, J. D., A New Hierarchy of Korteweg–deVries Equations, *Proc. Roy. Soc. London Ser. A* **351**, 407–422 (1976).
243. Satsuma, J. and Kaup, D. J., A Bäcklund Transformation for a Higher Order Korteweg–deVries Equation, *J. Phys. Soc. Japan* **43**, 692–697 (1977).
244. Dodd, R. K. and Gibbon, J. D., The Prolongation Structure of a Higher Order Korteweg–deVries Equation, *Proc. Roy. Soc. London Ser. A* **358**, 287–296 (1977).
245. Sawada, S. and Kotera, T., A Method for Finding N-soliton Solutions of the K.d.V Equation and K.d.V-like Equation, *Progr. Theoret. Phys.* **51**, 1355–1367 (1974).
246. Hirota, R. and Ramani, A., The Miura Transformations of Kaup's Equation and of Mikhailov's Equation, *Phys. Lett. A* **76**, 95–96 (1980).
247. Kaup, D. J., On the Inverse Scattering Problem for Cubic Eigenvalue Problems of the Class $\psi_{xxx} + 6Q\psi_x + 6R_\psi = \lambda\psi$, *Stud. Appl. Math.* **62**, 189–216 (1980).
248. Fordy, A. P. and Gibbons, J., Some Remarkable Nonlinear Transformations, *Phys. Lett. A* **75**, 325 (1980).
249. Leibbrandt, G., New Exact Solutions of the Classical Sine-Gordon Equation in $2 + 1$ and $3 + 1$ Dimensions, *Phys. Rev. Lett.* **41**, 435–438 (1978).
250. Christiansen, P. L., A Bäcklund Transformation for the $3 + 1$-dimensional Sine-Gordon Equation, *Proc. 8th Int. Conf. Non-linear Oscillations, Prague*, 1978.
251. Christiansen, P. L., Application of New Bäcklund Transformations for the $2 + 1$ and $3 + 1$-dimensional Sine-Gordon Equation, *Z. Angew. Math. Mech.* **60**, T242–T243 (1980).
252. Christiansen, P. L. and Olsen, O. H., Ring-shaped Quasi-soliton Solutions to the Two- and Three-dimensional Sine-Gordon Equation, *Phys. Scripta* **20**, 531–538 (1979).
253. Tenenblat, K. and Terng, C. L., A Higher Dimension Generalisation of the Sine-Gordon Equation and its Bäcklund Transformation, *Bull. (New Series) Amer. Math. Soc.* **1**, 589–593 (1979).
254. Popowicz, Z., Bäcklund Transformations for the Generalized Sine-Gordon Equation in $2 + 1$ and $3 + 1$ Dimensions, *Lett. Math. Phys.* **3**, 431–436 (1979).
255. Pohlmeyer, K., Integrable Hamiltonian Systems and Interactions through Quadratic

Constraints, *Comm. Math. Phys.* **46**, 207–221 (1976).
256. Kadomtsev, B. and Petviashvili, V. I., On the Stability of Solitary Waves in a Weakly Dispersing Media, *Soviet Phys. Dokl.* **15**, 539–541 (1970).
257. Chen, H. H., A Bäcklund Transformation in Two Dimensions, *J. Math. Phys.* **16**, 2382–2384 (1975).
258. Chen, H. H., Relation Between Bäcklund Transformations and Inverse Scattering Problems in "Lecture Notes in Mathematics 515, Bäcklund Transformations" (R. M. Miura ed.), pp. 241–252. Springer-Verlag, Berlin, 1976.
259. Chen, H. H., General Derivation of Bäcklund Transformations from Inverse Scattering Problems, *Phys. Rev. Lett.* **33**, 925–928 (1974).
260. Newell, A. C., The Interrelation Between Bäcklund Transformations and the Inverse Scattering Transform *in* "Lecture Notes in Mathematics 515 Bäcklund Transformations" (R. M. Miura, ed.) pp. 227–240. Springer-Verlag, Berlin, 1976.
261. Dryuma, V., Analytic Solution of the Two-dimensional Korteweg-deVries (KdV) Equation, *JETP Lett. Engl. Transl.* **19**, 387–388 (1974).
262. Satsuma, J., N-soliton Solution of the Two-dimensional Korteweg-deVries Equation, *J. Phys. Soc. Japan* **40**, 286–290 (1976).
263. Satsuma, J. and Ablowitz, M. J., Two-dimensional Lumps in Nonlinear Dispersive Systems, *J. Math. Phys.* **20**, 1496–1501 (1979).
264. Ablowitz, M. J. and Haberman, R., Nonlinear Evolution Equations—Two and Three Dimensions, *Phys. Rev. Lett.* **35**, 1185–1188 (1975).
265. Anderson, R. L., Barut, A. O. and Raczka, R., Bäcklund Transformations and New Solutions of Nonlinear Wave Equations in Four-dimensional Space–Time, *Lett. Math. Phys.* **3**, 351–358 (1979).
266. McCarthy, P. J., Existence of Strong Bäcklund Transformations in Four or More Dimensions and Generalization of a Family of Bäcklund Transformations, *Lett. Math. Phys.* **2**, 493–498 (1978).
267. Levi, D., Pilloni, L. and Santini, P. M., Bäcklund Transformations for Nonlinear Evolution Equations in $2 + 1$ Dimensions, *Phys. Lett. A* **81**, 419–423 (1981).
268. Kaup, D. J. and Newell, A. C., An Exact Solution for a Derivative Nonlinear Schrödinger Equation, *J. Math. Phys.* **19**, 798–801 (1978).
269. Kawata, T. and Inoue, H., Exact Solutions of the Derivative Nonlinear Schrödinger Equation under the Vanishing Conditions, *J. Phys. Soc. Japan* **44**, 1968–1976 (1978).
270. Nakamura, A. and Chen, H. H., Multi-soliton Solutions of a Derivative Nonlinear Schrödinger Equation, *J. Phys. Soc. Japan* **49**, 813–816 (1980).
271. Morris, H. C. and Dodd R. K., The Two Component Derivative Nonlinear Schrödinger Equation, *Phys. Scripta* **20**, 505–508 (1979).
272. Wadati, M., Konno, K., and Ichikawa, Y. H., A Generalization of the Inverse Scattering Method, *J. Phys. Soc. Japan* **46**, 1965–1966 (1979).
273. Wadati, M., Konno, K., and Ichikawa, Y. H., New Integrable Nonlinear Evolution Equations, *J. Phys. Soc. Japan* **47**, 1698–1700 (1979).
274. Benjamin, T. B., Internal Waves of Permanent Form in Fluids of Great Depth, *J. Fluid Mech.* **29**, 559–592 (1967).
275. Matsuno, Y., Bilinearization of Nonlinear Evolution Equations: IV. Higher Order Benjamin–Ono Equations, *J. Phys. Soc. Japan* **49**, 1584–1592 (1980).
276. Ono, H., Algebraic Solitary Waves in Stratified Fluids, *J. Phys. Soc. Japan* **39**, 1082–1091 (1975).
277. Nakamura, A., Bäcklund Transform and Conservation Laws of the Benjamin-Ono Equation, *J. Phys. Soc. Japan* **47**, 1335–1340 (1979).
278. Nakamura, A., N-periodic Wave and N-soliton Solutions of the Modified Benjamin–Ono Equation, *J. Phys. Soc. Japan* **47**, 2045–2046 (1979).

279. Matsuno, Y., Exact Multi-soliton Solution of the Benjamin–Ono Equation, *J. Phys. A* **12**, 619–621 (1979).
280. Satsuma, J. and Ishimori, Y., Periodic Wave and Rational Soliton Solutions of the Benjamin-Ono Equation, *J. Phys. Soc. Japan* **46**, 681–687 (1979).
281. Bock, T. L. and Kruskal, M. D., A Two-parameter Miura Transformation and the Benjamin–Ono Equation, *Phys. Lett. A* **74**, 173–176 (1979).
282. Joseph, R. I., Solitary Waves in a Finite Depth Fluid, *J. Phys. A* **10**, L 225–L 227 (1977).
283. Kubota, T., Ko, D. R. S. and Dobbs, D., Weakly Nonlinear, Long Internal Gravity Waves in Stratified Fluids of Finite Depth, *J. Hydronaut.* **12**, 157–165 (1978).
284. Matsuno, Y., Exact Multi-soliton Solution for Nonlinear Waves in a Stratified Fluid of Finite Depth, *Phys. Lett. A* **74**, 233–235 (1979).
285. Nakamura, A. and Matsuno, Y., Exact One- and Two-periodic Wave Solutions of Fluids of Finite Depth, *J. Phys. Soc. Japan* **48**, 653–657 (1980).
286. Nakamura, A., Exact N-soliton Solution of the Modified Finite Depth Fluid Equation, *J. Phys. Soc. Japan* **47**, 2043–2044 (1979).
287. Chen, H. H., Hirota, R., and Lee, Y. C., Inverse Scattering Problem for Internal Waves with Finite Fluid Depth, *Phys. Lett. A* **75**, 254–256 (1980).
288. Satsuma, J., Ablowitz, M. J., and Kodama, Y., On an Internal Wave Equation Describing a Stratified Fluid with Finite Depth, *Phys. Lett. A* **73**, 283–286 (1979).
289. Toda, M., Vibration of a Chain with Nonlinear Interaction, *J. Phys. Soc. Japan* **22**, 431–436 (1967).
290. Hirota, R., Exact N-soliton Solution of a Nonlinear Lumped Network Equation, *J. Phys. Soc. Japan* **35**, 286–288 (1973).
291. Flaschka, H., On the Toda Lattice: II. Inverse-scattering Solution. *Progr. Theoret. Phys.* **51**, 703–716 (1974).
292. Chen, H. H. and Liu, C. S., Bäcklund Transformation Solutions of the Toda Lattice Equation, *J. Math. Phys.* **16**, 1428–1430, (1975).
293. Wadati, M. and Toda, M., Bäcklund Transformation for the Exponential Lattice, *J. Phys. Soc. Japan* **39**, 1196–1203 (1975).
294. Wadati, M., Wave Propagation in Nonlinear Lattice: I. *J. Phys. Soc. Japan* **38**, 673–680 (1975).
295. Konno, K. and Sanuki, H., Bäcklund Transformation for Equation of Motion for Nonlinear Lattice under Weak Dislocation Potential. *J. Phys. Soc. Japan* **39**, 22–24 (1975).
296. Orfanidis, S. J., Discrete Sine-Gordon Equations, *Phys. Rev. D* **18**, 3822–3827 (1978).
297. Orfanidis, S. J., Sine-Gordon Equation and Nonlinear σ Model on a Lattice, *Phys. Rev. D* **18**, 3828–3832 (1978).
298. Hirota, R., Nonlinear Partial Difference Equations: I. A Difference Analogue of the Korteweg–deVries Equation, *J. Phys. Soc. Japan* **43**, 1424–1433 (1977).
299. Hirota, R., Nonlinear Partial Difference Equations: III. Discrete Sine-Gordon Equation, *J. Phys. Soc. Japan* **43**, 2079–2086 (1977).
300. Konno, K., Kameyama, W., and Sanuki, H., Effect of Weak Dislocation Potential on Nonlinear Wave Propagation in Anharmonic Crystal, *J. Phys. Soc. Japan* **37**, 171–176 (1974).
301. Hirota, R. and Satsuma, J., Nonlinear Evolution Equations Generated from the Bäcklund Transformation for the Toda Lattice, *Prog. Theoret. Phys.* **55**, 2037–2038 (1976).
302. Hirota, R. and Satsuma, J., A Variety of Nonlinear Network Equations Generated from the Bäcklund Transformation for the Toda Lattice, *Prog. Theoret. Phys. Suppl.* **59**, 64–100 (1976).

303. Hirota, R. and Satsuma, J., N-soliton Solutions of Nonlinear Network Equations Describing a Volterra System, *J. Phys. Soc. Japan* **40**, 891–900 (1976).
304. Hirota, R., Exact N-soliton Solution of Nonlinear Lumped Self-dual Network Equations, *J. Phys. Soc. Japan* **35**, 289–294 (1973).
305. Hirota, R., Nonlinear Partial Difference Equations: II. Discrete-time Toda Equation, *J. Phys. Soc. Japan* **43**, 2074–2078 (1977).
306. Hirota, R., Nonlinear Partial Difference Equations: IV. Bäcklund Transformation for the Discrete-time Toda Equation, *J. Phys. Soc. Japan* **45**, 321–332 (1978).
307. Hirota, R., Nonlinear Partial Difference Equations: V. Nonlinear Equations Reducible to Linear Equations, *J. Phys. Soc. Japan* **46** 312–319 (1979).
308. Kingston, J. G. and Rogers, C., On Nonlinear Equations Amenable to the Inverse Scattering Method, *Canad. J. Phys.* **53**, 58–61 (1975).
309. Calogero, F., Nonlinear Evolution Equations Solvable by the Inverse Spectral Transform, *in* "Lecture Notes in Physics 80, Proceedings, Mathematical Problems in Theoretical Physics, Rome," (G. Dell'Antonio, S. Doplicher, and G. Jona–Lasinio, eds.), pp. 235–269. Springer-Verlag, Berlin, 1978.
310. Wadati, M., Sanuki, H. and Konno, K., Relationships Among Inverse Method, Bäcklund Transformation and an Infinite Number of Conservation Laws, *Prog. Theoret. Phys.* **53**, 418–436 (1975).
311. Crum, M. M., Associated Sturm–Liouville Systems, *Quart. J. Math. Oxford* **6**, 121–127 (1955).
312. Satsuma, J., A Wronskian Representation of N-soliton Solutions of Nonlinear Evolution Equations, *J. Phys. Soc. Japan* **46**, 359–360 (1979).
313. Beltrami, E., "Opere Matematiche," Vol. 3, 349–382. Milano, Hoepli, 1911.
314. Bers, L. and Gelbart, A., On a Class of Differential Equations in Mechanics of Continua, *Quart. Appl. Math.* **1**, 168–188 (1943).
315. Bauer, K. W. and Ruscheweyh, S., Differential Operators for Partial Differential Equations and Function Theoretic Applications, *in* "Lecture Notes in Mathematics 791." Springer-Verlag, Berlin, 1980.
316. Arndt, W., "Die Torsion von Wellen mit Achsensymmetrischen Bohrungen und Hohlräumen." Dissertation, Göttingen 1916.
317. Weinstein, A., Transsonic Flow and Generalized Potential Theory. *Proc. Aeroballist. Res. Symposia, Naval Ordnance Laboratory*, 73–82 (1949).
318. Weinstein, A., Discontinuous Integrals and Generalized Potential Theory, *Trans. Amer. Math. Soc.* **63**, 342–354 (1948).
319. Parsons, D. H., Irrotational Flow of a Liquid with Axial Symmetry. *Proc. Edinburgh Math. Soc.* **13**, 201–204 (1963).
320. Rogers, C., Bäcklund Transformations and Invariance Properties in Axially Symmetric Flow, *Ann. Soc. Sci. Bruxelles*, **86**, 211–219 (1972).
321. Rogers, C. and Kingston, J. G., Application of Bäcklund Transformations to the Stokes–Beltrami Equations, *J. Austral. Math. Soc.* **15**, 179–189 (1973).
322. Burns, J. C., The Iterated Equation of Generalized Axially Symmetric Potential Theory: I–III, *J. Austral. Math. Soc.* **7**, 263–300 (1967); IV, Ibid **9**, 153–160 (1969); V, Ibid **11**, 129–141 (1970); VI, Ibid **18**, 318–327 (1974).
323. Payne, L. E. and Pell, W. H., The Stokes Flow Problem for a Class of Axially Symmetric Bodies, *J. Fluid Mech.* **7**, 529–549 (1960).
324. Pell, W. H. and Payne, L. E., The Stokes Flow About a Spindle, *Quart. Appl. Math.* **18**, 257–262 (1960).
325. Pell, W. H. and Payne, L. E., On Stokes Flow about a Torus, *Mathematika* **7**, 78–92 (1960).

326. Rogers, C. and Gladwell, G. M. L., On Punch, Crack, and Torsion Problems Linked by Bäcklund Transformations. Univ. of Waterloo, Dept. Applied Mathematics preprint.
327. Sneddon, I. N. and Lowengrub, M., "Crack Problems in the Classical Theory of Elasticity." Wiley, New York, 1969.
328. Weinstein, A., On Cracks and Dislocations in Shafts under Torsion, *Quart. Appl. Math.* **10**, 77–81 (1952).
329. Reissner, E. and Sagoci, H. F., Forced Torsional Oscillations of an Elastic Half-space: I. *J. Appl. Phys.* **15**, 652–654 (1944).
330. Boussinesq, J., "Applications des Potentials à L'étude de L'equilibre et des Mouvement des Solides Élastiques." Paris, 1885.
331. Sneddon, I. N., "Fourier Transforms." McGraw-Hill, New York, 1951.
332. Gladwell, G. M. L., "Contact Problems in the Classical Theory of Elasticity." Sijthoff and Noordhoff, Winchester, Maine, 1980.
333. Beltrami, E., Sulla Teoria della Funzioni Potenziali Simmetriche, *Atti. Accad. Sci. Ist. Bologna* **2**, 461–498 (1881).
334. Watson, G. N., "Theory of Bessel Functions." Cambridge Univ. Press, London and New York, 1944.
335. Reissner, E., Freie und Erzwungene Torsionsschwingungen des Elastischen Halbräumes, *Ingr.-Arch.* **8**, 229–245 (1937).
336. Sneddon, I. N., "Mixed Boundary Value Problems in Potential Theory." North-Holland Publ., Amsterdam, 1966.
337. Hobson, E. W., "The Theory of Spherical and Ellipsoidal Harmonics." Cambridge Univ. Press, London and New York, 1931.
338. Bergman, S., "Integral Operators in the Theory of Partial Differential Equations." Springer-Verlag, Berlin and New York, 1968.
339. Kreyszig, E., Function Theoretic Integral Operator Methods for Partial Differential Equations, personal communication.
340. von Mises, R. and Schiffer, M., On Bergman's Integration Method in Two-dimensional Compressible Fluid Flow *in* "Advances in Applied Mathematics," Vol. I. Academic Press, New York, 1948.
341. Alferov, V. D. and Ryashentsev, V. I., One Method for Solving the Plane Problem of Pressure Filtration to Boreholes in Nonhomogeneous Strata, *Izv. Akad. Nauk SSSR Meh. Zidk. Gaza* **1**, 71–77 (1973).
342. Sneddon, I. N. and Elliot, H. A., The Opening of a Griffith Crack under Internal Pressure, *Quart. Appl. Math.* **4**, 229–267 (1946).
343. Busbridge, I. W., Dual Integral Equations, *Proc. London Math. Soc.* **44**, 115–129 (1938).
344. Friedlander, F. G., Simple Progressive Solutions of the Wave Equation, *Proc. Cambridge Philos. Soc.* **43**, 360–373 (1946).
345. Rogers, C. Moodie, T. B., and Clements, D. L., Radial Propagation of Rotary Shear Waves in an Initially Stressed Neo-Hookean Material, *J. Mécanique*, **15**, 595–614 (1976).
346. Moodie, T. B., Rogers, C., and Clements, D. L., Radial Propagation of Axial Shear Waves in an Incompressible Elastic Material under Finite Deformation, *Internat. J. Engrg. Sci.* **14**, 585–603 (1976).
347. Moodie, T. B., Rogers, C., and Clements, D. L., Large Wavelength Pulse Propagation in Curved Elastic Rods, *J. Acoust. Soc. Amer.* **59**, 557–563 (1976).
348. Moodie, T. B., On the Propagation, Reflection, and Transmission of Transient Cylindrical Shear Waves in Nonhomogeneous Four-parameter Viscoelastic Media. *Bull. Austral. Math. Soc.* **8**, 397–411 (1973).
349. Rogers, C., Clements, D. L., and Moodie, T. B., Les Ondes de Cisaillement à Symétrie Sphérique ou Cylindrique pour un Matériau Visco-élastique Non-homogène et Isotropique, *Utilitas Math.* **10**, 167–177 (1976).

REFERENCES

350. Barclay, D. W., Moodie, T. B., and Rogers, C., Cylindrical Impact Waves in Inhomogeneous Viscoelastic Media, *Acta Mech.* **29**, 93–117 (1978).
351. Neuber, H., "Kerbspannungslehre." Springer-Verlag, Berlin and New York, 1958.
352. Neuber, H., Theory of Stress Concentration for Shear-strained Prismatical Bodies with Arbitrary Nonlinear Stress–strain Law, *Trans. ASME Ser. E, J. Appl. Mech.* **28**, 544–550 (1961).
353. Sokolovsky, V. V., Longitudinal Displacement of Plastic Mass between Non-circular Cylinders, *Prikl. Math. Meh.* **23**, 732–739 (1959).
354. Sih, G. C. and MacDonald, B., Effect of Material Nonlinearity on Crack Propagation, *Internat. J. Engrg. Sci.* **12**, 61–77 (1974).
355. Dobrovol'sky, V. L., On Antiplane Deformation for Materials Which Do Not Obey Hooke's Law, *Internat. J. Solids and Structures* **1**, 195–205 (1965).
356. Shapiro, A. H., "The Dynamics and Thermodynamics of Compressible Fluid Flow," Vol. I. Ronald Press, New York, 1953.
357. Payne, D. A., Bäcklund Transformations in Several Variables, *J. Math. Phys.* **21**, 1593–1602 (1980).
358. Kinnersley, W., Symmetries of the Stationary Einstein–Maxwell Field Equations: I, *J. Math. Phys.* **18**, 1529–1537 (1977).
359. Kinnersley, W. and Chitre, D. M., Symmetries of the Stationary Einstein–Maxwell Field Equations: II, *J. Math. Phys.* **18**, 1538–1542 (1977).
360. Kinnersley, W. and Chitre, D. M., Symmetries of the Stationary Einstein–Maxwell Field Equations: III, *J. Math. Phys.* **19**, 1926–1931 (1978).
361. Kinnersley, W. and Chitre, D. M., Symmetries of the Stationary Einstein–Maxwell Field Equations: IV. Transformations which Preserve Asymptotic Flatness, *J. Math. Phys.* **19**, 2037–2042 (1978).
362. Hoenselaers, C., Symmetries of the Stationary Einstein–Maxwell Field Equations: V, *J. Math. Phys.* **20**, 2526–2529 (1979).
363. Hoenselaers, C., Kinnersley, W., and Xanthopoulos, B. C., Symmetries of the Stationary Einstein–Maxwell Field Equations: VI, *J. Math. Phys.* **20**, 2530–2536 (1979).
364. Kinnersley, W., Generation of Stationary Einstein–Maxwell Fields, *J. Math. Phys.* **14**, 651–653 (1973).
365. Hoenselaers, C., On Generation of Solutions of Einstein's Equations, *J. Math. Phys.* **17**, 1264–1267 (1976).
366. Chitre, D. M., Characterization of Certain Stationary Solutions of Einstein's Equations, *J. Math. Phys.* **19**, 1625–1626 (1978).
367. Kinnersley, W. and Chitre, D. M., Group Transformation that Generates the Kerr and Tomimatsu–Sato Metrics, *Phys. Rev. Lett.* **40**, 1608–1609 (1978).
368. Hoenselaers, C., Kinnersley, W., and Xanthopoulos, B. C., Generation of Asymptotically Flat, Stationary Space-times with Any Number of Parameters, *Phys. Rev. Lett.* **42**, 481–482 (1979).
369. Geroch, R., A Method for Generating Solutions of Einstein's Equations, *J. Math. Phys.* **12**, 918–924 (1971).
370. Geroch, R., A Method for Generating New Solutions of Einstein's Equations: II, *J. Math. Phys.* **13**, 394–404 (1972).
371. Cosgrove, C. M., Ph.D Thesis, University of Sydney, 1979.
372. Cosgrove, C. M., New Family of Exact Stationary Axisymmetric Gravitational Fields Generalising the Tomimatsu–Sato Solutions, *J. Phys. A* **10**, 1481–1524 (1977).
373. Cosgrove, C. M., Limits of the Generalised Tomimatsu–Sato Gravitational Fields, *J. Phys. A* **10**, 2093–2105 (1977).
374. Cosgrove, C. M., A New Formulation of the Field Equations for the Stationary Axisymmetric Vacuum Gravitational Field: I. General Theory, *J. Phys. A* **11**, 2389–2404 (1978).

375. Cosgrove, C. M., A New Formulation of the Field Equations for the Stationary Axisymmetric Vacuum Gravitational Field: II. Separable Solutions, *J. Phys. A* **11**, 2405–2430 (1978).
376. Harrison, B. K., Bäcklund Transformation for the Ernst Equation of General Relativity, *Phys. Rev. Lett.* **41**, 1197–1200 (1978).
377. Belinskii, V. A. and Zakharov, V. E., Integration of the Einstein Equations by Means of the Inverse Scattering Problem Technique and Construction of Exact Soliton Solutions, *Soviet Phys. JETP* **48**, 985–994 (1978).
378. Neugebauer, G., Bäcklund Transformations of Axially Symmetric Stationary Gravitational Fields, *J. Phys. A* **12**, L67–L70 (1979).
379. Cosgrove, C. M., "The Second Marcel Grossmann Meeting on the Recent Developments of General Relativity, Proceeding." Trieste, Italy, 1979.
380. Cosgrove, C. M., Relationships between the Group-theoretic and Soliton-theoretic Techniques for Generating Stationary Axisymmetric Gravitational Solutions, *J. Math. Phys.* **21**, 2417–2447 (1980).
381. Neugebauer, G. and Kramer, D., Generation of the Kerr-NUT Solution from Flat Space-time by Bäcklund Transformations: preprint, Friedrich–Schiller Universität, Jena, 1979.
382. Kramer, D. and Neugebauer, G., The Superposition of Two Kerr Solutions, *Phys. Lett. A* **75**, 259–261 (1980).
383. Neugebauer, G., Recursive Calculation of Axially Symmetric Stationary Einstein fields, *J. Phys. A* **13**, 1737–1740 (1980).
384. Neugebauer, G., A General Integral of the Axially Symmetric Stationary Einstein Equations, *J. Phys. A* **13**, L19–L21 (1980).
385. Herlt, E., Static and Stationary Axially Symmetric Gravitational Fields of Bounded Sources: I. Solutions Obtainable from the Van Stockhum Metric, *Gen. Relativity Gravitation*, **9**, 711–719 (1978).
386. Herlt, E., Static and Stationary Axially Symmetric Gravitational Fields of Bounded Sources: II. Solutions Obtainable from Weyl's Class, *Gen. Relativity Gravitation*, **11**, 337–342 (1979).
387. Kodama, Y. and Wadati, M., A Canonical Transformation for the Sine-Gordon Equation, *Prog. Theor. Phys.* **56**, 342–343 (1976).
388. Kodama, Y. and Wadati, M., Theory of Canonical Transformations for Nonlinear Evolution Equations: I, *Prog. Theor. Phys.* **56**, 1740–1755 (1976).
389. Kodama, Y., Theory of Canonical Transformations for Nonlinear Evolution Equations: II, *Prog. Theor. Phys.* **57**, 1900–1916 (1977).
390. Dodd, R. K. and Bullough, R. K., The Generalized Marchenko Equation and the Canonical Structure of the AKNS–ZS Inverse Method, *Phys. Scripta* **20**, 514–530 (1979).
391. Sasaki, R., Canonical Structure of Bäcklund Transformations, *Phys. Lett. A* **78**, 7–10 (1980).
392. Leibbrandt, G., Morf, R., and Wong, S., Solutions of the Sine-Gordon Equation in Higher Dimensions, *J. Math. Phys.* **21**, 1613–1624 (1980).
393. Case, K. M., Bäcklund Transformations in Four-dimensional Space-time, *Lett. Math. Phys.* **4**, 87–92 (1980).
394. Wilson, W. and Swamy, N. V. V. J., Note on Bäcklund Transformations, Dirac Factorization and the Sine-Gordon Equation: II, *Nuovo Cimento A* **56**, 44–50 (1980).
395. Rund, H., Variational Problems and Bäcklund Transformations Associated With the Sine-Gordon and Korteweg–deVries Equations and Their Extensions, *in* "Lecture Notes in Mathematics 515 Bäcklund Transformations" (R. M. Miura, ed.), pp. 199–226. Springer-Verlag, Berlin, 1976.

396. 't Hooft, G., Symmetry Breaking through Bell–Jackiw Anomalies, *Phys. Rev. Lett.* **37**, 8–11 (1976).
397. Ward, R. S., On Self-dual Gauge Fields, *Phys. Lett. A* **61**, 81–82 (1977).
398. Atiyah, M. F. and Ward, R. S., Instantons and Algebraic Geometry, *Comm. Math. Phys.* **55**, 117–124 (1977).
399. Corrigan, E., Fairlie, D. B., Yates, R. G., and Goddard, P., The Construction of Self-dual Solutions to SU(2) Gauge Theory, *Comm. Math. Phys.* **58**, 223–240 (1978).
400. Yang, C. N., Condition of Self Duality for $SU(2)$ Gauge Fields on Euclidean Four-dimensional Space, *Phys. Rev. Lett.* **38**, 1377–1379 (1977).
401. Pohlmeyer, K., On the Lagrangian Theory of Anti-self-dual Fields in Four-dimensional Euclidean Space, *Comm. Math. Phys.* **72**, 37–47 (1980).
402. Belavin, A. A. and Zakharov, V. E., Yang–Mills Equations as Inverse Scattering Problem, *Phys. Lett. B* **73**, 53–57 (1978).
403. Daniel, M. and Viallet, C. M., The Geometric Setting of Gauge Theories of the Yang–Mills Type, *Rev. Modern Phys.* **52**, 175–197 (1980).
404. Brihaye, Y. and Nuyts, J., Invariance Properties of Yang Equations and Applications, *J. Math. Phys.* **21**, 909–912 (1980).
405. Yih, C. S., "Fluid Mechanics," McGraw-Hill, New York, 1969.
406. Christiansen, P. L., On Bäcklund Transformations and Solutions to the 2+1 and 3+1-dimensional Sine-Gordon Equation, *in* "Proceedings 4th Scheveningen Conference on Differential Equations." Springer-Verlag, Berlin and New York, 1979.
407. Shadwick, W. F., "The Bäcklund Problem: Symmetries and Conservation Laws for Some Nonlinear Differential Equations." Ph.D. Thesis, Univ. of London, 1979.
408. Crampin, M., Solitons and $\mathfrak{SL}(2, \mathbb{R})$, *Phys. Lett. A* **66**, 170–172 (1978).
409. Wahlquist, H. D. and Estabrook, F. B., Prolongation Structures of Nonlinear Evolution Equations, *J. Math. Phys.* **16**, 1–7 (1975).
410. Estabrook, F. B. and Wahlquist, H. D., Prolongation Structures of Nonlinear Evolution Equations: II, *J. Math. Phys.* **17**, 1293–1297 (1976).
411. Dodd, R. K. and Gibbon, J. D., The Prolongation Structures of a Class of Nonlinear Evolution Equations, *Proc. Roy. Soc. London Ser. A* **359**, 411–433 (1978).
412. Morris, H. C., A Prolongation Structure for the AKNS System and Its Generalization, *J. Math. Phys.* **18**, 533–536 (1977).
413. Matsuno, Y., The Bäcklund Transformations of the Higher-Order Korteweg–deVries Equations, *Phys. Lett. A* **77**, 100–102 (1980).
414. Matsuno, Y., Interaction of the Benjamin–Ono Solitons, *J. Phys. A* **13**, 1519–1536 (1980).
415. Nakamura, A. and Hirota, R., Second Modified KdV Equation and its Exact Multi-soliton Solution, *J. Phys. Soc. Japan* **48**, 1755–1762 (1980).
416. Ehlers, F. E., and Carrier, G. F., "Methods of Linearization in Compressible Flow," Monograph IV, Part II, Hodograph Method, Air Material Command Report F-TR-1180 B-ND 1948.
417. Kaplan, C., The Flow of a Compressible Fluid Past a Curved Surface, NACA Report 768 1943.
418. Lin, C. C., On an Extension of the Kármán–Tsien Method to Two-dimensional Subsonic Flows with Circulation Around Closed Profiles, *Quart. Appl. Math.* **4**, 291–297 (1946).
419. Costello, G. R., "Method of Designing Airfoils with Prescribed Velocity Distributions in Compressible Potential Flows." NACA Tech. Note 1913, 1949.
420. Poritsky, H., Polygonal Approximation Method in the Hodograph Plane, *J. Appl. Mech.* **16**, 123–133 (1949).
421. Busemann, A., Hodographenmethode der Gasdynamik, *Z. Angew. Math. Mech.* **17**, 73–79 (1937).

422. Carrier, G. F., Elbows for Accelerated Flows, *J. Appl. Mech.* **14**, 108–112 (1947).
423. Berryman, J. G., Evolution of a Stable Profile for a Class of Nonlinear Diffusion Equations with Fixed Boundaries, *J. Math. Phys.* **18**, 2108–2115 (1977).
424. Berryman, J. G. and Holland, C. J., Evolution of a Stable Profile for a Class of Nonlinear Diffusion Equations: II, *J. Math. Phys.* **19**, 2476–2480 (1978).
425. Ames, W. F., "Nonlinear Partial Differential Equations in Engineering." Academic Press, New York, 1965.
426. Aronson, D. G., Regularity Properties of Flows Through Porous Media, *Soc. Ind. Appl. Math. J. Appl. Math.* **17**, 461–467 (1969).
427. Aronson, D. G., Regularity Properties of Flows Through Porous Media: the Interface, *Arch. Rational Mech. Anal.* **37**, 1–10 (1970).
428. Knerr, B. F., The Porous Medium Equation in One Dimension, *Trans. Amer. Math. Soc.* **234**, 381–415 (1977).
429. Bluman, G. W. and Cole, J. D., "Similarity Methods for Differential Equations." Springer-Verlag, Berlin and New York, 1974.
430. Rosen, G., Nonlinear Heat Conduction in Solid H_2, *Phys. Rev. B* **19**, 2398–2399 (1979).
431. Bluman, G. W. and Kumei, S., On the Remarkable Nonlinear Diffusion Equation

$$\partial/\partial x[a(u+b)^{-2}\partial u/\partial x] - \partial u/\partial t = 0,$$

J. Math. Phys. **21**, 1019–1023 (1980).
432. Berryman, J. G., Slow Diffusion on the Line, *J. Math. Phys.* **21**, 1326–1331 (1980).
433. Boiti, M. and Pempinelli, F., Nonlinear Schrödinger Equation, Bäcklund Transformations and Painlevé Transcendents, *Nuovo Cimento B* **59** 40–58 (1980).
434. Bianchi, L, "Lezioni Sulla Teoria delle Funzioni di Variabile Complessa e della Funzioni Ellittiche." Pisa, 1916.
435. Boiti, M. and Pempinelli, F., Similarity Solutions of the Korteweg–deVries Equation, *Nuovo Cimento B* **51**, 70–78 (1979).
436. Boiti, M. and Pempinelli, F., Similarity Solutions and Bäcklund Transformations of the Boussinesq Equation, *Nuovo Cimento B* **56**, 148–156 (1980).
437. Boiti, M. and Pempinelli, F., Similarity Solutions of the KdV Equation: Bäcklund Transformations and Painlevé Transcendents, *in* "Lecture Notes in Physics 120" (M. Boiti, F. Pempinelli, and G. Soliani, eds.). Springer-Verlag, Berlin, 1980.
438. Chudnovsky, G. V., The Inverse Scattering Problem and Application to Arithmetics, Approximation Theory and Transcendental Numbers, *in* "Lecture Notes in Physics, 120" (M. Boiti, F. Pempinelli, and G. Soliani, eds.). Springer-Verlag, Berlin, 1980.
439. Anderson, R. L. and Turner, J. W., A Type of Bäcklund-like Invariance Transformation for a Class of Second-order Ordinary Differential Equations, *Lett. Math. Phys.* **1**, 37–42 (1975).
440. Airault, H., Rational Solutions of Painlevé Equations, *Stud. Appl. Math.* **61**, 31–53 (1979).
441. Motz, H., Pavlenko, V. P., and Weiland, J., Acceleration and Slowing Down of Nonlinear Wave Packets in a Weakly Nonuniform Plasma, *Phys. Lett. A* **76**, 131–133 (1980).
442. Gardner, R. B., Constructing Bäcklund Transformations in Partial Differential Equations and Geometry, *in* "Lecture Notes in Pure and Applied Mathematics 48" (C. I. Byrnes, ed.) Dekker, New York, 1979.
443. Watanabe, Y., Bäcklund Transformation as a Continuous Two-parameter Mapping, *J. Phys. Soc. Japan* **41**, 727–728 (1976).
444. Toda, M., Vibration of a Chain with Nonlinear Interaction, *J. Phys. Soc. Japan* **33**, 431–436 (1967).
445. Toda, M. and Wadati, M., A Canonical Transformation for the Exponential Lattice, *J. Phys. Soc. Japan* **39**, 1204–1211 (1975).

446. McCarthy, P. J., Some Linear Bäcklund Transformations, *Lett. Math. Phys.* **4**, 39–43 (1980).
447. Steudel, H., Noether's Theorem and the Conservation Laws of the Korteweg–deVries Equation, *Ann. Physik.* **32**, 445–455 (1975).
448. Steudel, H., Ableitung einer Kontinuierlichen Mannigfaltigkeit von Erhaltungssätzen der Modifizierten Korteweg–deVries Gleichung nach dem Noetherschen Satz, *Ann. Physik.* **32**, 459–465 (1975).
449. Shadwick, W. F., Noether's Theorem and Steudel's Conserved Currents for the Sine-Gordon Equation, *Lett. Math. Phys.* **4**, 241–248 (1980).
450. Shadwick, W. F., The Hamilton-Cartan Formalism for rth Order Lagrangians and the Integrability of the Korteweg–deVries and Modified Korteweg–deVries Equations, *Lett. Math. Phys.* **5**, 137–141 (1981).

Author Index

Numbers in parentheses are reference numbers and indicate that an author's work is referred to although the name is not cited in the text. Numbers in italics show the page on which the complete reference is listed.

A

Ablowitz, M. J., 6(288), 65, 91, 104(288), *307, 313, 314*
Adamson, I. T., 127, *307*
Airault, H., 69(440), *320*
Alferov, V. D., 255, *316*
Ames, W. F., 9(425), 18, 168(425), *309, 320*
Anderson, R. L., 13, 69(439), 82, *307, 313, 320*
Arndt, W., 11, 239, 245, *315*
Aronson, D. G., 9(426,427), 168(426,427), *320*
Asano, N., 62(209), *311*
Askar, A., 198(95), 230(95), *306*
Atiyah, M.F., 84, *319*
Atkinson, C., 262(100), *306*

B

Bäcklund, A. V., 1, 12, 13, *302*
Barclay, D. W., 262(350), *317*
Barnard, T. W., 3, 22, 28, 29, *303*
Barut, A. O., 82(265), *313*
Bateman, H., 9, 38, 151, 153(53), 154, *304, 309*

Bauer, K. W., 239, 254, *315*
Bean, C. P., 17(16), *303*
Belavin, A. A., 84, *319*
Belinskii, V. A., 6, 115, *318*
Bell, J. F., 207, *308*
Beltrami, E., 239, 249, *315, 316*
Benjamin, T. B., 6(274), 93, *313*
Benney, D. J., 62, *311*
Berezin, Y. A., 43, *310*
Bergman, S., 254, *316*
Berryman, J. G., 9(423, 424), 168(423, 424), 171, *320*
Bers, L., 239, *315*
Bespalov, V. I., 62(205), *310*
Bianchi, L., 69, *307, 320*
Blackstock, D. T., 38, *309*
Bluman, G. W., 168(429), 170(431), *320*
Bock, T. L., 101, *314*
Boiti, M., 69(435–437), *320*
Born, M., 232, *308*
Boussinesq, J., 69, 249, *312, 316*
Brihaye, Y., 92, *319*
Broer, L. J. F., 225, *308*

Brown, G. H., 16(164), *309*
Bruschi, M., *304*
Bullough, R. K., 16, *307, 319*
Burgers, J. M., 38, 41, *307, 309*
Burns, J. C., 241, *315*
Busbridge, I. W., 260, *316*
Busemann, A., 162(421), *319*

C

Calogero, F., 65, *304, 315*
Carrier, G. F., 162(416, 422), *319, 320*
Cartan, E., *304*
Case, K. M., 82, *304, 318*
Castell, S. P., 9(62), 10(83), 163(62), *305*
Caudrey, P. J., 5, 79, 80, 82, *312*
Cekirge, H. M., 10, 11, 180(88), 190, 191, 195, 197, 198(95), 202, 204, 205(85), 206(85), 207, 208, 209, 211(85), 212(85), 214, 216, 217(85), 230(95), *306, 308*
Chaplygin, S. A., 8, *304*
Chen, H. H., 6(292), 62(219), 64, 83, 104, 105, 113, *311, 313, 314*
Chitre, D. M., 115(359–361, 366, 367), *317*
Chiu, S. C., *304*
Christiansen, P. L., 82, 84, 86, *312, 319*
Chu, C.-w., 42, *304*
Chu, F. Y. F., 43(185), *310*
Chudnovsky, G. V., 69(438), *320*
Clairin, J., 12, *307*
Clements, D. L., 11(97), 180(87), 261(101), 262(93, 100, 102, 345–347, 349), 267, *306, 316*
Coburn, N., 10(68), 160, 182, *305*
Cole, J. D., 39, 40, 168(429), *304, 320*
Coleman, S., 17(23), *303*
Corrigan, E., 84, 92, *319*
Cosgrove, C. M., 7, 115(372–375, 379), *317, 318*
Costello, G. R., 162(419), *319*
Courant, R., 219, *308*
Crampin, M., 142, *319*
Critescu, N., 203, *308*
Crum, M. M., 4, 66, *315*

D

Daikoku, K., 56, 114, *310*
Daniel, M., 84(403), *319*
Darboux, G., *307*

Davey, A., 62, *311*
de Blois, R. W., 17(16), *303*
de Gennes, P. G., 62(215), *311*
De Martini, F., 226, *308*
de Vries, G., 3, 43, *303*
Dobbs, D., 6(283), 101(283), *314*
Dobrovol'sky, V. L., 268, 271, 272, *317*
Dodd, R. K., 5(242), 16, 79(242), 80(242), 81, 82(242), 146, *307, 312, 313, 319*
Dombrovskii, G. A., 10, 179, *305*
Donato, A., 230, *308*
Donth, H., 16(10, 162), *302, 309*
Döring, W., 17(165), *309*
Dragos, L., 10, 162(74), *305*
Dryuma, V., 83, *313*

E

Ehlers, F. E., 162(416), *319*
Eisenhart, L. P., 8, 18, 144, *309*
Elliot, H. A., 260, *316*
Enz, U., 16(13), 17(166), *303, 309*
Eringen, A. C., 198, *308*
Estabrook, F. B., 4, 51, 54, 55, 57, 140, *304, 319*

F

Fairlie, D. B., 84(399), 92(399), *319*
Fergason, J. L., 16(164), *309*
Fermi, E., 43, *310*
Flanders, H., 126, 277, 278(122), 280, *307*
Flaschka, H., 6(291), 105, *314*
Fordy, A. P., 5(248), 82, *312*
Forsyth, A. R., 2(107), 15, 16(107), *307*
Frankl', F. L., 182, *307*
Frenkel, J., 16(160), *309*
Fried, B. D., 62(212), *311*
Friedlander, F. G., 262(344), *316*
Friedrichs, K. O., 219, *308*
Fusco, D., 230, *308*

G

Gardner, C. S., 3, 4(37), 43, 124, *303, 304*
Gardner, R. B., *320*
Gelbart, A., 239, *315*
Gerdzhikov, V. S., 69, *311*

Geroch, R., 115(370), *317*
Gibbon, J. D., 5(242), 17, 79(242), 80(242), 81, 82(242), 146, *309, 312, 319*
Gibbons, J., 5(248), *312*
Gibbs, H. M., 24, 26, *307*
Gladwell, G. M. L., 249, *316*
Glauert, H., 162(126), *307*
Goddard, P., 84(399), 92(399), *319*
Goldsworthy, F. A., 189, *308*
Goursat, E., 12, 16, *306, 307*
Grad, H., 10, 183, *305*
Grimshaw, R., 64, 78, *311, 312*
Gustafson, T. K., 226(153), *308*

H

Haar, A., 8, 151, *304*
Haberman, R., *313*
Harrison, B. K., 6, 115, 150, *318*
Hasegawa, A., 62(211), *311*
Hasimoto, H., 62, *311*
Hayes, W.D., 189, *308*
Herlt, E., 122, *318*
Herrmann, R., 56, *310*
Hirota, R., 5(246), 6(287, 290), 56, 59, 69, 73, 74, 75, 77, 78, 79, 81, 83, 90, 91, 104(287), 105(299), 106(299), 111, 113, 114, 115(305–307), *307, 310, 312, 314, 315, 319*
Hobson, E. W., 246(337), 251, *316*
Hoenselaers, C., 115(362, 363, 365, 368), *317*
Holland, C. J., 9(424), 168(424), *320*
Hopf, E., 39, *304*
Hopf, F. A., *303*

I

Ibragimov, N. H., 13, *307*
Ichikawa, Y. H., 62(212–214), 65(272, 273), *311, 313*
Imamura, T., 62(213), 311
Inoue, H., *313*
Ishimori, Y., 98, 101, *314*
Iur'ev, I. M., 183, *308*

J

James, I. N., 17(173), *309*
Jeffrey, A., 10, 43, 225, *305, 308, 310*

Johnson, H. H., *307*
Johnson, W. J., 16(159), *309*
Joseph, R. I., 6(282), 101(282), *314*
Josephson, B. D., 16, 30, *302*

K

Kadomtsev, B., 5, 90, *313*
Kakutani, T., 4(203), 43, 57, *310*
Kameyama, W., 105(300), 106(300), 108(300), 109(300), 110(300), *314*
Kaplan, C., 162(417), *319*
Karal, F. C., Jr., 262(90), *306*
Karpman, V. I., 38, 43, 62(210), *309, 310, 311*
Kaup, D. J., 5, 65(121), 78, 80, 81, *307, 312, 313*
Kawata, T., *313*
Kazakia, J. Y., 11, 193, 197(142), 198, 202, 217(142), 222(142), 223, 224, 225(142), 226, 229, 230, 231, 233, 236, 237(94), *306, 308*
Keller, J. B., 262(90, 91), *306*
Kelley, P. L., 62(206), 226(153), *308, 311*
Khristianovich, S. A., 182, *307*
Kingston, J. G., 9(62, 63), 65(308), 163(62), 186, 240, *305, 308, 315*
Kinnersley, W., 115(358–361), 363, 364, 367, 368), *317*
Knerr, B. F., 9(428), 168(428), *320*
Ko, D. R. S., 6(283), 101(283), *314*
Köchendörfer, A., 16(10, 161, 162), *302, 309*
Kodama, Y., 6(288), 104(288), xi(387), xi(388), xi(389), *314, 318*
Konno, K., 8(310), 65(272, 273), 66(310), 68, 105, 106(300), 108, 109, 144(310), *311, 313, 314, 315*
Konopelchenko, B. G., 65, *311*
Kontoteva, T., 16(160), *309*
Korobeinikov, V. P. 225, *308*
Korteweg, D. J., 3, 43, *303*
Kotera, T., 5, 79, 82, *312*
Kramer, D., 119, 122(382), *318*
Kreyszig, E., 254, *316*
Kruskal, E. M., 62(210), *311*
Kruskal, M. D., 4(37), 43, 55, *303, 304, 310, 314*
Kubota, T., 6(283), 101(283), *314*
Kulik, I. O., 16(157), *309*
Kulish, P. P., 69, *311*
Kumei, S., 168, 170(431), *320*

L

Ladik, J. F., *304*
Lake, B. M., 62, 63, *311*
Lamb, G. L., Jr., 2, 4, 13, 16, 17, 22, 26, 27, 44, 51, 54, 58, 59, 60, 145, *303*
Lax, P. D., 79, *312*
Lebwohl, P., 16(7), *302*
Lee, Y. C., 6(287), 104(287), *314*
Leibbrandt, G., 3, 29, 34, 35, 82, 84, 86(392), 87, 88, 89, *303, 312, 318*
Leibovich, S., 43(186), 43, *310*
Levi, D., 68(267), *304, 313*
Lewis, R. M., 262(91), *306*
Lie, S., 12, 13, *306, 307*
Lighthill, M. J., 38, 39, *309*
Lin, C. C., 162(418), *319*
Liu, C. S., 6(292), 62(219), 64, 105, 113, *311, 314*
Loewner, C., 7, 9, 16, 139, 152, 171, 172, *305*
Lonngren, K., 43(187), 55, *310*
Lowengrub, M., 316, 248(327), 253, *316*
Luneburg, R. K., 262(92), *306*

M

MacDonald, B., 268, *317*
McCarthy, P. J., 175, *313, 321*
McLaughlin, D. W., 18, 19, 22, 43(185), *309, 310*
Maki, K., 17(167), *309*
Martin, M. H., 151(138, 172), 153, 188, *308, 309*
Matsuno, Y., 5, 80, 93, 98, 102, 103, *313, 314, 319*
Miura, R. M., 4, 43, *303, 304, 310*
Mizushima, Y., 56(200), 114(200), *310*
Moodie, T. B., 262(93, 345–350), *306, 316, 317*
Morf, R., 82(392), 86(392), 87(392), 89(392), *318*
Morikawa, G. K., 3, 43, *303*
Moroz, I. M., 17(173), *309*
Morris, H. C., 17(24), 150, *303, 313, 319*
Mortell, M. P. 11, 196, 198, *308*
Motz, H., 62(220), 65, *311, 320*
Movsesian, L. A., 10(77), 185, *305*
Müller, W. 179, *307*

N

Nakamura, A., 93, 98, 99(278), 100, 101, 103, 104, 111, *313, 314, 319*
Nariboli, G. A., 43, *310*
Neuber, H., 262(351, 352), 263, 265, 268(352), *317*
Neugebauer, G., 6, 115, 119, 122(382–384), *318*
Newell, A. C., 65(121), *307, 313*
Nikol'skii, A. A., 10(75), 185, *305*
Nuyts, J., 92, *319*

O

Olsen, O. H., 83, *312*
Ono, H., 4(203), 6(276), 57, 62, 93, *310, 311, 313*
Orfanidis, S. J., 105(297), *314*
Osborn, R. A., 10(66), 179(66), *305*
Owczarek, J. A., 195, *309*

P

Parker, D. F., 42, *310*
Parsons, D. H., 239, *315*
Pasta, J., 43, *310*
Pavlenko, V. P., 62(220), 65(441), *311, 320*
Payne, D. A., 124, *317*
Payne, L. E., 242(325), 243(99), 244, 247, 249, 250, 251, 253, *306, 315*
Pell, W. H., 242(325), *315*
Pempinelli, F., 69(435–437), *320*
Peregrine, D. H., 50, *310*
Pérès, J., 10, *305*
Petviashvili, V. I., 5, 90, *313*
Phillips, O. M., 44, *310*
Pilloni, L., 68(267), *313*
Pirani, F. A. E., 7(48), 8, 123(48), 124, 134, 137(47, 48), 138, 143(48), 146, 147(48), 150, 290, *304*
Pohlmeyer, K., 83, 84, 92, 143(255), *312, 319*
Popowicz, Z., 5, 83, *312*
Poritsky, H., 162(420), *319*
Power, G., 9(59, 60), 10, 154, 155, 157, 160, 179, *304, 305, 307*
Prim, R. C., 9(57, 58), *304*

R

Raczka, R., 82(265), *313*
Ragnisco, O., *304*
Ramani, A., 5(246), 81, *312*
Reissner, E., 248, 250, *316*
Rivlin, R. S., 225, *308*
Robinson, D. C., 7(47, 48), 8(48), 123(48), 124(48), 134, 137(47, 48), 138(48), 143(48), 146(48), 147(48), 150(48), 290(48), *304*
Rogers, C., 8(50), 9(61–63), 10(66, 83), 11(97), 65(308), 157, 163, 179(66), 180(87, 88), 183, 185(132), 186, 188, 198, 230(95), 240, 243(326), 261(101), 262(93, 100, 102, 345–347, 349, 350), 267, *304, 305, 306, 308, 315, 316, 317*
Rosen, G., 9, 168, 169(430), 170, *320*
Rosen, N., 16(163), *309*
Rosenstock, H. B., 16(163), *309*
Roskes, G., 62, *311*
Rubinstein, J., 16(14), *303*
Ruscheweyh, S., 239, 254, *315*
Ryashentsev, V. I., 255, *316*
Rykov, V. A., 10(80), *305*

S

Sachdev, P. L., 42, *309*
Sagoci, H. F., 248, *316*
Santini, P. M., 68(267), *313*
Sanuki, H., 8(310), 66(310), 105(300), 106(300), 108(300), 109(300), 110(300), 144(310), *314, 315*
Sasaki, R., *318*
Satsuma, J., 5, 6, 56, 59(201), 66, 69, 73, 75, 78, 79, 80, 83, 90, 91, 98, 101(280), 104, 113, 114, *310, 312, 313, 314, 315*
Sauer, R., 10, 179, *305, 307*
Sawada, S., 5, 79, 82, *312*
Schiffer, M., 255, *316*
Scott, A. C., 16(8, 9, 158, 159), 18, 19, 22, 43(185, 187), 55, *302, 303, 309, 310*
Scott-Russell, J., 43, *310*
Scully, M. O., *303*
Sedov, L. I., 10, *306*
Seebass, A., 43(186), *310*
Seebass, R., 183, 184, 185, *308*
Seeger, A., 16(10, 161, 162), *302, 309*
Segur, H., 65(121), *307*
Seymor, B. R., 11, 196, 198, *308*
Shabat, A. B., *307*
Shadwick, W. F., 7(48, 407), 8(48, 407), 9(63), 18, 123(48, 407), 124(48), 137(48), 138(48), 143(48), 144, 146(48), 147(48), 148, 150(48), 290(48), 297, *304, 305, 309, 319, 321*
Shapiro, A. H., 160, 269, *317*
Shen, M. C., 43, *310*
Shimizu, K., 62(214), *311*
Sih, G. C., 268, *317*
Skyrme, T. H. R., 16(11, 12), *302, 303*
Slusher, R. E., 24, 26, *307*
Smith, P. 9(59, 60), 154, 155, 157, 160, 186, *304, 305, 307, 308*
Sneddon, I. N., 244, 246(336), 248(327), 249, 250, 253, 260, *316*
Sokolovsky, V. V., 262, 268, *317*
Steudel, H., *321*
Stewartson, K., 62, *311*
Su, C. H., 43, *304*
Suhubi, E. S., 198, *308*
Swamy, N. V. V. J., *318*

T

Talanov, V. I., 62(205, 207), *310, 311*
Tamada, K., 182, 184, *307, 308*
Tamama, T., 56(200), 114(200), *310*
Taniuti, T., 10, 43, 62(208, 209, 213), *305, 310, 311*
Tappert, F., 43, 62(211), *310, 311*
Tasso, H., 42, *309*
Teichmann, J., 42, *309*
Tenenblat, K., 82, *312*
Terng, C. L., 82, *312*
t'Hooft, G., 84(396), *319*
Toda, M., 6(293), 105, 111, 113, (444), *314, 320*
Tomilov, E. D., 10(76), 185, *305*
Tomotika, S., 182, *307*
Townes, C. H., 226(153), *308*
Tsien, H. S., 9, 151, 160, 162, *304*
Tsuneto, T., 17(167), *309*
Tsuzuki, T., 62, *311*
Turner, J. W., 69(439), *320*

U

Ulam, S., 43, *310*
Ursell, F., 69(237), *312*
Ustinov, M. D., 10(78, 79), 185, 189, *305*

V

van Wijngaarden, L., 43, *310*
Varley, E. 10, 11, 190, 191, 193, 195, 197(142), 202, 204, 205(85), 206(85), 207, 208, 209, 211(85), 212(85), 214, 216, 217(85, 142), 222(142), 223, 224, 225(142), *306, 308*
Varma, C. M., 43, *310*
Venkataraman, R., 11, 198, 225, 226, 229, 230, 231, 233, 236, 237(94), *306, 308*
Viallet, C. M., 84(403), *319*
von Kármán, T., 10, *305*
von Mises, R., 9(55), 160, 255, *304, 316*

W

Wadati, M., 6(293), 8(310), 58, 59, 65(272, 273), 66, 68, 105, 110, 111, 113, 144, (387), (388), *307, 311, 313, 314, 315, 318, 320*
Wahlquist, H. D., 4, 51, 54, 55, 57, 140, 145, *304, 319*
Ward, R. S., 84(397), *319*
Washimi, H., 43, 62(208), *303, 311*
Watanabe, Y., (443), *320*
Watson, G. N., 249, *316*
Wei, C. C., 43, *310*
Weiland, J., 62(220), 65(441), *311, 320*
Weinstein, A., 11(328), 239, 245, 252, 253, *306, 315, 316*
Whitham, G. B., 38, 39, 41, 42, 44, 49, 50, *303*
Wilson, W., *318*
Wolf, E., 232, *308*
Wong, S., 82(392), 86(392), 87(392), 89(392), *318*

X

Xanthopoulos, B. C., 115(368), *317*

Y

Yajima, N., 62(209), *311*
Yang, C. N., 84(400), *319*
Yates, R. G., 84(399), 92(399), *319*
Yih, C. S., 152, *319*
Yuen, H. C., 62, 63, *311*

Z

Zabusky, N. J., 4(33), 43, 55, 57, *303, 310*
Zakharov, V. E., 6, 62, 64, 84, 115, *307, 311, 319*

Subject Index

A

Abel integral equation, 260
AKNS system, 65–69, 106–107
 and Bäcklund maps, 141, 150
Alfvén
 number, 184
 speed, 184
Amplitude dispersion, 50
Antiplane
 contact problems in linear elastostatics, 261–262
 crack problems in linear elastostatics, 257–261
 deformation of nonlinear elastic media, 262–273
Asymptotic wave front expansions, 262
Attenuation dispersion, 48

B

Bäcklund maps, 134–145
 and Bäcklund transformations, 139–143
 and constraints, 142–143
 and the Korteweg–deVries equation, 140–141
 and the Loewner transformations, 138–139
 and the sine-Gordon equation, 137–138, 141, 142
 composition with symmetries, 299–301
 ordinary, 137
Bäcklund transformations
 bilinear operator formulation of, 5, 69–82, 90–91, 92–104, 113–115
 classical, 2, 4, 15
 in continuum mechanics, 8–11
 in higher dimensions, 5, 82–92
 in nonlinear heat conduction, 168–171
 jet-bundle formulation, 7–8, 123–150
 of the Benjamin–Ono equation, 95–96
 of the Boussinesq equation, 70–71
 of the Burgers' equation, 3, 38–39
 of the cubic Schrödinger equation, 4, 68
 of the Davey–Stewartson equation, 68
 of the diffusion equation, 175
 of the Ernst equation, 6, 7, 117–118

of the Haar-type, qv
of the Kadomtsev–Petviashvili equation, 90
of the Korteweg–deVries equation, 4, 51–54, 60–62, 67, 74, 79–82, 140–141, 175
of the Lax hierarchy, 80
of the Liouville equation, 13–14
of the Loewner type, qv
of the modified Korteweg–deVries equation, 58–59, 60–62
of the reciprocal type, qv
of the sine-Gordon equation, 16–18, 31–32, 84–85
of the Toda Lattice equation, 6, 111
of the Wadati Lattice equation, 111
of the Yang equations, 92, 143
Bell's Law, 202, 207–208, 209
Benjamin–Ono equation, 6, 92–101
  Bäcklund transformation of, 95–96
  bilinear representation of, 93–95
  modified, 99–101
  permutability theorem for, 97–98
Bergman
  integral operators, 11, 255
  series, 254–257, 262
Bianchi diagram
  for the Benjamin–Ono equation, 96
  for the Boussinesq equation, 72, 73
  for the elliptic sine-Gordon equation, 37, 38
  for the Ernst equation, 119, 120
  for the 1+1 sine-Gordon equation, 19, 22, 25, 27, 28
  for the 3+1 sine-Gordon equation, 87
  for the Toda Lattice equation, 112
Bianchi–Lie transformation, 2
Boussinesq
  equation, 5, 69–73
  problem, 248–250
  system, 50
Burgers' equation, 38
  and reduction to the heat equation, 38–42

C

Canonical projection maps, 125
Cauchy–Riemann equations,
  and Bergmann series, 254
  and the Stokes–Beltrami system, 239
  in nonlinear elastostatics, 262, 265, 269
  in subsonic gasdynamics, 161–162
Characteristics,
  in gasdynamics, 192
  in nonlinear dielectrics, 230
  in nonlinear elastodynamics, 211
Chen's method, 83
Clairin's method, 16
Cole–Hopf transformation, 3, 38–39
  and the solution of initial value problems, 39–42
Conformal mapping, 239
Conservation Laws, 8, 92
Contact
  forms, 127, 128
  modules, 127, 128
  problems, 253–273
  structures, 126–128
  transformations, 13
Crack problems, 250–253
  antiplane, 257–262
Cross section maps, 125
Cubic Schrödinger equation
  and AKNS system, 66, 68
  and deep water gravity waves, 62–64
  and Painlevé transcendents, 69
  as a canonical form, 64–65
  Bäcklund transformation of, 4, 68
  symmetries, 69

D

Davey–Stewartson equation
  and deep water gravity waves, 63–64
  Bäcklund transformation of, 68
Deformation
  Cauchy tensor, 198
  Green tensor, 198
  Law, 263
  uniaxial, 199
Diffeomorphism, 163, 294
Differential-difference equations, 105–115
  Konno–Sanuki transformation, 105–107
  Self-dual nonlinear networks, 114
  Toda–Lattice, 111–115
  Volterra system, 114
  Wadati Lattice, 111
Differential ideal, 289
Differential p-forms, 277–279
  on jet bundles, 286–288
Dirichlet problem, 252
Dispersion
  amplitude, 50

SUBJECT INDEX

attenuation, 48
  pure, 48
  relation, 48
Dombrovskii transformation, 179–180
Drag function, 153

E

Einstein equations, 115–116
Elastodynamics
  inhomogeneous linear, 262
  nonlinear, 10–11, 197–225
Elastostatics
  Bäcklund transformations in, 11, 242–273
  equilibrium equations in, 242, 243, 253, 263
Electromagnetic wave propagation, 11, 225–237
Electrostatic potential, 245, 249, 252
Elliptic sine-Gordon equation, 3, 31–38
Equilibrium equations, 242, 243, 253, 263
Ernst equation
  Neugebauer transformations of, 117–118
  permutability theorem for, 118–119
Extension, 198
Exterior
  derivative, 279–282
  differential systems, 288, 290–291
  ideal, 287
  product, 277

F

Fermi–Pasta–Ulam problem, 43
Fibered products, 132–134
Frankl' approximation, 182

G

Gasdynamics
  Lagrangian formulation, 189
  Loewner-type transformations in, 171–183, 189–196
  Martin's equation in, 153, 188
  reciprocal relations in, 151–162, 185–189
  subsonic, 155, 160–162, 172, 178–180
  supersonic, 155, 160, 172, 182–183
  transsonic, 172, 180–182
Generalized axially symmetric potential theory, 239, 241

General relativity
  Bäcklund transformations in, 115–122
Gravity waves, 46
Griffith crack, 258, 260

H

Haar transformation, 8, 9, 10, 155
Hard elastic materials, 203, 204, 208, 209
  ideally, 202, 203, 206, 207, 210
Heat
  classical linear equation, 39, 168, 170, 175
  nonlinear conduction, 9, 168–171
Hirota bilinear operators
  basic properties, A1
  formulation of Bäcklund transformations, 5, 69–82, 90–91, 92–104, 113–115
Hodograph
  systems, 161, 172, 176–183, 184, 190, 200, 227, 265
  transformation, 160–161, 184, 190, 200, 227, 264
Hooke's Law, 208, 212, 242, 245, 262, 263, 270
Hypersonic flow, 189

I

Interior product, 283–284
Inverse scattering, 65–69

J

Jet bundles
  and the formulation of Bäcklund transformations, 7–8, 123–150
  fibered products of, 132–134
  partial differential equations on, 129–132
Joseph's equation, 6, 101–104
  Bäcklund transformation of, 102–103
  bilinear representation of, 102
  modified, 103–104
Josephson junctions
  generation of fluxons, 32–38
  magnetic flux along, 29–31

K

Kadomtsev–Petviashvili equation, 5

Bäcklund transformation of, 90
permutability theorem for, 75, 90–91
Kármán–Tsien approximation, 9, 159, 160, 161, 179, 185, 268, 269
Kaup's water wave equation, 78–79
k-equivalent maps, 124
Kerr–NUT metric, 119–122
k-jet bundles, 124
Klein–Gordon equation, 130
Konno–Sanuki transformation, 106–107
Korteweg–deVries equation
  and Bäcklund maps, 140–141
  and the AKNS system, 66, 67
  as a canonical form, 43, 78
  Bäcklund tranformations of, 4, 51–54, 60–62, 67, 74, 140–141
  bilinear formulation for, 73
  higher dimensional, 5, 75, 83, 90–91
  higher order, 5, 79–82
  in water wave theory, 3, 43–50
  linearized, 175
  modified, 4, 57–62, 66, 68, 75
  permutability theorem for, 54–55, 74–75
  soliton solutions of, 55–57, 75–77

L

Lagrangian equations
  in gasdynamics, 189
  in nonlinear elasticity, 199
  reduction to canonical form, 190–191, 200–201
Lax
  hierarchy, 5, 79, 80
  system, 83
Lie
  Bäcklund tangent transformations, 13
  derivative, 284–285
  groups, 142
  theorem, 144–45
Lift function, 153
Liouville's equation, 13–14
Local
  approximation and model laws, 210–211, 230
  impedance, 214
  reflection coefficient, 214, 232
  transmission coefficient, 232
Loewner transformations

  and Bäcklund maps, 138–139
  and Stokes–Beltrami systems, 240–241
  and termination of Bergman series, 256–257
  in gasdynamics, 171–183, 189–196
  in linear elastostatics, 253
  in magnetogasdynamics, 183–185
  in nonlinear dielectric theory, 227–230
  in nonlinear elastodynamics, 200–201
  in nonlinear elastostatics, 265

M

Magnetogasdynamics
  Loewner transformations in, 183–185
  reciprocal transformations in, 162–168
  sub-Alfvénic regions, 184
  super-Alfvénic regions, 184
Martin's equation
  in nonsteady gasdynamics, 188
  in steady gasdynamics, 153
Material
  points, 198
  velocity, 199
Maxwell's equations, 226
Miura's transformation, 60–62
  for higher order Korteweg–deVries equations, 82
Mixed boundary value problems, 248, 250–252, 258
Model Laws
  in gasdynamics, 179, 191
  in nonlinear dielectrics, 229–230
  in nonlinear elastodynamics, 202–211
  in nonlinear elastostatics, 266–268
Mode III displacement, 253, 258
Monge–Ampère equations
  Bäcklund transformations of, 15–16, 38–39, 154–155, 188–189
  Burgers' equation, 38
  Martin's equations, 153, 188
Monomial, 278

N

Nakamura transformation, 99–101
Neuber equations, 262
Neugebauer transformations, 117–118
Neumann problem, 252

SUBJECT INDEX

N-periodic waves, 101
N-waves, 41, 42

O

Oblate spheroidal coordinates, 246, 250
$O(3)$ nonlinear σ model, 5, 83

P

Painlevé transcendents, 5, 69
Particle displacement, 198
Penny shaped crack, 248, 250–252
Permutability theorems
  for lattice equations, 107, 111
  for the Benjamin–Ono equation, 97–98
  for the Boussinesq equation, 71–73
  for the cubic Schrödinger equation, 68–69
  for the elliptic sine-Gordon equation, 34–38
  for the Ernst equation, 118–119
  for the Kadomtsev–Petviashvili equation, 75, 90–91
  for the Korteweg–deVries equation, 54–55, 74–75
  for the modified Korteweg–deVries equation, 59, 75
  for the 1+1 sine-Gordon equation, 19–22
  for the 3+1 sine-Gordon equation, 86–87
  for the Toda Lattice equation, 112–113
Persey curves, 271
Phase, 48
Pohlmeyer's transformation, 143
Point transformations, 294–296
Prim Law, 155
Projection maps, 133
Prolongation, 131
Pull-back maps, 282–283
Punch problems, 246–248

R

Reciprocal transformations, 9, 10, 151
  in gasdynamics, 151–162, 185–189
  in magnetogasdynamics, 162–168
Reissner–Sagoci problem, 248
Riccati equations
  and the AKNS system, 67
  in gasdynamics, 177
  in nonlinear elasticity, 201
Riemann invariants
  in gasdynamics, 192
  in nonlinear dielectrics, 230
  in nonlinear elasticity, 201, 211
Rosen–Bluman–Kumei transformation, 169–171

S

Saint–Venant compatibility conditions, 242
Sawada–Kotera equation, 79–82
Schrödinger equation
  cubic, 4, 62–66, 68–69
  linear, 66
Self-dual nonlinear networks, 114
Shallow water waves
  approximation, 48
  nonlinear equations, 49
Shock formation
  in nonlinear dielectrics, 237
  in nonlinear elasticity, 221–222
Shock tube
  reflection of a centered wave in, 189–196
Signal
  functions, 212
  speed, 190
Similarity solutions, 5
Sine-Gordon equation
  and Bäcklund maps, 137–138, 141, 142
  and the AKNS system, 66, 68, 141
  as a submanifold of a jet bundle, 129–130
  Bäcklund transformations of, 16–18, 31–32, 84–85
  elliptic, 3, 31–38
  group of symmetries of, 297
  higher dimensional, 82, 84–89
  one dimensional, 2, 16–29
  permutability theorems for, 19–22, 34–38, 86–87
  theorem of Lie, 144–145
  Wahlquist–Estabrook procedure for, 148–150
Soft elastic materials, 203, 208, 209
  ideally, 203, 205, 206, 209, 210
Sokolovsky Law, 262, 263, 267, 268–270
Soliton
  solutions of higher order sine-Gordon equations, 85–89
  solutions of lattice equations, 107–110

solutions of the Korteweg–deVries equation, 55–57
solutions of the modified Korteweg–deVries equation, 59
solutions of the sine-Gordon equation, 22–29
Solution manifold, 289
Source maps, 125
Standard basis, 127
Stokes
   flow, 242
   relation, 63
Stokes–Beltrami systems
   generalized, 238–239, 245
   in continuum mechanics, 11
   in elastostatics, 244, 245
Strain
   Eulerian tensor, 198
   Lagrangian tensor, 198
Stress function, 264, 271, 273
Sturm–Liouville systems, 4, 66
Symmetries
   and the cubic Schrödinger equation, 69
   composition with Bäcklund maps, 299–301
   of exterior systems, 297–298
   of systems of differential equations, 143–145, 297

T

Tangent spaces, 276–277
Target maps, 125
Toda Lattice, 111–115
   associated lattices, 113–114
   Bäcklund transformation for, 111
   permutability theorem for, 112–113
Torsion problems, 239, 245, 246–253
Total derivative operators, 131
Traction, 199
Tricomi equation
   and the Stokes–Beltrami system, 239
   as a canonical form in transsonic flow, 180–182
Tschaplygin–Molenbroek
   system, 265
   transformation, 264

U

Ultrashort optical pulses, 2–3, 22–29
   generation of $2N\pi$ pulses, 24–29
Ustinov's transformation, 189

V

Vector fields, 276–277
   characteristic, 289
Viscoelastodynamics, 262
Volterra system, 114

W

Wadati Lattice, 110–111
   Bäcklund transformation for, 111
Wahlquist–Estabrook
   procedure, 145–150
   transformation, 4, 51–54, 67
Warping function, 264, 271, 272
Waves
   classical equation as canonical form, 182, 190, 201, 227
   deep water gravity, 63–64
   dispersive, 48
   frequency, 48
   group velocity, 48
   in gasdynamics, 189–196
   in nonlinear dielectric media, 225–237
   in nonlinear elasticity, 197–225
   in weakly inhomogeneous plasma, 64
   length, 48
   number, 48
   N-wave, 41, 42
   period, 48
   periodic, 42
   reflection, 191–196, 211–225, 230–237
   slowly varying solitary, 64
   steepness, 48
Wedge product, 277
Weinstein's correspondence principle, 11, 239, 241, 244, 245
   application in elastostatics, 247–248, 249, 251, 252, 253, 257
   generation as a Bäcklund transformation, 240–241
   iterated, 241
Wronskian's, 66

Y

Yang equations, 83–84
   Bäcklund transformation of, 92, 143
   one-parameter family of Bäcklund maps for, 145